I0049652

A Journey from
Raw Materials to Ceramics

by

Fatih ŞEN

Sen Research Group, Department of Biochemistry, Kutahya Dumlupinar University, Turkey

Published by **Materials Research Forum LLC**
Millersville, PA 17551, USA

Published as part of the book series
Materials Research Foundations
Volume 184 (2025)
ISSN 2471-8890 (Print)
ISSN 2471-8904 (Online)

Print ISBN 978-1-64490-382-7
eBook ISBN 978-1-64490-383-4

Distributed worldwide by

Materials Research Forum LLC
105 Springdale Lane
Millersville, PA 17551
USA
https://mrforum.com

Manufactured in the United States of America
10 9 8 7 6 5 4 3 2 1

Table of Contents

Chapter 1

Introduction and Definition of Ceramics

Gulfem Binal[1]*, Okan Odabas[2], Kazim Erden Karaoglanli[3], Yasin Ozgurluk[4],
Abdullah Cahit Karaoglanli[5], Cumhur Eren Isık[6], Iskender Isık[7], Fatih Sen[8]*

[1]Bartin University, Faculty of Engineering, Architecture and Design, Department of Metallurgical and Materials Engineering, Bartin, Türkiye

[2]Bartin University, Faculty of Engineering, Architecture and Design, Department of Mechanical Engineering, Bartin, Türkiye

[3]Bartin University, Faculty of Sciences, Department of Biotechnology, Bartin, Türkiye

[4]Bartin University, Vocational School of Health Services, Program of Medical Services and Techniques, Bartin, Türkiye

[5]Bartin University, Faculty of Engineering, Architecture and Design, Department of Metallurgical and Materials Engineering, Bartin, Türkiye

[6]Department of Handicrafts, Kutahya Fine Arts Vocational School, Kutahya Dumlupınar University, Evliya Çelebi Campus, 43000 Kutahya, Türkiye

[7]Department of Materials Science & Engineering, Faculty of Engineering, Kutahya Dumlupınar University, Evliya Çelebi Campus, 43100 Kutahya, Türkiye

[8]Sen Research Group, Department of Biochemistry, Kutahya Dumlupinar University, Kutahya 43000, Türkiye

gbinal@bartin.edu.tr, fatihsen1980@gmail.com

Abstract

Materials are defined as anything with mass, and materials can be divided into three groups: ceramics, metals, and polymers. Ceramics is the subject of this introductory chapter. Materials made of ceramics have been produced for thousands of years, almost as long as human civilization. The word "ceramics" is generally used to refer to a broad category of materials, most of which are inorganic and need high temperatures to be processed or manufactured. These materials are typically separated into two groups: advanced ceramics and traditional ceramics. This chapter briefly discusses the definitions, processing technology, structure, properties, classification, and applications of ceramic materials.

Keywords

Ceramics, Definitions, Processing, Structure, Properties, Applications

Materials Research Forum LLC
https://doi.org/21741/9781644903834

1. Introduction

Humans have utilized ceramics for many centuries. The earliest ceramic artifacts discovered date back more than 20.000 years. Although clay served as the foundation for all ancient ceramics, its characteristics differed widely based on firing temperature and composition [1]. Ceramics are hard materials composed of minerals that go through physicochemical reactions like clay and are fired at high temperatures after being shaped [2]. In many different fields, like the construction, energy, and environmental industries, they are frequently utilized [3]. The Greek words keramikos, which means "of pottery" or "for pottery," and keramos, which means "potter's clay, tile, pottery," are the sources of the word "ceramic." It is a human-made "artificial stone" that blends the four fundamental elements defined by the ancient Greeks: earth or clay, water, fire, and air [4]. From the ancient art and technique of ceramics to a vast array of materials connected to cutting-edge and contemporary technology, ceramics have evolved over a long time.

These days, a wide variety of materials and goods are included in the term "ceramic," ranging from various aspects of everyday life products to materials used in industry [5]. Because ceramics are among the oldest and most diverse groups of materials, they have been defined in a variety of ways. These definitions include the classification of materials according to their chemical composition (silicates, oxides, and non-oxides), their mechanical and physical properties, or their applications (construction materials, functional materials, and high-temperature materials). Inorganic, nonmetallic materials are the most commonly used and the most straightforward definition of ceramics [6]. It is challenging to describe ceramics nowadays because a variety of products made by producing and firing inorganic substances other than clay are included in this category. The optimum definition, however, is any product that is shaped and made from nonmetallic, inorganic raw materials and that, through firing, is changed from an incoherent powder into a solid item with a partially crystalline and vitreous structure. Ceramics can be defined in two groups: traditional ceramics and technical ceramics. Products made from natural oxides, such as tiles, bathroom goods, bricks, tableware, and various refractories, are included in the first group.

The second group includes products such as bioceramics, electronic ceramics, technical and electrical porcelains, special refractories, and catalysts [7]. According to Kingry et al., ceramics are the art and science of manufacturing and utilizing solid objects that are mostly made of nonmetallic inorganic materials. This term encompasses not only materials like pottery, enamels, porcelain, glass, and cement but also ferroelectrics, glass ceramics, nonmetallic magnetic materials, manufactured single crystals, and a wide range of other items that were not around for a while or may not even exist now [8]. The American Ceramic Society defines ceramics as inorganic, non-metallic materials with a crystalline structure. These materials are composed of compounds formed between metallic and non-metallic elements like silicon and nitrogen (Si_3N_4), calcium and oxygen (CaO), and aluminum and oxygen (Al_2O_3) [9]. Some ceramics, a large family of materials, are described below. Porcelain, terracotta, stoneware, and earthenware are all included in the general term "pottery." "All fired ceramic wares that contain clay when formed, except technical, structural, and refractory products," is the definition of pottery according to ASTM Standard C 242-01: "Standard Terminology of Ceramic Whitewares and Related Products". Whiteware is a type of ceramic product that is fired and is composed of a glazed or unglazed body. It is generally white in color and has a fine texture. Whiteware includes many product classifications, such as porcelain, tile, earthenware, china, and semivitreous ware. Whiteware is manufactured using natural raw materials, with clay constituting the predominant component. Throughout history, bodies of stoneware, porcelain, earthenware, and china have been differentiated by virtue of their compositions and firing temperatures. Another characteristic of porcelains that sets them apart is the absence of open porosity in the fired body [10].

Glass can be defined as a non-crystalline solid, a randomly structured material, or a supercooled liquid. Generally speaking, "glass" refers to an amorphous solid with non-directional features that is transparent, hard, and rigid at ambient temperature, as well as having the ability to function as plastic at higher temperatures [11]. Glass in thin layers, known as glaze or enamel, depending on whether the substrate it is deposited on is porous (such as porcelain) or smooth (such as metal), has long been highly regarded [12]. A thin layer of glassy material fused by firing to the ceramic body's surface is called a ceramic glaze. The primary distinction between glaze and glass is the glaze's high surface-to-volume ratio and its ability to interact with both the atmosphere and the ceramic body.

Glazes are open systems, with their chemical composition changing during firing mostly because of their contact with ceramics and the volatilization of certain compounds, whereas glass is essentially a closed system, with its chemical composition not changing appreciably during firing. Element diffusion occurs through the glaze towards the body and through the body towards the glaze because of the interaction between the ceramic body and the glaze. As a result of diffusion, glazes are not uniform and exhibit gradients in chemical composition throughout the thickness of the glaze [13]. Porcelain enamel is an inorganic glassy coating that is applied to a metal or ceramic substrate. It is made by covering a prefabricated substrate with an appropriate powder coating, either wet or dry, and then firing it for a short period of time to melt the coating and smooth the surface. The selection of the substrate is based on special needs, including cost, strength, stiffness, and thermal characteristics. Enamel can be developed specifically for aesthetics, thermal shock resistance, chemical resistance, ease of cleaning, physical durability, or a combination of these parameters. In essence, enamel is a glass that softens at a low temperature of 510–530 °C. Therefore, it is mostly utilized in heating systems and kitchenware with low operating temperatures [14]. Ceramic materials produced through controlled glass nucleation and crystallization are known as glass ceramics. Glasses are melted, shaped, and thermally transformed into a ceramic material that is mostly crystalline. The fundamental principle of controlled internal crystallization is effective nucleation, which allows the formation of fine, randomly oriented grains with no microcracks, voids, or other porosity [15].

Refractories are ceramics used in the construction of kilns and other high-temperature equipment. They are also utilized in the iron and steel sectors, as well as in the cement, ceramic, and glass kiln industries. There are various types of refractories, such as high-alumina, fireclay, and basic refractories [16]. Cement is a ceramic material composed of clay, limestone, and other minor additives. The finely ground raw materials are combined with water to make a slurry, which is then sintered in a cement kiln (1400–1600 °C), cooled and pulverized, and then a little gypsum is added to retard setting. There are various kinds of cement, such as refractory and portland cement [16]. Advanced ceramics refer to the development of ceramic materials that include distinct and enhanced qualities compared to traditional ceramics. Advanced ceramics are ceramics with outstanding functionality. In other words, advanced ceramics are ceramic materials designed to achieve the desired characteristics using composition control and functional production methods. There are various terms used to describe this sort of ceramic, such as advanced, engineered, fine, technical, and so on. It is commonly referred to as 'advanced' or 'technical' ceramics in American literature, whereas it is referred to as 'fine' ceramics in Japanese literature. In Europe, the phrase "technical" ceramics is more frequently utilized. Advanced ceramics generally have a crystalline structure. Advanced ceramics, as a key component of modern technology, currently play an important function in industries like communication, energy, transportation, environment, and so on [17]. Nonmetallic inorganic advanced ceramics can be utilized as components for 'structural' design restriction materials. Wear resistance, high temperature stability and strength, corrosion resistance, and chemical inertness are just a few of the beneficial characteristics that these materials display. Furthermore, they are typically manufactured from synthetic powders, which can be adjusted for applications and production procedures. These materials can also be named technical ceramics, engineering ceramics, or high-tech ceramics.

Traditional ceramics, on the other hand, are manufactured from minerals that arise naturally and are utilized for items with less stringent design requirements, like whitewares, tiles, refractories, and bricks [18]. 'Bioceramics' refers to ceramics used for the healing and rebuilding of diseased or damaged parts of the body. Bioceramics are mostly utilized for the treatment of hard tissues, including teeth, bones, and joints. Hydroxyapatite, alumina, bioactive glasses, titania, glass-ceramics, calcium aluminates, and tricalcium phosphate are some examples of bioceramics [19]. Electronic ceramics, also known as functional ceramics, are materials with electronic properties that enable them to be utilized for insulating, conductive, capacitive, sensor, electro-optical, resistive, and magnetic applications [20].

2. Processing Technology

Because earlier ceramic products were produced from naturally existing materials like clay minerals and ceramics have been associated with clay. From the past to the present, the same raw materials have been used as the main ingredients in the production of traditional ceramics such as whitewares, glass, construction materials, and so on [21]. Figure 1 illustrates the procedures involved in ceramic manufacture. Raw materials used for ceramic production can be categorized as synthetic or natural resources. After quarrying, the natural, and hence less expensive, raw materials are mechanically and physically processed. Commonly used raw materials include quartz sands and rocks, kaolins, clays, and feldspars; also, zircon sands, magnesia, and baddeleyite can be used. Al_2O_3 and SiC are examples of synthetic raw materials. Specific unique methods, comprising multiple chemical reactions, are required for their manufacturing. The selection of ceramic raw materials is determined by their geographical proximity and accessibility, with the aim of minimizing transportation expenses. Additionally, their suitability for specific functions in the manufacturing process of ceramics, such as acting as polymers, inerts, or fluxes, is taken into consideration [22], [23]. Fluxing materials are utilized to facilitate the creation of a liquid phase during the firing process. The effectiveness of a fluxing agent is determined by the quantity of liquid phase generated at the temperature of firing and its viscosity. Fluxes play a crucial role in determining the characteristics of the final ceramic product [24]. Ceramic powders have been produced using both chemical and physical processes.

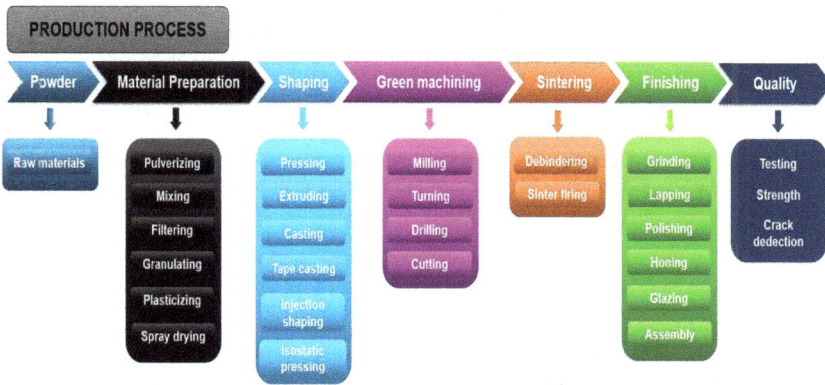

Figure 1. *The procedures involved ceramic manufacture.*

Using methods for ceramic forming, ceramic powders can be formed into variously shaped green ceramics, which are then sintered into consolidated ceramics. However, to create ceramics that are too expensive or difficult to make through the powder-forming and sintering approaches, several ceramic processing methods have been designed and are still being worked on. While some ceramic forming procedures, such as casting and extrusion, are similar to those used for the fabrication of metal and polymer components, the processing pathways for ceramics are far broader than those for metallic and polymer goods [25]. Synthesize precursor powder for conventional ceramics, various powders must be crushed, ground, separated from impurities, and blended in accordance with the necessary composition. Milling or comminution are other terms for size reduction achieved by crushing and grinding.

Raw material milling is followed by batching of milled components in predetermined amounts. To achieve homogeneity, batching is followed by mixing and then shaping to give the material its shape. The formed body is dried before firing, and the fired body becomes consolidated material [26]. The shaping step of the processing involves compacting and forming a ceramic powder with high porosity and low mechanical characteristics into the desired shape after it has been appropriately prepared through washing procedures, homogenization, and so on. The most common extant forming procedures are porous mold or tape molding, axial and isostatic dry pressing, and injection molding or extrusion. According to the kind of product to be produced, the time and expense constraints in the manufacturing systems, and other factors, each of these forming techniques is applied. High packing density and low porosity are often the goals of the forming operations (porosity will be reduced entirely or partially during the sintering process). To prevent microstructural flaws, which would reduce the final material's qualities, the packaging should be as homogenous as possible, in addition to having a high degree of coordination. Several variables, including the distribution and size of the particles, the degree of agglomeration, and the flow of powder throughout the forming process, must be regulated in order to achieve this.

Other unconventional techniques exist to produce green parts; these include hot pressing, the sol-gel method, deposition of molten particles (SHS), and deposition in the vapor phase (CVD) [27]. The ceramic pattern is frequently machined after it has been formed in order to smooth out any seams and rough surfaces or to alter its shape. Green ceramics are machined using a variety of techniques, such as laminating multilayer ceramics, blanking and punching to cut the shape and produce holes or cavities, and surface grinding to smooth surfaces. Typically, thermal processing is the final stage in the production of ceramic parts. At this stage, the loosely bound particle body formed through the shaping process is subjected to heat treatment to create a solid body with the specific qualities required for its intended application. Thermal processing involves sequential operations of drying and binder removal, sintering, and vitrification. The primary objective of the drying process in ceramics is to eliminate moisture from the malleable ceramic material prior to subjecting it to elevated temperatures during firing. Typically, the process of removing water through drying is conducted at temperatures equal to or below 100 °C and may require up to 24 hours for a large ceramic component. Many organic binders can be eliminated from ceramic components by subjecting them to temperatures ranging from 200 °C to 300 °C; however, specific hydrocarbon residues may necessitate exposure to even higher temperatures. Sintering is the solid-state diffusion process that causes small particles of a substance to bond together. During the thermal treatment process in ceramic manufacture, the porous compact undergoes a transformation and becomes a dense and coherent material. Sintering is a frequently utilized method for manufacturing ceramic forms composed of materials such as titanates, beryllia, alumina, and ferrites.

The final characteristics of the ceramic products are determined by a series of reactions and transformations that occur in clay minerals and supplementary minerals such as feldspar, quartz, dolomite, and calcite throughout the firing stage of traditional ceramic materials [28]. Through the process of ceramic firing, those crystalline structures, when they surpass their stability limitations, slightly decompose, and at the same time, others occur [29]. During the firing process, traditional sintered ceramic bodies, such as porcelain or stoneware, generate considerable amounts of glassy phases that are still prominent in the body after sintering. Those glassy phases allow for densification through a process known as viscous flow sintering. Therefore, all traditional sintered ceramics must have the capacity to create appropriate glassy phases with a viscosity that is high enough to avoid the ware from deformation and low enough to permit rapid sintering. The resulting product is composed of a glass matrix that contains residual (quartz and feldspars) or newly formed (mullite) crystalline phases [30]. Glazes are frequently applied to ceramic items.

Glazes are thin silicate coatings that resemble glass and have both functional and artistic importance. Pre-melted fruits, or inorganic materials, are the raw materials used to make glazes. The prepared glaze powders are applied to the pre-fired or unfired body with different techniques. The melting temperature of the applied glaze is lower than the body on which it is applied. Glazed bodies are then subjected to the firing process. If over-glaze decoration is used, the products undergo an extra firing at a lower temperature. The ceramic industry uses raw materials, water, auxiliary materials, and energy in its manufacturing process. Depending on the manufacturing procedures used, there are differences in the environmental impact of ceramic materials. It pollutes the environment by emitting particles (dust), soot, and gases (CO_x, NO_x, SO_x, heavy metals, and chlorinated and fluorinated inorganic compounds). It contaminates water by releasing insoluble particles and other inorganic materials like heavy metals. It also has an impact on soil because it releases waste formed throughout production, like used plaster molds, sludge, solid waste (ash), used sorption and adsorption materials, broken pieces, and packaging waste. A further aspect to take into account is the high energy consumption involved in the manufacturing of ceramic materials, particularly during the drying and high-temperature firing parts that define the process and require the use of electricity, natural gas, butane, propane, coal, biomass, biogas, and other fuels [31]. In any industrial sector, the manufacturing of ceramics requires a significant quantity of energy. A significant portion of the overall production costs in the ceramics industry is attributed to energy costs [32].

Table 1 summarizes the main stages in ceramic manufacturing as described in Section 2. It starts with the selection of raw materials, which can be natural (such as clay, kaolin, feldspar, and quartz) or synthetic (such as alumina and silicon carbide), chosen according to their availability, cost, and role in the process.

Table 1. Summary of ceramic processing technology.

Stage / Technique	Description	Purpose	Examples of Materials / Techniques
Raw Material Selection	Choosing suitable natural or synthetic materials for production	Ensure required properties and reduce cost	Natural: clay, kaolin, feldspar, quartz, zircon sands / Synthetic: alumina (Al_2O_3), silicon carbide (SiC)
Raw Material Pre-processing	Crushing, grinding, removing impurities, blending in precise ratios	Achieve homogeneity and prepare for forming	Mechanical milling, sieving, magnetic separation
Conventional Forming	Giving the green body its initial shape while maintaining desired porosity	Obtain a near-net shape before sintering	Uniaxial or isostatic dry pressing, tape casting, slip casting, injection molding, extrusion
Non-conventional Forming	Special methods for complex or high-cost shapes	Provide higher precision or advanced properties	Hot pressing, sol-gel method, self-propagating high-temperature synthesis (SHS), chemical vapor deposition (CVD)
Mechanical Processing After Forming	Adjusting shape and removing defects before firing	Improve dimensions and surface quality	Cutting, drilling, grinding, multilayer lamination
6. Thermal Processing	Drying, binder removal, sintering, and vitrification	Convert the green body into a dense, coherent product	Drying $\leq100°C$, binder removal 200–300°C, sintering (alumina, ferrites, titanates), vitrification for glass phase formation
Glazing	Applying a functional or decorative glassy coating	Enhance wear resistance, waterproofing, and appearance	Low-melting silicate coatings, spraying, dipping
Environmental Impact	Emission of pollutants and generation of solid/liquid waste	Reduce environmental footprint and improve sustainability	Emissions: COx, NOx, SOx, heavy metals; Waste: used plaster molds, sludge, broken pieces

Materials Research Forum LLC
https://doi.org/21741/9781644903834

3. Structure and Properties

Ceramic materials are inorganic since they are not based on carbon. Ceramic materials are also two- or more-atom compounds. Many materials used in ceramics are oxides, including magnesia (MgO), zirconia (ZrO_2), alumina (Al_2O_3), and silica (SiO_2) [33], [34]. In addition, they can be found in a wide range of other compounds, including carbides like SiC, nitrides like Si_3N_4, and silicates like $MgSiO_3$ and $ZrSiO_4$. Materials made from ceramics are hence nonmetallic inorganic compounds, like nitrides, carbides, and oxides [35]. The most flexible category of materials is ceramic materials. This adaptability stems from the chemical bonds, which are mostly composed of strong ionic and covalent bonds in varying ratios. Several certain characteristics of ceramic materials are determined by the bonding, including limited ductility, high hardness and fragilities paired with tenacities, high wear strength, high modulus, and relatively high fusion temperatures. Electrons in ceramic materials interact to form chemical bonds. Due to their lack of conduction electrons, ceramics are good electrical insulators [36]. The silicates comprise the most extensive category of ceramic materials. This comprises clays, micas, and talc, which are regarded as being constructed using silicon tetrahedrons as building blocks. Although predominantly composed of silicon tetrahedrons, unlike these materials, the structure of glass is irregular or random.

This discrepancy causes uneven binding strengths in glass, leading to melting "ranges" as opposed to precise melting points. It is demonstrated that glassy materials can thus develop in a viscous state at high temperatures. Other nonsilicate oxide crystals are formed by using the anion as the point of reference, and they include close-packed oxygens, like body-centered cubic, face-centered cubic, and hexagonal close-packed structures, in which the cations occupy fourfold (tetrahedra) and sixfold (octahedra) interstitial sites. Ferromagnetic and ferroelectric materials have special characteristics that are associated with their closely packed crystal structures [37], [38]. Ceramics are the most complex of all inorganic materials created by humans. Ceramics are typically composed of a glassy phase together with one or more crystalline phases, with pores at the grain boundaries [39]. Aluminosilicate compounds and crystalline residual minerals, primarily quartz, are the crystalline phases in traditional ceramics. Typically, these are mullites with the chemical formulas $Al_4 [Al (Si_2Al) O_{13}]$ or $Al_4O_3 [AlSi_2O_3]_2$. Mullite typically has an acicular habit. Mullite crystals in their protruding shape provide a micro-reinforced framework. The resulting structure can withstand tension stresses more effectively. It is beneficial to the strength indices during stretching and bending. Bending strength is critical for structural ceramics. The improved bending strength consequently increases the frost resistance of the final materials. In the meantime, because the mullite crystals are not densely packed, the final materials' density does not rise. This determines the ceramic material's thermal efficiency. The amorphous phase is formed during the sintering of ceramics and contains glass-forming oxides. Additionally, some clay impurities and metakaolin compounds produced when kaolinite is heated may be found in X-ray amorphous components. The type and content of clay determine the composition of the glass phase that occurs after firing. In the ceramic material, the glass phase's volume could approach 60%. Compression strength is achieved by means of the glass phase. The ceramics exhibit slight elastic deformations due to the presence of the glass phase, which are subsequently followed by brittle fractures. Overfiring causes a significant rise in density and firing shrinkage. This is problematic when producing efficient wall tiles [40]. The structure of ceramics, glasses, or glass-ceramics determines their distinctive characteristics. Ceramics contain a wide range of grain sizes, morphologies, grain boundaries, and phase boundaries, all of which influence their processing and characteristics.

Consequently, the ceramic microstructure has a significant impact on the final ceramic properties, including hardness and density. They have a wide range of adjustable characteristics, comprising but not limited to mechanical strength, electrical conductivity, corrosion resistance, surface functionality (surface energy, biocompatibility, etc.), and optical features [41]. These qualities can be controlled through the control of the raw materials, production methods, and sintering conditions. Mechanical parameters are significant features of ceramic materials. Every ceramic product is subjected to mechanical stress throughout technological procedures such as drying, firing, and actual use. Two of the most significant physical characteristics of ceramic materials are flexural strength and Young's modulus, which are included in theoretical models and computations pertaining to the acceptable loading of ceramic products. They are also essential in determining the maximum sintering rate, together with a coefficient of thermal conductivity and a coefficient of thermal expansion [42]. To maximize the resistance to mechanical abrasion, thermal shock, and chemicals, one of the most crucial physicochemical qualities to comprehend is the pore structure.

Ceramic production typically accepts porosity as a defect that might decrease the quality of the product. The regulation of pore structure provides beneficial features to ceramic products, like a high chemical and mechanical resistance, low thermal conductivity, or a decrease in the ceramic weight [43]. The intrinsic brittleness of ceramic materials restricts their ability to withstand mechanical damage and presents an important limitation to their extensive engineering uses. One way to solve this problem is to include pores in the structure as well. Remarkably, this greatly expands their functions because porous ceramics supply a special set of characteristics. Certain characteristics that are scarcely found in their dense counterparts like large specific surface area, low density, strong thermal shock resistance, high toughness, low dielectric constant, and good thermal insulation ability open up a whole new world of applications in numerous industrial applications [44]. The operational performance of the porous ceramic material is influenced by various parameters associated with its pore structure, including pore shape, size, distribution, and connection. For example, the pores can provide insulation at high temperatures, trap contaminants in a filter, promote tissue growth in a bio-scaffold, or serve as the structural framework for reinforcement in a ceramic-metal composite. These materials can be developed to have the desired characteristics for a specific use by adjusting and improving the processing procedures and controlling the microstructure [45].

4. Classification of Ceramics and Their Applications

Ceramics is a broad field that deals with many different materials. There have been attempts in the past few decades to split the field into two categories: advanced ceramics and traditional ceramics (Figure 2). However, the term advanced has not gained widespread acceptability, and other terms like technical, engineering, fine, and special are frequently used. Materials used since ancient civilizations are closely related to those in traditional ceramics. These include structural clay products, pottery, and clay-based refractories, which we can also include in the category of glasses, concretes, and cement. The ceramics business still largely consists of traditional ceramics, but in recent years, attention has been drawn to advanced ceramics, which are ceramics that have been developed since about the 1950s, with very few exceptions. Advanced ceramics comprise those used for structural applications at both ambient and high temperatures (structural ceramics) and those for magnetic, optical, electrical, and electronic applications (functional ceramics) [46].

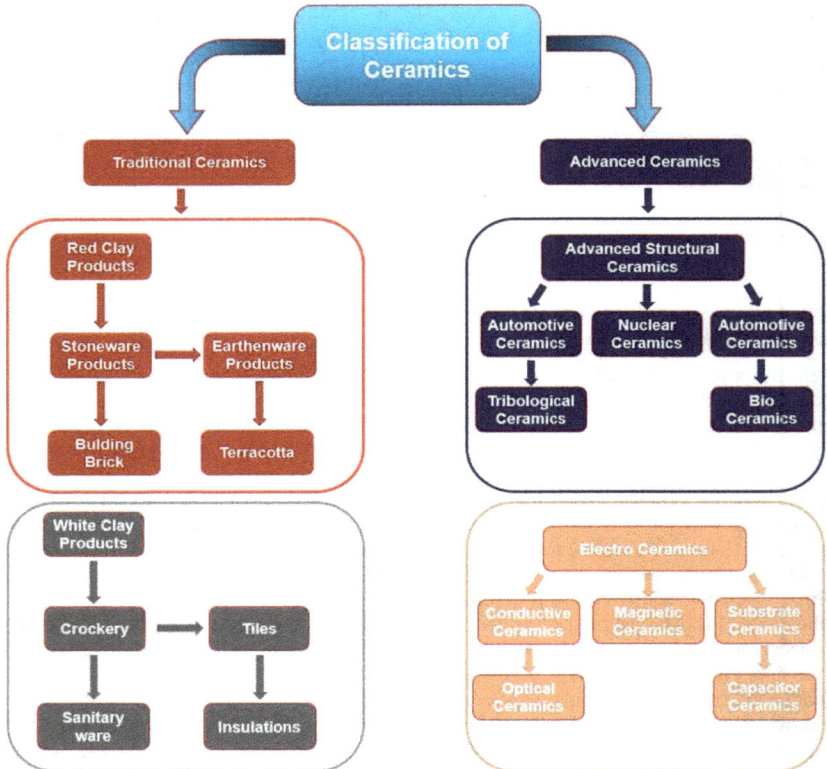

Figure 2. Classification of ceramics [43].

Ceramics encompass a broad category of materials that possess a vast array of applications. Beyond their utilization in building materials like cement, bricks, and wall or floor tiles, they also have applications in a multitude of other industries, including advanced manufacturing, dental, and biomedical applications. Dental ceramics, for instance, are widely recognized for their exceptional aesthetic qualities, which make them ideal for use in dental applications [47]. Ceramics are essential in advanced manufacturing methods such as 3D printing of advanced ceramics for bone replacement and biomedical applications [48]. In addition, ceramics are utilized for manufacturing porous materials for heat insulation and filtration, demonstrating their adaptability and potential for a range of industrial uses [49].

Abrasive ceramic materials play an important role in the grinding, lapping, and polishing of items needing tight dimensional tolerances. Flint, corundum, silicon carbide (carborundum), and diamond dust are examples of abrasives. In order to cut through the substance, the abrasive grains must be sufficiently hard. They are frequently attached to sandpaper or grinding wheels, although they can also be utilized loose or embedded within waxes and pastes [50]. Ceramics, like Si_3N_4, are important structural materials due to their ability to modify properties based on microstructure. Ceramics are utilized in a wide range of fields, such as armor, microelectronics, and micro mechatronics [51], [52].

Additionally, they find applications in cutting-edge materials like Si-B-C-N ceramics, which provide valuable understanding of the correlation between the structure and properties of ceramics [53]. Moreover, ceramics have a vital function in electrical applications, particularly glass ceramics, which exhibit advantageous electrical characteristics [54]. The advancement of strong and durable glass ceramics has given rise to new possibilities in growing industries, emphasizing the ongoing progression and ingenuity in ceramic materials [55]. Because ceramic composite materials can withstand high temperatures and harsh environmental conditions, they have been used as thermal barriers in the aviation industry [56].

The thermal barrier coating (TBC) system is made up of multiple layers, including a ceramic topcoat (TC), a metallic bond coat (BC), a superalloy substrate, and a thermally grown oxide (TGO) layer (Figure 3). The fundamental function of TC is to lower the surface temperature of the superalloy substrate. The two primary methods used to manufacture the TC are atmospheric plasma spraying (APS) and electron beam physical vapor deposition (EB-PVD). Yttria-stabilized zirconia (YSZ) has been widely used in the fabrication of TC layers because of its exceptional thermal compatibility, high fracture toughness, low thermal conductivity, and high thermal shock resistance [57]. Alternative ceramic materials (e.g., $Gd_2Zr_2O_7$, $Sm_2Zr_2O_7$, $Nd_2Zr_2O_7$, and $La_2Zr_2O_7$, rare earth zirconates with pyrochlore structure or fluorite-type structure) can be used in TBC production [58]. The application of ceramic coatings on engine components has been researched to improve engine efficiency, reduce pollutants, and optimize combustion characteristics [59]. The effects of applying ceramic coatings on engine performance and emission characteristics have also been studied to produce cleaner production [60].

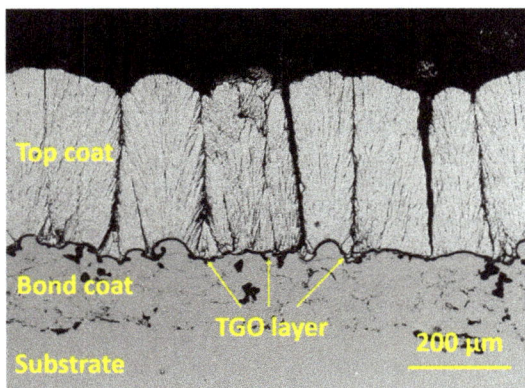

Figure 3. Cross-sectional SEM image of YSZ TBC system deposited by EB-PVD. The TGO layer is formed during use at high temperatures [61], reprinted with permission from MDPI.

Magnetic ceramic materials (such as lead ferrite) contain ferromagnetic hysteresis loops. A hysteresis loop found in certain magnetic ceramics is extremely like the square loop that is usually preferred for use in electronic computer memory circuits. In rocket and missile development, nitride ceramics are used in the rocket throat and nose cones due to their good erosion resistance and resistance to severe temperatures. Another use of ceramics is in the nuclear energy sector. Uranium ceramics are now an essential component of nuclear reactor technology [62]. The development of high-entropy ceramics for potential application as core materials in nuclear reactors requires a balance between irradiation resistance and thermal conductivity.

Table 2 summarizes the main classifications of ceramics and their applications as described in Section 4. It illustrates the difference between traditional ceramics used since ancient times and advanced ceramics that have emerged since the mid-20th century, highlighting areas of use such as structural, medical, and military applications. The table shows typical materials for each category and their role in industries ranging from construction to aerospace and nuclear energy.

Table 2. *Classification of ceramics and their applications.*

Category	Description	Example Applications	Example Materials
Traditional Ceramics	Materials used since ancient times; mainly clay-based	Bricks, tiles, pottery, cement, refractories	Clay, kaolin, feldspar, quartz
Advanced Ceramics	Developed mainly since the 1950s; higher performance	Structural: armor, microelectronics, aerospace; Functional: magnetic, optical, electrical devices	Si_3N_4, Si-B-C-N, glass ceramics
Dental & Biomedical Ceramics	High aesthetics, biocompatibility	Dental restorations, bone replacement (3D printing)	Dental porcelain, bio-ceramics
Porous Ceramics	Lightweight, high thermal insulation, filtration	Heat insulators, industrial filters	Alumina foams, silica-based materials
Abrasive Ceramics	Extremely hard, for shaping and finishing	Grinding, polishing, cutting	Flint, corundum, SiC, diamond dust
Thermal Barrier Coating Ceramics (TBCs)	Multi-layer coatings for heat protection in engines/aerospace	Protect turbine blades, improve efficiency	YSZ, $Gd_2Zr_2O_7$, $Sm_2Zr_2O_7$, rare-earth zirconates
Magnetic Ceramics	Exhibit ferromagnetic hysteresis	Memory circuits, electronics	Lead ferrite, ferrites
Nitride Ceramics	High erosion & temperature resistance	Rocket nozzles, nose cones	Si_3N_4, BN
Nuclear Ceramics	Withstand radiation, high temperatures	Nuclear reactor cores, fuel elements	Uranium ceramics, high-entropy ceramics

5. Conclusion

Difficulties, including poor mechanical properties of ceramic-matrix self-lubricating composites and high friction coefficients, have restricted the use of ceramics in tribological fields. The development of high-performance, self-lubricating ceramic composites with laminated-graded structures, on the other hand, shows promise in overcoming these restrictions. Additionally, the electrical discharge machining (EDM) of engineered ceramics has been impeded by their hard and brittle nature, which makes typical processing methods challenging. Nevertheless, EDM has been discovered as a frequently utilized nonconventional material removal procedure for machining engineered ceramics, provided they are electrically conductive. The advancement of novel ceramic materials necessitates an integrated examination of crystal structure and electrophysical characteristics, underscoring the multidisciplinary essence of ceramic research and the requirement for a thorough comprehension of these materials. Additionally, ceramics have been utilized in innovative composite materials, including ceramic coatings, which are derived from conventional ceramic substances, illustrating the versatility of ceramics in advancing technology. The life cycle sustainability assessment (LCSA) has been used to benchmark the environmental performance of the ceramic industry, demonstrating the growing interest. The potential for sustainable practices within the industry has been highlighted by the exploration of the use of ceramic brick waste as a raw material for the manufacturing of restoration in the sustainability of ceramic production. This demonstrates how promoting environmental sustainability is becoming increasingly crucial to the ceramics industry. Additionally, research into using discarded ceramic bricks as a raw material for restoration material production has brought attention to the industry's potential for sustainable practice. Research into sustainable and environmentally friendly sources of ceramic materials is still going on, as evidenced by the use of alternative materials such as rice husk ash (RHA) as a silica precursor in ceramic manufacturing. Additionally, ceramics' special qualities, such as their capacity to insulate electrical current, make them appropriate for uses, like in silicon nitride-based carbon-ceramic composites, where they demonstrate remarkable refractoriness and durability.

In conclusion, ceramics encompass a broad spectrum of materials with diverse applications, ranging from traditional construction materials to advanced biomedical and manufacturing technologies. Their distinctive characteristics and capacity for customization according to microstructure render them essential in contemporary technical and medical progress. Nanotechnology can enhance the characteristics of ceramic materials. Nanotechnological applications can enhance material durability, optimize surface qualities, and facilitate more efficient utilization across diverse sectors. Current research and development in this domain persist in broadening the possible uses of ceramics, establishing them as a crucial focus in materials science and engineering.

References

[1] Bergstrom, L. (2001). Colloidal processing of ceramics. Handbook of applied surface and colloid chemistry, 1, 201-217.

[2] Niu, C. and Zhang, M. (2022). Using image feature extraction to identification of ancient ceramics based on partial differential equation. Advances in Mathematical Physics, 2022, 1-12. https://doi.org/10.1155/2022/3276776

[3] Boccaccini, A. R., Singh, M., Marra, J., Dogan, F., Lin, H. T., & Watanabe, T. (Eds.). (2010). Ceramics for environmental and energy applications. Ceramic Transactions Volume: 217. John Wiley & Sons. https://doi.org/10.1002/9780470909874

[4] Kolb, C.C. (2022). Ceramics. In: Gilbert, A.S., Goldberg, P., Mandel, R.D., Aldeias, V. (eds) Encyclopedia of Geoarchaeology. Encyclopedia of Earth Sciences Series. Springer, Cham. https://doi.org/10.1007/978-3-030-44600-0_32-1

[5] Hennicke, H. W., & Hesse, A. (1991). Traditional Ceramics. In Concise Encyclopedia of Advanced Ceramic Materials (pp. 488-494). Pergamon. https://doi.org/10.1016/B978-0-08-034720-2.50135-0

[6] Warlimont, H. (2018). Ceramics. In: Warlimont, H., Martienssen, W. (eds) Springer Handbook of Materials Data. Springer Handbooks. Springer, Cham. https://doi.org/10.1007/978-3-319-69743-7_17

[7] Nastro, V., Vuono, D., Guzzo, M., Niceforo, G., Bruno, I., & De Luca, P. (2006). Characterisation of raw materials for production of ceramics. Journal of Thermal Analysis and Calorimetry, 84(1), 181–184. https://doi.org/10.1007/s10973-005-7206-6

[8] Kingery W.D., Kent Bowen H., Uhlmann D.R., Introduction to Ceramics, 2. Edition, 1976, John Wiley & Sons

[9] Babu, P. J., Alla, R. K., Alluri, V. R., Datla, S. R., & Konakanchi, A. (2015). Dental ceramics: Part I–An overview of composition, structure and properties. Am J Mater Eng Technol, 3(1), 13-18. DOI:10.12691/materials-3-1-3

[10] ASTM (American Society for Testing and Materials), 2015. ASTM C242-15, Standard Terminology of Ceramic Whitewares and Related Products. West Conshohocken: ASTM International. www.astm.org.

[11] Carty, W. M., & Senapati, U. (1998). Porcelain—raw materials, processing, phase evolution, and mechanical behavior. Journal of the American Ceramic Society, 81(1), 3-20. https://doi.org/10.1111/j.1151-2916.1998.tb02290.x

[12] Mooney, J. F. (1996). Ceramics & glass. In Industrial Minerals and Their Uses (pp. 459-481). William Andrew Publishing.

[13] Colomban, P. (2021). Glazes and Enamels. Encyclopedia of Glass Science, Technology, History, and Culture, 1309–1325. doi:10.1002/9781118801017.ch10.6

[14] Pradell, T., & Molera, J. (2020). Ceramic technology. How to characterise ceramic glazes. Archaeological and Anthropological Sciences, 12(8), 189. https://doi.org/10.1007/s12520-020-01136-9

[15] Bachar, A., Mabrouk, A., Meneses, D. D. S., Veron, E., Sadallah, Y., & Echegut, P. (2018). Study of the firing type on the microstructure and color aspect of ceramic enamels. Journal of Alloys and Compounds, 735, 2479-2485. https://doi.org/10.1016/j.jallcom.2017.11.364

[16] Holand, W., & Beall, G. H. (2019). Glass-ceramic technology 3rd Edition. John Wiley & Sons. DOI:10.1002/9781119423737

[17] Eke, H. O. (2007). Development of ceramics technology education: its roles, problems and solutions. A Journal of Knowledge Review, 15(3), 39-44.

[18] Singh, K., Kaur, M., & Kumar, A. (2021). Progress in Advanced Ceramics: Energy and Environmental Perspective. Advanced Ceramics for Energy and Environmental Applications; Kumar, A., Ed, 1-12.

[19] Lewis, G. (1991). Applications for traditional ceramics. In: Engineered Materials Handbook Ceramics and Glasses. ASM International, Materials Park, OH, Vol. 4, Sect. 12, pp. 893–4

[20] Schoenung, J. M. (2001). Structural Ceramics. Encyclopedia of Materials: Science and Technology, 8921–8926. doi:10.1016/b0-08-043152-6/01605-3

[21] Boccaccini, A. R. (2005). Ceramics. In Biomaterials, artificial organs and tissue engineering (pp. 26-36). Woodhead Publishing.

[22] Buchanan, R. C. [1991] Electrical/electronic applications for advanced ceramics. In: Engineered Materials Handbook Ceramics and Glasses. ASM International, Materials Park, OH, Vol.4, Sect. 15, pp. 1105–6

[23] Ribeiro, M.J., Tulyaganov, D. (2021). Traditional Ceramics Manufacturing. In: Baino, F., Tomalino, M., Tulyaganov, D. (eds) Ceramics, Glass and Glass-Ceramics. PoliTO Springer Series. Springer, Cham. https://doi.org/10.1007/978-3-030-85776-9_3

[24] Dias, F. G., Segadães, A. M., Perottoni, C. A., & Cruz, R. C. (2017). Assessment of the fluxing potential of igneous rocks in the traditional ceramics industry. Ceramics International, 43(18), 16149-16158. https://doi.org/10.1016/j.ceramint.2017.08.190ç

[25] Zinkle, S. J., & Kinoshita, C. (1997). Defect production in ceramics. Journal of Nuclear Materials, 251, 200-217.

[26] [26] Wang, S. F., Zhang, J., Luo, D. W., Gu, F., Tang, D. Y., Dong, Z. L., ... & Kong, L. B. (2013). Transparent ceramics: Processing, materials and applications. *Progress in solid state chemistry*, *41*(1-2), 20-54

[27] Reed, J. S. (1995). Principles of ceramics processing. 2nd Edition. John Wiley & Sons, Inc. (US).

[28] Xiao, P. (2023). "Frontiers in ceramics" grand challenges. Frontiers in Ceramics, 1, 1137377.

[29] Maitra, S. (2013). A Brief Description of the Processing of Ceramics. Published by: Indian Institute Of Ceramics, 22(1), 16.

[30] Eliche-Quesada, D., Pérez-Villarejo, L., & Sánchez-Soto, P. J. (2019). Introductory Chapter: Ceramic Materials-Synthesis, Characterization, Applications and Recycling. Ceramic Materials-Synthesis, Characterization, Applications and Recycling, 1.

[31] Timellini, G. et al. (2016). Ceramic Industry Air Quality. Emissions Into the Atmosphere From Ceramic Tile Processes, pp. 707–729.

[32] Ewsuk, K. G. (1999). Ceramic processing (No. SAND99-3035J). Sandia National Lab.(SNL-NM), Albuquerque, NM (United States); Sandia National Lab.(SNL-CA), Livermore, CA (United States).

[33] Smith, W. F., & Hashemi, J. (2019). Ceramics, In Foundations of materials science and engineering. (pp. 615). Mcgraw-Hill Publishing.

[34] Montero, M. A., Jordán, M. M., Hernández-Crespo, M. S., & Sanfeliu, T. (2009). The use of sewage sludge and marble residues in the manufacture of ceramic tile bodies. Applied Clay Science, 46(4), 404-408. https://doi.org/10.1016/j.clay.2009.10.013

[35] Jordán, M. M., Boix, A., Sanfeliu, T., & De la Fuente, C. (1999). Firing transformations of cretaceous clays used in the manufacturing of ceramic tiles. Applied Clay Science, 14(4), 225-234. https://doi.org/10.1016/S0169-1317(98)00052-0

[36] Venturelli, C., & Paganelli, M. (2014). Study of Residual Stresses in Traditional Ceramics. Key Engineering Materials, 608, 14–20. https://doi.org/10.4028/www.scientific.net/KEM.608.14

[37] García-González, J., Rodríguez-Robles, D., Juan-Valdés, A., Morán-del Pozo, J. M., & Guerra-Romero, M. I. (2015). Ceramic ware waste as coarse aggregate for structural concrete production. Environmental technology, 36(23), 3050-3059. https://doi.org/10.1080/09593330.2014.951076

[38] Hader, M. A., Darabseh, T. T., & AlOthman, H. A. (2011). Exergy Analysis of Ceramic Production in Jordan. Jordan Journal of Mechanical & Industrial Engineering, 5(6).

[39] Farid, S. B. (2018). Overview. Bioceramics: for materials science and engineering. Woodhead Publishing. Page 1.

[40] Phillips, G. (2012). A concise introduction to ceramics. Springer Science & Business Media. https://doi.org/10.1007/978-94-011-6973-8

[41] Bakunov, V.S., Belyakov, A.V. (2000). Creep and Structure of Ceramics. Inorganic Materials 36, 1297–1301. https://doi.org/10.1023/A:1026658404494

[42] Vladimirov, K. D., Saraeva, P. I., & Chernykh, T. N. (2019, December). Influence of process parameters of firing clay materials on phase composition and structure of ceramics. In IOP Conference Series: Materials Science and Engineering (Vol. 687, No. 2, p. 022001). IOP Publishing. DOI 10.1088/1757-899X/687/2/022001

[43] Fraser, R., & Girtan, M. (2023). A Selective Review of Ceramic, Glass and Glass–Ceramic Protective Coatings: General Properties and Specific Characteristics for Solar Cell Applications. Materials, 16(11), 3906. https://doi.org/10.3390/ma16113906

[44] Youngovim, O. M. M. T. I., & Keramiki, M. P. T. (2011). Relationship between mechanical strength and Young's modulus in traditional ceramics. Mater. Tehnol, 45, 375-378.

[45] Coulon, A., Cohen, M., & Pillet, G. (2023). Light-weighing traditional ceramics by porosity control and consequences on mechanical strength. Ceramics International. https://doi.org/10.1016/j.ceramint.2023.11.269

[46] Chen, Y., Wang, N., Ola, O., Xia, Y., & Zhu, Y. (2021). Porous ceramics: Light in weight but heavy in energy and environment technologies. Materials Science and Engineering: R: Reports, 143, 100589. https://doi.org/10.1016/j.mser.2020.100589

[47] Hammel, E. C., Ighodaro, O. R., & Okoli, O. I. (2014). Processing and properties of advanced porous ceramics: An application based review. Ceramics International, 40(10), 15351-15370. https://doi.org/10.1016/j.ceramint.2014.06.095

[48] Gilboa, A., Karasik, A., Sharon, I., & Smilansky, U. (2004). Towards computerized typology and classification of ceramics. *Journal of Archaeological Science*, *31*(6), 681-694. https://doi.org/10.1016/j.jas.2003.10.013

[49] Kurian, M., & Thankachan, S. (2023). Introduction: Ceramics classification and applications. In *Ceramic catalysts* (pp. 1-17). Elsevier. https://doi.org/10.1016/B978-0-323-85746-8.00009-6

[50] Rahaman, M. N. (2017). Ceramic processing. 2nd Edition. CRC press.

[51] Vasiliu, R. and Porojan, L. (2019). The influence of oral environment on the optical properties of heat- pressed ceramics. Materiale Plastice, 56(1), 271-276. https://doi.org/10.37358/mp.19.1.5164

[52] Withell, A., Diegel, O., Grupp, I., & Reay, S. (2011). Porous ceramic filters through 3d printing., 313-318. https://doi.org/10.1201/b11341-50

[53] Korjakins, A., Radina, L., & Bajare, D. (2015). Fabrication of porous ceramics as clay/glass composite. Environment Technology Resources Proceedings of the International Scientific and Practical Conference, 1, 72. https://doi.org/10.17770/etr2015vol1.201

[54] Fischer, T. (2009). Materials science for engineering students. Elsevier, Burlington, MA, pp. 17-23.

[55] Walker, L., Marotto, V., Rafiee, M., Koratkar, N., & Corral, E. (2011). Toughening in graphene ceramic composites. Acs Nano, 5(4), 3182-3190. https://doi.org/10.1021/nn200319d

[56] Chen, W., Rajendran, A., Song, B., & Nie, X. (2007). Dynamic fracture of ceramics in armor applications. Journal of the American Ceramic Society, 90(4), 1005-1018. https://doi.org/10.1111/j.1551-2916.2007.01515.x

[57] Bochenek, D., Niemiec, P., Adamczyk, M., Skulski, R., Zachariasz, R., Wodecka-Duś, B., … & Machnik, Z. (2017). The multicomponent pzt-type ceramics for micromechatronic applications. Archives of Metallurgy and Materials, 62(2), 667-672. https://doi.org/10.1515/amm-2017-0099

[58] Zhang, P., Jia, D., Yang, Z., Duan, X., & Zhou, Y. (2012). Progress of a novel non-oxide si-b-c-n ceramic and its matrix composites. Journal of Advanced Ceramics, 1(3), 157-178. https://doi.org/10.1007/s40145-012-0017-x

[59] Montedo, O., Alves, I., Faller, C., Bertan, F., Piva, D., & Piva, R. (2015). Evaluation of electrical properties of glass-ceramics obtained from mill scale. Materials Research Bulletin, 72, 90-97. https://doi.org/10.1016/j.materresbull.2015.07.040

[60] Fu, Q., Beall, G., Smith, C., Kohli, J., Youngman, R., Wheaton, B., … & Gulbiten, O. (2016). Strong, tough glass-ceramics for emerging markets. International Journal of Applied Glass Science, 7(4), 486-491. https://doi.org/10.1111/ijag.12247

[61] Vasile, B., Bîrcă, A., Surdu, V., Neacsu, I., & Nicoara, A. (2020). Ceramic composite materials obtained by electron-beam physical vapor deposition used as thermal barriers in the aerospace industry. Nanomaterials, 10(2), 370. https://doi.org/10.3390/nano10020370

[62] Karaoglanli, A. C., Ozgurluk, Y., Gulec, A., Ozkan, D., & Binal, G. (2023). Effect of coating degradation on the hot corrosion behavior of yttria-stabilized zirconia (YSZ) and blast furnace slag (BFS) coatings. Surface and Coatings Technology, 473, 130000. https://doi.org/10.1016/j.surfcoat.2023.130000

[63] Ozgurluk, Y., Doleker, K. M., & Karaoglanli, A. C. (2018). Hot corrosion behavior of YSZ, Gd2Zr2O7 and YSZ/Gd2Zr2O7 thermal barrier coatings exposed to molten sulfate and vanadate salt. Applied Surface Science, 438, 96-113. https://doi.org/10.1016/j.apsusc.2017.09.047

[64] Anandavelu, K. and Mahabubadsha, A. (2022). Review of ceramics coating on the engine components. Revista De Chimie, 73(3), 38-44. https://doi.org/10.37358/rc.22.3.8533

[65] Odabaş, O. (2020). Investigation of the effect of calcium-alumina-magnesium-silicate (CMAS) infiltration on the microstructural characteristics of thermal barrier coatings (TBCs). [Master's thesis, Bartın University].

[66] Abbas, S. and Elayaperumal, A. (2019). Experimental investigation on the effect of ceramic coating on engine performance and emission characteristics for cleaner production. Journal of Cleaner Production, 214, 506-513. https://doi.org/10.1016/j.jclepro.2018.12.040

[67] Ossai, A. A. (2001). Ceramic technology-Its role in other technologies. The Nigerian Academic Forum Vol. 1 No. 2.

[68] Xiang, H., Xing, Y., Dai, F., Wang, H., Su, L., Liu, M., … & Zhou, Y. (2021). High-entropy ceramics: present status, challenges, and a look forward. Journal of Advanced Ceramics, 10(3), 385-441. https://doi.org/10.1007/s40145-021-0477-y

[69] Zhang, Y., Su, Y., Fang, Y., Qi, Y., & Hu, L. (2016). High-performance self-lubricating ceramic composites with laminated-graded structure. https://doi.org/10.5772/62538

[70] Li, X., Xiong, J., Yuan, X., & Jiu-chen, F. (2011). Mechanism research on the electrical discharge wed machining of engineering ceramics materials.. https://doi.org/10.1109/mec.2011.6025988

[71] Ji, R., Y, L., Diao, R., Zhang, Y., Wang, F., Cai, B., ... & Xu, C. (2014). Experimental research on electrical discharge machining characteristics of engineering ceramics with different electrical resistivities. The International Journal of Advanced Manufacturing Technology, 75(9-12), 1743-1750. https://doi.org/10.1007/s00170-014-6258-9

[72] Duarte, A., Silva, E., Melo, E., Pereira, F., & Sombra, A. (2020). Characterization of the bfo-bzn-based electroceramic composite for application as dra. Cerâmica, 66(377), 1-6. https://doi.org/10.1590/0366-69132020663772796

[73] Zhang, Y., Qi, B., & Sun, X. (2017). The application of ceramic technology in spherical pipe joints. Proceedings of the Institution of Mechanical Engineers Part J Journal of Engineering Tribology, 231(8), 1078-1088. https://doi.org/10.1177/1350650117692660

[74] Ferrari, A., Volpi, L., Pini, M., Siligardi, C., Muiña, F., & Settembre-Blundo, D. (2019). Building a sustainability benchmarking framework of ceramic tiles based on life cycle sustainability assessment (lcsa). Resources, 8(1), 11. https://doi.org/10.3390/resources8010011

[75] Kamilovich, B. (2020). Waste from ceramic bricks, as a raw material for the production of restoration materials. International Journal of Emerging Trends in Engineering Research, 8(8), 4390-4393. https://doi.org/10.30534/ijeter/2020/56882020

[76] Milak, P., Souza, M., Bom, C., Mantas, P., Raupp-Pereira, F., & Oliveira, A. (2017). Valorization of rice husk ash in ceramic bricks. Cerâmica, 63(368), 490-493. https://doi.org/10.1590/0366-69132017633682137

[77] Fényi, B., Hegman, N., Wéber, F., Arató, P., & Balázsi, C. (2007). Dc conductivity of silicon nitride based carbon-ceramic composites. Processing and Application of Ceramics, 1(1-2), 57-61. https://doi.org/10.2298/pac0702057f

Chapter 2

History of Ceramics: Ancient Ceramics of Archaeometric Characterization

Cumhur Eren Isık[1*], Hussein Elaibi[2], Farah Mutlag[2,3], Irem Turk[2],
Iskender Isık[4], Fatih Sen[2*]

[1]Department of Handicrafts, Kutahya Fine Arts Vocational School, Kutahya Dumlupınar University, Evliya Çelebi Campus, 43000 Kutahya, Türkiye

[2]Sen Research Group, Department of Biochemistry, Kutahya Dumlupinar University, Kutahya 43000, Türkiye

[3]Ministry of Education, Karbala Education Directorate, Karbala, 56001, Iraq

[4]Department of Materials Science & Engineering, Faculty of Engineering, Kutahya Dumlupınar University, Evliya Çelebi Campus, 43100 Kutahya, Türkiye

eren84tr@gmail.com, fatihsen1980@gmail.com

Abstract

As is known, ceramic materials have been used to meet different needs in people's daily lives since the middle of the Neolithic Period. In this respect, ceramic findings will provide guiding information about the lifestyles and production technologies of civilizations. For this purpose, in addition to the classifications made for ceramics according to their form, design, and ornamentation features, ceramics can be characterized and classified in terms of the chemical and mineralogical contents through the advanced analysis methods used in ceramic archaeometry. Thanks to such studies, the production characteristics and origins of ceramics can be determined, and detailed information can be obtained regarding the social, cultural, economic, and artistic aspects of the period and societies to which the findings belong. This study presents a brief guide to be used as a draft during ceramic archaeometry studies. The working principles in ceramic archaeometry and the characterization techniques for ancient ceramics are presented in the form of sequential process steps to be followed. In particular, the determination of chemical composition, mineralogical content, microstructural-microchemical properties, and molecular structure is specified together with the analysis techniques, which are briefly discussed. Finally, the technical aspects of ceramic archaeometry are discussed in detail.

Keywords

Ancient Ceramics, Archaeometry, Ceramics Characterization

1. Introduction

The world has hosted many civilizations and, therefore, cultures throughout history. Accordingly, besides its natural heritage, a very rich accumulation has been formed in terms of cultural heritage. One of the regions where cultural richness is strongly felt is undoubtedly the Anatolian geography. Located in the north of Mesopotamia, which is mentioned as the cradle of civilizations, Anatolia has taken on the role of a natural bridge between the Asian and European continents, which differ in many ways. Having different climates and vegetation diversity, the Anatolian geography has been the scene of different lifestyles and cultures throughout history. Besides being a magnificent source containing thousands of years of information, this situation, which resulted in a rich cultural diversity, has made Anatolia a focal point for centuries, especially in the studies of historians and archaeologists. As a result of the archaeological excavations, it has been observed that the prehistoric civilizations in Anatolia have left numerous movable works (i.e., metal, ceramic, glass, stone, textile, etc.) together with architectural structures that reflect their lifestyles, socio-cultural structures, and art [1].

Requiring great care, time, and labor, archaeological excavations are carried out in many different parts of the world. The material group that predominated in these excavations was generally ceramics (beginning from the Neolithic Age with ceramics). The term "ceramic", as a material group, includes terracotta products such as brick, tile, glass, porcelain, and pottery. For this reason, the most comprehensive and short definition for ceramics would be "terracotta" (fired soil). It is known that the word "ceramic" is derived from the Greek word "Keramos", meaning "burnt material". Therefore, the definition of "terracotta" and the word "keramos" seem to be compatible with each other. In addition to this short definition, ceramics can be defined more comprehensively as the material obtained by shaping the compositions of inorganic materials after they gain a certain plasticity, subsequently drying and firing them until they reach sufficient hardness and strength. As can be seen, the production of ceramics consists of four main steps: i. raw material (recipe), ii: shaping, iii: drying, iv: firing. These four stages may also include additional processes within and/or between them (e.g., purification of raw materials, enrichment, pre-drying, application of slip/glaze, etc.). In terms of raw materials, the three raw materials that make up ceramics are clay, quartz, and feldspars. Clays are one of the indispensable raw materials for ceramics to take permanent shape, since they are naturally formed, consist of fine-grained minerals, become plastic when sufficient water is added, and retain their shape by drying or firing. Quartz and feldspar, as non-plastic raw materials, are effective in determining the physical properties of ceramics and the formation of vitrification, respectively [2], [3].

Ceramic materials have been used by humans for many different purposes in different periods. Especially, daily use items (e.g. kitchen and storage ware) are the most needed and, accordingly, the most useful materials in prehistoric times. In this context, it is seen that most of the items, such as cooking and kitchen wares, were ceramic. Owing to its soil-based character, ceramic products are permanent after use (non-disposable) and suffer little damage over time. As a fired material, ceramic can withstand almost all geological, biological, environmental, and climatic effects and thus maintains its existence for centuries and reflects the period in which it was manufactured. Therefore, ceramic finds are one of the most significant materials in archeology. Metal products, sculptures, and architectural structures have been the main artifacts that have survived until today and reflect the cultural, social, economic, and artistic aspects of the civilizations they belong to, likewise ceramics. Each of the material groups and architectural/artistic works belonging to different periods in history appears as cultural heritage and sheds light on history with its characteristic features [4], [5].

Ceramics, which started with the Neolithic Period pottery and have continuously developed in parallel with the lifestyle of human beings from the beginning of the ceramic period in Anatolia to the present, have constantly renewed and developed themselves. According to today's archaeological data, the Neolithic Age, dated to approximately between 10000 and 6000 BC, is one of the most substantial periods in human history. The Neolithic Age is the beginning of significant innovations for humanity, and this period is also an indication of the transition to settled life and the foundation of human life today. In the Neolithic Age, which is also called the "Polished Stone Age" or "New Stone Age", the lifestyles of the societies living in Anatolia began to show differences (for instance, the transition to agriculture after the improvements in climatic conditions). Therefore, in this age, it is seen that human beings have begun to possess a productive identity in a way. People who started to move to a living system that can overlap with today's village life took a step towards settled life in this age, and on the other hand, they took a step towards development with their social, economic, and cultural characteristics. One of the important developments in this context was the making of earthenware that gains strength by firing. The existence of terra cotta objects in Anatolia could be understood from finds dating back thousands of years. The earliest clay pots in Anatolia were found in the 7th millennium BC in Konya, Burdur, and Antalya regions. The earliest examples were monochrome, rough, and in simple forms. Afterwards, the later works in Çatalhöyük and Hacılar in the middle of the 6th millennium BC were more successful and became one of the first works of art produced by humanity in the field of ceramics. The multi-colored potteries in that period are truly eye-catching in their form and pattern [6–11].

2. Archaeometry

A general definition can be made for archaeometry as the characterization of historically valuable artifacts and materials with analytical techniques. The material groups examined may comprise different artifacts/products such as pottery, metal items, glass, skeleton, brick, tile, porcelain, mortar, plaster, rock paintings, and manuscripts which were recovered or unearthed during the archaeological excavations, salvage excavations, and/or conservation-restoration works. It can be deduced that the beginning of archaeometry dates to the beginning of the 19th century. The chemical analyses performed by MH Klaproth in the 1800s on some coins, glass, and Medieval sculptures can be shown among the first archaeometric studies [12,13]. Klaproth's analyses were followed by the studies of many researchers such as HH Coghlan, F Rathgen, JR Partington, and CH Desdh. With the projects carried out in this field, a great interest in archaeometry began to emerge [14,15]. The history of the ceramic period is illustrated in Figure 1.

In 1878, Baron De Geer investigated the remains of perennial plants in the lake and swamp sediments in Sweden and developed an absolute dating method called "Varv analysis," which is based on counting the sludge layers. In the 1920s, Yugoslav mathematician and astronomer Milutin Milankovitz had the assumption that spots in the solar system caused climate changes in the world, and revealed that the Ice Ages can be dated back to 600000 years through the mathematical calculation of these changes [16,17].

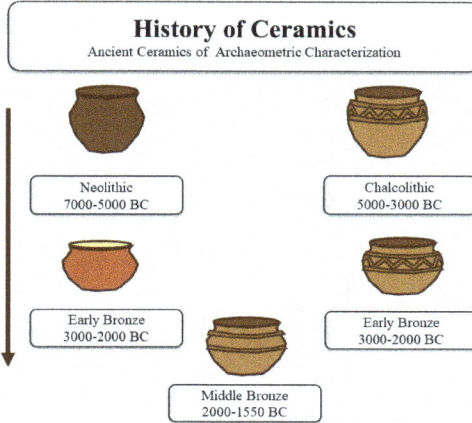

Figure 1. The history of the ceramic period.

In this context, one of the important steps has been the book called "Radiocarbon Dating", which offers a new absolute dating method to archeology by measuring the radioactive carbon 14 (C-14) found in organic materials that have expired [18]. This book had a great repercussion in archeology, because the fact that the ancient materials unearthed during the archaeological excavations carried out for many years could be dated with a new technique provided a substantial advantage. Material groups, which were previously considered only unilaterally, have now begun to be analyzed and interpreted from many different aspects, thanks to the multidisciplinary approach of archaeometry.

An interdisciplinary work plan forms the basis of an archaeometry investigation. In this type of research, which can be carried out by combining different fields of science, work teams are formed according to the characteristics of the material group examined. For instance, biologists, chemists, and paleontologists may take part in archaeometry research together with the excavation head in the examination of skull, fossil, and/or bone samples unearthed in an archaeological excavation. Civil engineers, architects, art historians, and restorers can work together in archaeometry studies, which aim to restore a historical architectural structure and/or to carry out other conservation-restoration applications. The working groups mentioned in these examples should also include archaeometrists who have made archaeometry their field of expertise (it may differ according to the type of material). In this context, a multidisciplinary approach covering materials, ceramics, mining and geological engineering, physics, chemistry, archeology, conservation-restoration of cultural properties, and art history would be appropriate in archaeometry investigations of ceramic finds. According to the type of find, scope of work, and the questions to be answered, experienced researchers from these disciplines will make important contributions to the examination of ceramics. In general, the term "ceramic

23

archaeometry" can be used for this type of work involving the characterization of ancient ceramics. Section 2 of Table 1 summarizes and illustrates the timeline of the development of archaeology.

Table 1. Timeline of Archaeometry Development.

Period/Year	Key Event/Development	Contributors
Early 19th Century	First chemical analyses of artifacts (coins, glass, sculptures).	MH Klaproth
1878	Development of "Varve Analysis" for absolute dating by counting sediment layers.	Baron De Geer
1920s	Theory linking climate changes to astronomical cycles, dating Ice Ages back to 600,000 years.	Milutin Milankovitch
Mid-20th Century	Introduction of Radiocarbon Dating (C-14) for organic materials revolutionized archaeology.	"Radiocarbon Dating" Book
Modern Era	Expansion into multidisciplinary research (chemistry, physics, geology, engineering, archaeology).	Collaborative research teams.

3. Working Principles in Ceramic Archaeometry

A predetermined work plan is required in the archaeometry examination of ancient ceramics. For that purpose, the current situation should be revealed, and the questions to be answered should be determined by considering the archaeological background of the excavation areas where the ceramics were unearthed, and the periods they belong to. In a characterization study of many ceramic finds, representative samples should be first selected, and the ceramics should be recorded through a documentation process. Then, the ceramics should be purified, prepared by analysis, and analyzed with the appropriate methods. At the last stage, the obtained data should be processed, interpreted, and recorded. The main lines of these steps, which reflect the working principle of ceramic archaeometry, can be expressed as follows [19–22].

3.1 Selection of representative samples

Budget, time, and labor are the most important parameters in the selection of representative samples from many ceramic finds. Budget and time can be used optimally by selecting representative samples that will reflect the ceramic groups among hundreds of ceramics that repeat themselves. In the archaeological excavation areas, ceramics belonging to different periods or cultures can be found. At this point, representative examples should be determined in line with the scope and priorities of the study, which would direct the research. The decisive factors in sample selection can be the features such as color, form, and type of part represented. It would be beneficial to make the selection of the representative samples together with the head of the excavation and/or an expert dealing with ancient ceramics [23–25].

3.2 Documentation

In the documentation process, macro definitions are made by taking photographs of ceramics. Some of the ceramics, particularly the ones in small sizes, could be completely ground to use in analyses requiring powders. Therefore, the documentation of ceramics is very important. During the documentation process, photographs of the ceramics are suggested to be taken so as to demonstrate the front, back, and profile. The color of the background to be used in photographing should be determined according to the hues of the ceramics. After creating an environment that would minimize the shadow formation, the ceramics should be photographed using a scale. Apart from photographing, macro definitions of ceramics are noted. For this purpose, the color and texture of the ceramics, as well as the slip, glaze, and decorations (if any), could be defined. Sample codes should be given to the ceramics in addition to their archaeological features (e.g., locus number, the period they belong to, date of the excavation, etc.) [23,26,27].

3.3 Purification

The purification process is of great importance, especially in preventing errors that may arise from impurities. Undesirable impurities can accumulate in ceramics that have been waiting underground for hundreds or thousands of years. Due to the porous structure of the earthenware, impurities such as water-soluble salts and secondary calcite can penetrate ceramics. Although these contents can be distinguished with some analysis techniques, the results may lead to unexpected inferences in some cases. In order to prevent such adverse conditions, ceramics are soaked in pure water for a while (24 or 48 hours, in general, depending on the amount and type of contamination/impurity) before the analysis. Thus, the salts dissolving in water could be removed from the body, and the impurities are softened. Sanding (i.e., use of SiC sandpaper) or acid washing of ceramics is not preferred. Unlike the ceramics found in archaeological excavations, it may not be appropriate to apply the cleaning process to the ceramic materials (glass, tile, etc.) to be examined in the context of conservation and restoration. Because in conservation-restoration studies, it is generally aimed to reveal the causes of deterioration, and for this purpose, materials are mostly analyzed in their initial state in the first place [28,29].

3.4 Preparation of samples

Analyses can be destructive or non-destructive. In general, portable analyzers are non-destructive, and such types of devices generally do not require sample preparation. Portable devices are the most suitable options for on-site examination of immovable cultural heritages, and the historical artifacts that cannot be taken out of the museums. If the analyses are applied in a lab, powder and bulk samples should be prepared for the destructive characterization methods. Powder samples can be prepared by taking random parts from the purified ceramics and grinding them in agate or porcelain mortars. It will be beneficial for this powder to pass through a 63-micron sieve. In the analyses requiring bulk samples (usually in microscopic examinations), a sample of appropriate size for the microscope should be taken from the ceramics. A pretreatment could be needed for some of the analyses (e.g., coating, cutting, and polishing processes may be mandatory for image acquisition in microscopes) [23,26].

3.5 Selection of the analyses and characterization process

The selection of analytical techniques for studying ancient ceramics should be based on the aim and scope of the research. For example, the "alloy" mode of a portable X-ray fluorescence (XRF) spectrometer can be effectively employed to characterize the composition of colored glasses or glazes. In the examination of materials such as mortar and plaster, simple spot tests may complement advanced instrumental analyses, providing essential information on aggregate particle size and distribution, as well as the presence of oil, protein, and soluble salts. The key considerations in determining the most appropriate analytical methods include the nature of the material, the specific properties to be investigated, and the constraints of budget and time [27,30].

Several priority questions guide the selection of analyses: What is the chemical composition of the ceramics? What is their mineralogical content? How are different ceramics related in terms of chemical and mineralogical characteristics (for classification purposes)? Is there a match between clay samples from the excavation site (or its vicinity) and the ceramics (for provenance determination)? What are the firing temperature ranges of the ceramics? What are the chemical and mineralogical characteristics of slip and glaze layers, if present? What types of colorants, if any, are used in the ceramics? and finally, what was the atmosphere condition during the firing process, reducing or oxidizing?

The answers to these questions will determine the analytical techniques most suitable for the study. Certain practical aspects must also be considered during the implementation of analyses. For large ceramic samples from which powder can be prepared, analyses requiring bulk samples should be performed first, followed by powder preparation for subsequent tests. If the analyses require sample coating, as in scanning electron microscopy, the surface should be cleaned prior to powder preparation when possible. Powder samples that can be analyzed directly without additives may be reused across different methods; however, when pellets are prepared by adding binders to the powder, the binder acts as an impurity, making such samples unsuitable for other tests that require pure powders. In these cases, analyses that require pure powders should be performed first, followed by those that can use powder–binder mixtures. Considering these factors, developing a scheduled characterization procedure at the outset of the study will ensure a more efficient, systematic, and reliable analytical process.

3.6 Interpretation of the results

Archaeometry studies require more budget as the number of samples increases. In case of a sufficient budget, the number of samples can be increased, or different analysis techniques can be added. However, what matters is the interpretation of the results. For instance, if it is assumed that 100 ceramics are analyzed with five different analytical techniques, 500 data points will be obtained after the study. These data can be in the form of spectra, curves, numerical tables, and photographs, depending on the type of analysis. Regardless of the number of samples and the type of analyses, the interpretation of the analysis data should be done somehow. Specific discussions can be held for each analysis during the interpretation of the results. For instance, in an analysis in which the chemical composition is revealed, it can be determined what the ceramics contain chemically (oxide/element) and how many groups they create in this sense, or microstructural properties of the ceramics can be identified through electron microscopy. In addition to this, considering the period ceramics belong to and the culture they reflect, it can be revealed whether there is a change or an interaction between the ceramics by comparing the data (if any) from previous archaeometry studies. Especially, the statistical approaches play an active role in the interpretation of the results. Statistical analyses are the guides in demonstrating [31–34] the relations of ceramics both among themselves and with others. Section 3 of Table 2 summarizes and illustrates the **step-by-step guide to ceramic characterization in archaeometry.**

Table 2. Step-by-step guide to ceramic characterization in archaeometry.

Step	Key Procedures	Considerations & Techniques
1. Sample Selection	- Select representative samples from ceramic finds.	- Consider budget, time, and labor constraints. - Prioritize based on color, form, and typology. - Collaborate with archaeologists/ceramic experts.
2. Documentation	- Photograph ceramics (front, back, profile). - Record macro-features (color, texture, slip/glaze, decorations). - Assign sample codes.	- Use neutral backgrounds and proper lighting. - Include a scale in photos. - Note archaeological context (locus, period, excavation date).
3. Purification	- Soak ceramics in pure water (24–48 hrs) to remove soluble salts/contaminants. - Avoid sanding/acid washing.	- Not applicable for conservation studies (e.g., tiles, glass) where the original state is critical.
4. Sample Preparation	- **Non-destructive**: Portable analyzers (e.g., pXRF). - **Destructive**: - Powder samples (agate mortar, <63µm sieve). - Bulk samples (cut/polished for microscopy).	- For powders: Use pure samples first; binder-added pellets limit reuse. - Prioritize analyses requiring bulk samples before powdering.
5. Analysis Selection	Choose techniques based on research questions: - Chemical composition (XRF, SEM-EDS). - Mineralogy (XRD). - Microstructure (SEM, petrography). - Firing conditions (thermal analysis).	- Match techniques to material properties (e.g., "alloy" mode in pXRF for glazes). - Combine spot tests with advanced methods for mortars/plasters.
6. Data Interpretation	- Compare data (spectra, tables, images) statistically. - Correlate with archaeological context. - Identify compositional groups/provenance.	- Use multivariate statistics (PCA, clustering). - Cross-reference with existing archaeometric studies.

4. Characterization Techniques for Ancient Ceramics

From an archaeometry perspective, the analytical techniques used to study ancient ceramics are primarily chemical and mineralogical, since ceramic materials are composed of minerals such as clay, quartz, feldspars, carbonates, and others. Materials belonging to the ceramic group include pottery, glass, tiles, bricks, porcelain, and other similar products. The analytical methods applied in this field vary according to the study's aims. For determining chemical composition, techniques include X-ray fluorescence (XRF) spectroscopy, atomic absorption spectrometry (AAS), X-ray photoelectron spectroscopy (XPS) [35,36], laser-induced plasma spectroscopy (LIPS), proton-induced X-ray emission (PIXE) spectroscopy [37–40], neutron activation analysis (NAA), inductively coupled plasma mass spectrometry (ICP-MS) [41–43], inductively coupled plasma atomic emission spectroscopy (ICP-AES) [44–47], and proton-induced gamma-ray emission (PIGE) spectroscopy [48–50]. To determine mineralogical content, X-ray diffraction (XRD) [51–53] and petrography are commonly used. For studying microstructural and microchemical properties, scanning electron microscopy with energy-dispersive X-ray spectrometry (SEM-EDX) and transmission electron microscopy with energy-dispersive X-ray spectrometry (TEM-EDX) are applied [39,54–56]. Molecular structure analysis can be performed using Fourier transform infrared (FTIR) spectroscopy and Raman spectroscopy. Changes in materials during heat treatment can be analyzed through thermogravimetric (TG) analysis [57,58], differential thermal analysis (DTA), thermomechanical analysis (TMA), and differential scanning calorimetry (DSC). The characterization of ceramic materials requires a careful selection of these techniques, depending on the needs of the research. In archaeometry studies, the choice of analytical methods should be made in line with the study's objectives and resources. Among the methods listed, the most frequently used for ancient ceramics are XRF, XRD, petrography, SEM-EDX, FTIR, Raman, and TG-DTA, with the possibility of adding other techniques depending on the content, scope, and possibilities of the study.[59].

4.1 Determination of chemical composition

XRF analysis is mostly preferred for the identification of the chemical composition of ancient ceramics. XRF analysis can be performed in the laboratory or in situ. In other words, the XRF device is available both as a laboratory type and as a portable one. It can be deduced that the laboratory type XRF device is more sensitive. Scanning of basic elements (as % wt. and ppm) reflecting the content of ceramic raw materials can be carried out with this type of XRF. Some of the portable XRF devices could only scan the elements with the atomic number above 11; thus, sodium, which is a significant element in some of the ancient ceramic artifacts, cannot be detected through such devices. For instance, if the presence of sodium cannot be detected in the examination of antique glasses, it will not be possible to discover whether the glass is soda-lime-silica type. On the other hand, the portable XRF device makes an important contribution to the examination of immovable cultural assets, and the artifacts that cannot be removed outside the museum. Portable XRF devices, which have a relatively fast analysis time and do not require sample preparation, are frequently preferred, especially in conservation-restoration studies. The main features of XRF analysis in ceramic archaeometry could be listed as follows [60–63].

The chemical composition, which is one of the most important parameters reflecting the main characteristics of ancient ceramics, can be identified by XRF. In this context, it would be assumed that XRF is essential in ceramic archaeometry in most cases, since the priority in the characterization of ancient ceramics is mostly to reveal the chemical content.

In light of the chemical composition, preliminary data are collected which would be directive in the prediction of the mineralogical compositions creating ceramics, and thus the raw materials. For instance, the amount of silicon and aluminum in the ceramics mostly refers to quartz and clay, respectively. The high calcium content is mostly assigned to calcite, while the presence of magnesium in high amounts could mostly point to dolomite. High sodium and potassium contents mostly indicate feldspar and/or plagioclase. At this point, the lowest/highest levels and the ratios of the elements or oxides should be considered. The ultimate presence of such minerals can only be verified through XRD. What is meant to be described here is to find the source of the detected elements. Although mineral/phase analysis is done through XRD, XRF data also form an important infrastructure about the content of the raw material and make a great contribution to the interpretation of XRD results.

The relations among a great number of ceramics can be revealed. For this purpose, ceramics are classified with a statistical approach. Ceramics can be divided into groups by hierarchical clustering analysis and principal component analysis. Thus, ceramics with different contents, if any, could be differentiated from the others. This provides significant data, especially for some of the archaeological questions. For instance, it can be thought that ceramics unearthed from the same locus but with different chemical compositions could have diverse raw material sources. This may suggest that those ceramics would have been produced with various raw materials or that they belong to a different workshop. In addition to the main elements, the distribution of trace elements can also classify ceramics. Strontium (Sr) and zirconium (Zr) are the prominent trace elements in ancient ceramics. In ceramics with colorants and/or glazes, contents such as copper (Cu), cobalt (Co), lead (Pb), arsenic (As), manganese (Mn), and tin (Sn) are the important ones. Also, the amounts of alkali and alkaline earth elements should be considered in glazes and glass.

An evaluation can be made about whether the ceramics were locally produced or imported. However, for this type of evaluation, soil/clay samples should be taken from the mound area and/or its immediate vicinity where the ceramics were recovered, and they should be analyzed with the same device. In this manner, the chemical compositions of ancient ceramics and soil samples can be determined and compared. Similarities or differences between the samples will give clues about the origin of production. This type of work is called "provenance". In provenance studies, if no clay/soil samples are available, a general evaluation can be made through the geological maps of the region (if any) where the ceramics were found. However, such an approach may be more limited than the other.

4.2 Identification of mineralogical content

The fact that ceramics are produced from soil turns them into a mineralogical material, and their soil-based structure protects the ceramics from adverse conditions in a sense. In this context, it can be said that the ceramics excavated after being underground for thousands of years do not undergo much degradation. One of the important criteria here is undoubtedly that ceramics are materials fired at high temperatures. The firing process can be seen as the last step to reach the final product for ceramics. Firing conditions are very effective and important in the transformation of ceramics from a raw state to a mature material/product. The main ceramic firing techniques used in ancient periods can be summarized as *i: bonfiring, ii: pit firing, iii: kiln firing*. The common purpose of these firing techniques is the heat treatment of shaped and dried ceramics to gain permanent strength. As a result of the changes and advances in ages and civilizations, different firing environments have been designed, and developments have been seen in ceramic production, as in everything else. In the end, regardless of the production technology, ceramics have been produced in a way and used, then managed to reach the present day. The determination of minerals and phases plays a significant role in revealing the production technologies of ceramics by minimizing all the mentioned possibilities above. This type of information is of great archaeological importance [64], [65], [66].

The main technique used to specify the minerals in ceramics is XRD analysis. This technique reveals the mineral/phase assemblages in the ceramic matrix. Decomposition of the existing minerals and the formation of new minerals/phases occur at certain temperatures. So, the mineral/phases detected by XRD analysis provide data indicating the temperatures to which the material was exposed. There are prominent differences between the firing of ceramics at low temperatures and high temperatures in terms of production technology. The determination of different firing temperature ranges among ceramics belonging to a certain culture, civilization, and period may indicate a different production technology and, therefore, a different production site. At this point, the provenance study will also support possible predictions. It would be unrealistic to achieve definitive results by performing XRF or XRD analysis alone. In ceramic archaeometry, the following can be revealed by XRD.

Another technique frequently used to determine the available mineral content in ceramics is petrography. In this analysis, a thin section sample is taken from the material and examined under a polarized microscope. The paste structure and content of ceramics can be determined in this way. In this context, porosity (% by volume), aggregate (% by volume), grain size (i.e., fine, medium, coarse), mineral content, and rock origin can be specified. Besides, additives such as ceramic/brick fracture in the ceramic can be seen under the polarizing microscope. The impurities, which are generally called "grog", can be clay lumps and rock fragments in addition to ceramic/brick fragments. These are the "temper materials" which are subsequently added to the ceramic body to increase the strength of the product. Temper materials may also contain vegetables and animal additives (e.g., straw, twigs, leaves, oil, blood, hair, etc.). Organic additives, which can also be seen in mortars, can act as binders in ceramics. Images can also be taken during the petrographic investigation. The colored microphotographs can provide information about the firing atmosphere of the ceramics (reducing/oxidizing), which plays an important role in understanding the ceramic firing technology [67], [68].

4.3 Determination of microstructural and microchemical properties

In addition to the macro-observation of the ceramic body, examinations on a micro scale can also be made. The most used method for this purpose is electron microscopy, which can be a scanning or transmission type. It can be thought that a more detailed study can be done with the transmission electron microscope (TEM), which is known to work more sensitively. However, it can be assumed from the archaeometry literature that scanning electron microscopy (SEM) is most preferred. The budget of the research, working time or the deadline of the study, the availability of the analysis device, and the number of samples are the primary key factors in selecting which type of microscope will be used. In general, TEM analyses are more expensive than SEM, and this is decisive in creating the budget of the study/project. However, depending on the number of samples, TEM by itself may be preferred in case of a comprehensive investigation of a particular ceramic group (e.g., glazed, colored tiles or glasses) for a specific purpose (e.g., restoration, conservation, determination of deterioration, etc.).

According to the working principle of electron microscopes, the sample surface must be conductive in order to take even images. The ceramic fragments are usually coated with gold or palladium, also with carbon in some cases. In electron microscopes, there are decisive parameters such as separation power, magnification, and contrast, all of which need to be set carefully. During the examination of interfaces in ceramics with secondary layers such as slip and glaze, it is expected that the microscope has sufficient separation power and magnification, as well as an appropriate contrast setting. The indistinguishable blur of the images will make it difficult to determine the body, slip, and/or glaze layers. If the layers can be identified, then the interactions between the body and other layers will also be revealed. Such points are important in determining the under-glaze, over-glaze, and colored glaze techniques. The fact that the electron microscope gives a colorless image indicates the necessity of an expert in adjusting the contrast degree in black and white images [69], [70].

Another point to be noted is that the magnification values should be the same so that the ceramic groups in the study can be compared among themselves. Different magnification values for the ceramics would make it difficult to make a comparison regarding the microstructural features of the samples. Therefore, in electron microscopy examinations, it is first recommended to take a representative analysis for a ceramic, to set a standard accordingly, and subsequently to examine all ceramics under the same conditions. Such a work plan will also minimize possible problems and obstacles. In addition to the magnification, separation power, and contrast settings of the microscope, the working distance, which refers to the distance between the sample holder and the lens, should be carefully optimized and should be studied at close values, although not the same. Depending on the thickness of the samples, the working distance of the device may vary in the examinations carried out on broken ceramic pieces.

Chemical characterization can also be carried out during electron microscopy analysis. For this purpose, chemical characterization can be performed through an EDX (energy dispersive X-ray) spectrometer integrated into the microscope, and the analysis can be carried out on a selected point, line, or area on the SEM image. In some cases (e.g., insufficient budget, time), SEM-EDX can be solely preferred instead of XRF in archaeometry studies of ceramics. This may allow the user to get information regarding both the microstructure and chemical composition of the ceramics at once. Namely, the general chemical composition of the ceramics can be determined through EDX by scanning at low magnifications, and then the microchemical contents of the samples at higher magnifications. As mentioned earlier, standardization is important for these applications. In addition, it is possible to see the distribution of the elements in color on the image through the elemental mapping on black-and-white SEM images. This feature can be found in advanced electron microscopes. In ceramic archaeometry, the following can be revealed by SEM-EDX [69], [70].

4.4 Determination of molecular structure

FTIR and Raman analyses, mostly as complementary techniques in ceramic archaeometry, are used to identify the molecular structure. In the FTIR analysis, the sample absorbs the infrared radiation transmitted to it, and the molecular content is determined by the bond vibrations that occur accordingly. FTIR analysis is generally performed in the wavelength range of 400-2000cm⁻¹, which is called the *"fingerprint region"* for the ceramics. This range can be expanded upon the aim and scope of the research. The powder sample is used in FTIR analysis, so the purification and powder preparation stages of the ceramics are very important in this sense. Clay minerals, quartz, feldspar, plagioclase, carbonates, iron minerals, and high-temperature minerals/phases in ancient ceramics can be detected by the FTIR technique thanks to their characteristic band values. Possible overlapping of some band values indicates that more than one mineral may be seen on FTIR spectra. In such cases, XRF, XRD, and/or petrography data of the ceramics would help interpret the results more reliably. One of the important types of information provided by the FTIR analysis is to determine whether calcite in the ceramic matrix is primary or not. Primary calcite normally has bands in the range of ca 1415-1430 cm⁻¹ in FTIR spectra, but an increase in the band value to ca 1440-1450 cm⁻¹ indicates secondary calcite, which may occur due to the burial conditions [71], [72].

Raman analysis specifies the crystal lattices and molecular vibrations of the sample, depending on the phase content, crystal structure, composition, and bond type of the ceramic [73]. This technique is generally used in the examination of colored ceramic finds. In this context, Raman analysis can be preferred especially for the determination of colorants in colored tiles, glasses, and glazed ceramics. This technique can be performed destructively or non-destructively. However, as with other techniques, Raman analysis needs to be applied by experts working particularly on ancient materials. As is known, the analysis techniques mentioned in this book chapter are already used by different disciplines, but there are some key points in the examination of historical findings and ancient artifacts. Consequently, it is substantial to have experts who know the characteristics of the materials and can analyze them accordingly. In

Raman analysis, if the material radiates, no data can be obtained. Therefore, the analysis should be performed after choosing a laser beam with the appropriate wavelength. Thanks to the microscope of the Raman device (if any), images can be taken of the sample. These images can be magnified, and the molecular structure of any point on the image can be revealed by sending the laser beam to that area. Since such studies will require a long working time, representative samples can be selected from the ceramic set, or the research schedule can be updated accordingly [74,75].

Table 3. *Characterization techniques for ancient ceramics.*

Analysis Type	Techniques	Purpose/Applications	Key Considerations
Chemical Composition	- XRF (lab/portable) - AAS, ICP-MS, ICP-AES, PIXE, NAA, LIPS, PIGE	- Quantify major/trace elements (e.g., Si, Al, Ca, Na, Cu, Co). - Classify ceramics (statistics). - Provenance studies (compare with local clays).	- Portable XRF cannot detect Na ($Z < 11$). - Lab XRF is more precise. - Combine with XRD for mineral context.
Mineralogical Content	- XRD - Petrography (polarized microscopy)	- Identify phases (quartz, calcite, feldspars). - Estimate firing temperatures. - Detect temper materials (grog, organics).	- Thin sections required for petrography. - XRD reveals primary vs. secondary calcite.
Microstructural/ Microchemical	- SEM-EDX - TEM-EDX (rare)	- Study microstructure (porosity, layer interfaces). - Map elemental distribution (e.g., glaze-body interaction).	- Sample coating (Au, C) needed. - Standardize magnification/contrast. - TEM is costly but detailed.
Molecular Structure	- FTIR - Raman spectroscopy	- Identify bonds (e.g., clay minerals, carbonates). - Detect colorants (glazes, tiles). - Distinguish primary vs. secondary calcite (FTIR: 1415–1450 cm^{-1}).	- FTIR requires powder samples. - Raman needs laser wavelength optimization.
Thermal Analysis	- TG/DTA - DSC, TMA	- Track weight loss (TG) and reactions (DTA). - Detect decomposition (e.g., calcite at 700–800°C). - Infer firing conditions.	- Secondary calcite shows lower decomposition temperatures (~700°C). - Heating rate: 10–20°C/min.

4.5 Thermal analysis

Thermal analysis methods, which are not essential for ancient ceramics but provide useful information when applied, are generally based on tracking the changes that occur by heating the sample from room temperature to higher temperatures. In this context, two techniques frequently used in ceramic archaeometry are thermogravimetric (TG) analysis and differential thermal analysis (DTA). These analyses can be performed separately. If only the TG data is sufficient, TG analysis can be employed alone. However, in case of the need for both TG and DTA data, it is possible to perform them simultaneously. In TG-DTA analysis, ceramic powders are heated from room temperature to high temperatures such as 1000-1200°C at a heating rate of 10- 20 °C per minute, and the changes are recorded. In the diagrams obtained, the DTA curve shows the endothermic and exothermic reactions that occur with the increase in temperature, and the TG curve shows the alterations in weight that happen during the reactions. The changes indicating the presence of physical water, chemical water, organic matter, carbonates, and high temperature minerals (polymorphic transformations) in ceramics can be followed upon the TG-DTA curves [65], [76].

In addition to FTIR analysis, secondary calcite can also be detected by the TG-DTA technique. In general, if primary calcite is present in the ancient ceramic and has not decomposed, the presence of this type of calcite can be confirmed by the endothermic effect seen around 800 °C and the high amount of weight loss that occurred due to this effect. If secondary calcite is present, a poor endothermic effect at a temperature close to 700 °C rather than 800 °C will be seen, and the weight loss will be observed at negligible levels accordingly. If the only aim or need is to detect secondary calcite in ancient ceramics, one of the FTIR and TG-DTA analyses can be selected [72]. Section 4 of Table 3 summarizes and illustrates Characterization techniques for ancient ceramics.

5. Conclusion

In ceramic archaeometry studies, archaeological data should first be taken into account, such as the cultural context and the period to which the ceramic finds belong. When examining ceramic materials (including tiles, bricks, glass, and similar) from historical buildings, it is essential to identify the period characteristics of the building and the distinctive features of the materials. Before starting the analysis, the research questions to be addressed, such as dating, chemical classification, provenance determination, or reproduction for conservation and restoration, should be clearly defined. Once the main objectives are set, a detailed work plan should be created according to the scope of the study, the number of samples, and their sizes to ensure proper progress. The work schedule should also determine the sequence of analyses; for example, if samples are small, analyses requiring bulk samples (such as petrography or electron microscopy) should be performed first, followed by preparation of powder samples for subsequent analyses. If bulk samples are to be coated for microscopic investigation, a decision should be made at the outset regarding which analyses will be performed, since coating materials may appear as impurities in powder samples used for other analyses. Alternatively, uncoated analysis can be carried out, although this may reduce image quality in electron microscopy. After analyzing the entire sample set, results can be compared with archaeometric data from ceramics of the same period, region, or culture to interpret production technologies. If soil or clay samples from the find's region are available, provenance studies can provide valuable information on the production location. In cases where numerous ceramic finds are studied, statistical analyses such as hierarchical cluster analysis or principal component analysis can be applied to make complex datasets more understandable. By incorporating archaeometric data from similar ceramic groups into these statistical methods, similarities and differences between the ceramics can be comparatively revealed.

References

[1]   Işık, İ. (2018). Karakterizasyon ve Tarihlendirme Çalışmalarının Arkeometrik Yöntemlerle İncelenmesi. Cedrus. https://doi.org/10.13113/CEDRUS.201834

[2]   Eramo, G. et al. (2004). Late Roman cooking pottery from the Tavoliere area (Southern

Italy): raw materials and technological aspects. Journal of Cultural Heritage.
https://doi.org/10.1016/J.CULHER.2003.05.002

[3] Kumari, N. and Mohan, C. (2021). Basics of Clay Minerals and Their Characteristic
Properties. Clay and Clay Minerals. https://doi.org/10.5772/INTECHOPEN.97672

[4] Li, J. et al. (2013). The history, development, and future prospects for laser ceramics: A
review. International Journal of Refractory Metals and Hard Materials.
https://doi.org/10.1016/j.ijrmhm.2012.10.010

[5] Sánchez-Climent, Á. (2024). Materiality and Immateriality: Exploring Material Culture in
the Construction of Cultural Meanings. Global Journal of Cultural Studies.
https://doi.org/10.6000/2817-2310.2024.03.13

[6] Bayram, G. (2018). Güneydoğu Anadolu'nun Neolitik Çağ Totemleri Ve Ritüel Nesneleri.
Amisos.

[7] Üniversitesi, S.D. et al. (2016). Buzul Çağı'ndan İlk Çağ'a Tüketimin Tarihi. Süleyman
Demirel University Visionary Journal.

[8] Yaka, R. et al. (2021). Variable kinship patterns in Neolithic Anatolia revealed by ancient
genomes. Current Biology. https://doi.org/10.1016/J.CUB.2021.03.050

[9] Sevin, V. (2003). Anadolu Arkeolojisi , Der Yayınları .

[10] ÖZDEMİR, M. (2016). Neolitik Dönemde Güneydoğu Anadolu Bölgesi / The
Southeastern Anatolia Region in Neolithic Period.

[11] Efecan, S. 1985-author 185620 et al. (2011). Geç neolitik döneme ait Hacılar kazılarında
bulunmuş terracota figürlerin çağdaş yorumlarla biçimlendirilmesi = Forming with
contemporary interpretations of late neolithic terracotta figures found in excavaiions at
Hacılar /.

[12] Akgonul, S.O. et al. (2020). 100. Yılında Türk Arkeolojisi. Süleyman Demirel
Üniversitesi Fen-Edebiyat Fakültesi Sosyal Bilimler Dergisi.
https://doi.org/10.35237/sufesosbil.760779

[13] Caley, E.R. (1949). Klaproth as a pioneer in the chemical investigation of antiquities.
Journal of Chemical Education.
https://doi.org/10.1021/ED026P242/ASSET/ED026P242.FP.PNG_V03

[14] Rousaki, A. et al. (2019). Archaeological investigations (archaeometry). Physical Sciences
Reviews. https://doi.org/10.1515/psr-2017-0048

[15] Esin, U. (1969). Kuantitatif spektral analiz yardımıyla Anadolu'da başlangıcından Asur
kolonileri çağına kadar bakır ve tunç madenciliği : (metin, kataloglar, resim ve haritalar).,
Taş Matbaası, İstanbul :

[16] Piggott, S. (1965). An Introduction to Prehistoric Archeology by F. Hole and R. F. Heizer.
New York: Holt, Rinehart and Winston, 1965. 316 pp., 28 figs. 56s. Antiquity.
https://doi.org/10.1017/S0003598X00039788

[17] Jäger, E. (1979). Introduction to Geochronology, in Lectures in Isotope Geology, Springer
Berlin Heidelberg, Berlin, Heidelberg, pp. 1–12.

[18] Bronk Ramsey, C. (2008). Radiocarbon dating: Revolutions in understanding.
Archaeometry. https://doi.org/10.1111/j.1475-4754.2008.00394.x

[19] Eramo, G. and Mangone, A. (2019). Archaeometry of ceramic materials. Physical

Sciences Reviews.

[20] Ricci, G. (2017). Archaeometric studies of historical ceramic materials.

[21] Liritzis, I. et al. (2020). Archaeometry: An Overview. Scientific Culture.

[22] Maritan, L. (2019). Archaeo-ceramic 2.0: investigating ancient ceramics using modern technological approaches. Archaeological and Anthropological Sciences.

[23] Slavíček, K. (2023). Ceramic petrography and the study of artefacts using instrumental methods. Nepublikovaná disertační práce obhájená na Ústavu geologických věd Masarykovy univerzity.

[24] Bayazit, M. et al. (2021). Investigation of a diagnostic group of Bronze Age pottery (Dark Rimmed Orange Bowl ware) from the upper Tigris Valley, Turkey. X-Ray Spectrometry.

[25] Cohen, A.S. (2024). Potting communities and conservatism in the Purépecha empire at Angamuco, Michoacán, Mexico. Ancient Mesoamerica.

[26] Lozada-Mendieta, N. and Villagran, X.S. (2025). Ceramic archaeometric studies in the Amazon and Caribbean regions: A review. Archaeometry.

[27] Puntın, M.Categorızıng The Vısıble And Invısıble Attrıbutes Of Ceramıc Pastes: A Database For Archaeometry. Ceramıcs Matter: Of The European Meetıng On Ancıent Ceramıcs–Emac 2023 & 1st Emac School.

[28] Camara, C.A. et al. (2023). High-Lead Glazed Ceramic Production in Western Iberia (Gharb al-Andalus) between the 10th and Mid-13th Centuries: An Approach from the City of Évora (Portugal). Ceramics.

[29] Gómez-Martínez, S. et al. (2023). High-Lead Glazed Ceramic Production in Western Iberia (Gharb al-Andalus) between the 10th and mid-13th Centuries: An Approach from the City of Évora (Portugal).

[30] Cereda, S. and Fragnoli, P. (2021). Petrography and Micromorphology Face-to-Face: the Potential of Multivocality in the Study of Earth-Based Archaeological Materials. Interdisciplinaria Archaeologica, XII/1, https://doi.org/10.24916/iansa

[31] Tunca, B. et al. (2022). Chemically complex double solid solution MAX phase-based ceramics in the (Ti, Zr, Hf, V, Nb)-(Al, Sn)-C system. Materials Research Letters.

[32] Pilania, R.K. et al. (2023). Synthesis of the chemically durable glass-ceramic matrix for radioactive waste immobilisation. Ceramics International.

[33] Guan, H. et al. (2022). Microstructure and tribological behavior of Ti3C2Tx MXene reinforced chemically bonded silicate ceramic coatings. Ceramics International.

[34] Zhang, Y. et al. (2022). Microstructure, cytocompatibility, and chemical durability of chemically strengthened LAS (Li2O-Al2O3-SiO2) glass-ceramic materials. Journal of the European Ceramic Society.

[35] Leonel, G.J. et al. (2023). Compositional analysis of SiOC (H) powders: a comparison of X-ray photoelectron spectroscopy (XPS) and combustion analysis. Ceramics.

[36] Zafeiratos, S. (2023). Applications of X-ray photoelectron spectroscopy to catalytic studies: from routine analysis to cutting-edge surface characterization, World Scientific.

[37] Clough, R. et al. (2023). Atomic spectrometry update: review of advances in the analysis of metals, chemicals and materials. Journal of Analytical Atomic Spectrometry.

[38]   Zahoor, M. et al. (2021). Spectroscopic Analysis of Ancient Ceramic From Agror Valley Khyber Pakhtunkhwa, Pakistan.

[39]   Magdy, M. (2022). Analytical techniques for the preservation of cultural heritage: Frontiers in knowledge and application. Critical reviews in analytical chemistry.

[40]   Kradolfer, S. (2021). Sampling and Elemental Analysis in the World of Art and Archaeology.

[41]   Das, D.D. et al. (2024). Neutron activation analysis: An excellent nondestructive analytical technique for trace metal analysis. Critical Reviews in Analytical Chemistry.

[42]   Samanta, S.K. et al. (2021). Intercomparison studies of Instrumental Neutron Activation Analysis using singles and gamma–gamma coincidence spectrometry for trace element determination in sodalime glass and sediment matrices and utilization of coincidence method for rapid automobile glass forensics. Nuclear Instruments and Methods in Physics Research Section A: Accelerators, Spectrometers, Detectors and Associated Equipment.

[43]   Alsabbagh, A. et al. (2025). Comparison of neutron activation analysis, X-ray fluorescence spectrometry and inductively-coupled plasma mass spectrometry for the determination of rare earth element concentrations in Jordanian monazite ore. Analytical Methods.

[44]   Khan, S.R. et al. (2022). Inductively coupled plasma optical emission spectrometry (ICP-OES): a powerful analytical technique for elemental analysis. Food Analytical Methods.

[45]   Kakuk, M. et al. (2025). Inductively Coupled Plasma Optical Emission Spectroscopy (ICP-OES): Exploring Versatile Applications in Industrial and Analytical Fields. Periodica Polytechnica Chemical Engineering.

[46]   Szymczycha-Madeja, A. et al. (2021). Development and validation of an analytical method for determination of al, ca, cd, fe, mg and P in calcium-rich materials by ICP OES. Molecules.

[47]   Yang, F. et al. (2025). Determination of Platinum Group Metals in Catalysts by Inductively Coupled Plasma–Atomic Emission Spectrometry (ICP-AES) with Coprecipitation Separation and Enrichment. Analytical Letters.

[48]   Samanta, S.K. et al. (2022). The standardization and application of an external (in air) particle induced gamma emission (PIGE) method for the rapid and non-destructive quantification of light elements at major to trace concentrations in coal, bottom ash and coke samples. Journal of Analytical Atomic Spectrometry.

[49]   Ghosh, M. et al. (2022). Determination of impurities in graphite using proton induced gamma ray emission, total reflection X-ray fluorescence and instrumental neutron activation analysis. Analytical Chemistry Letters.

[50]   Khan, N. et al. (2024). Quantification of lithium in hyper-stoichiometric lithium titanate by external (in air) particle induced gamma-ray emission (PIGE) method. Journal of Radioanalytical and Nuclear Chemistry.

[51]   Rawat, K. et al. (2022). X-Ray fluorescence and comparison with other analytical methods (AAS, ICP-aes, la-ICP-ms, IC, LIBS, SEM-EDS, and XRD). X-Ray Fluorescence in Biological Sciences: Principles, Instrumentation, and Applications.

[52]   Shathi, A.S. et al. (2024). Iron removal from red clay using oxalic acid leaching for

enhanced ceramic industry applications. Heliyon.

[53] Rosiak, A. et al. (2025). Advanced Characterization of Archaeological Ceramics. Annual Review of Materials Research. https://doi.org/10.1146/annurev-matsci-080522-111043

[54] Sharma, V. et al. (2023). Chemical characterization of automobile windshield glass samples by nuclear and radio-analytical techniques namely SEM-EDX, ED-XRF, PIXE, PIGE, and INAA and potential of external (in air) PIGE and INAA in conjunction with chemometrics for glass forensics. Journal of Analytical Atomic Spectrometry.

[55] Magdy, M. (2023). X-ray techniques dedicated to materials characterization in cultural heritage. ChemistrySelect.

[56] Vereshchagina, T.A. et al. (2024). Zirconosilicate Sorbent Based on Waste Fly Ash Cenospheres for Cesium Immobilization in a Ceramic Form. Russian Journal of Inorganic Chemistry.

[57] Ceylan, İ. et al. (2021). Development of CaO-rich blast furnace slag containing fluorine mica-based glass ceramic coatings. Ceramics International.

[58] Rashad, A.M. et al. (2024). Valorization of ceramic waste powder for compressive strength and durability of fly ash geopolymer cement. Arabian Journal for Science and Engineering.

[59] Zendri, E. (2024). Archaeometric studies of historical ceramic materials.

[60] Pérez-Arantegui, J. et al. (2004). Some aspects of the characterization of decorations on ceramic glazes. Applied Physics A: Materials Science and Processing. https://doi.org/10.1007/S00339-004-2508-2/METRICS

[61] Eramo, G. and Mangone, A. (2019). Archaeometry of ceramic materials. Physical Sciences Reviews. https://doi.org/10.1515/psr-2018-0014

[62] Simsek, G. et al. (2019). On-site pXRF analysis of glaze composition and colouring agents of "Iznik" tiles at Edirne mosques (15th and 16th-centuries). Ceramics International. https://doi.org/10.1016/J.CERAMINT.2018.09.213

[63] Simsek, G. et al. (2019). On-site pXRF analysis of body, glaze and colouring agents of the tiles at the excavation site of Iznik kilns. Journal of the European Ceramic Society. https://doi.org/10.1016/J.JEURCERAMSOC.2019.01.050

[64] Bertolino, S.R. et al. (2009). X-ray techniques applied to surface paintings of ceramic pottery pieces from Aguada Culture (Catamarca, Argentina). X-Ray Spectrometry. https://doi.org/10.1002/XRS.1124;PAGEGROUP:STRING:PUBLICATION

[65] Ion, R.M. et al. (2010). Thermal analysis of Romanian ancient ceramics. Journal of Thermal Analysis and Calorimetry. https://doi.org/10.1007/s10973-009-0226-x

[66] van der Weerd, J. et al. (2004). Identification of black pigments on prehistoric Southwest American potsherds by infrared and Raman microscopy. Journal of Archaeological Science. https://doi.org/10.1016/J.JAS.2004.03.008

[67] Petrography, C. and Middleton, A. (2016). Ceramic Petrography. Revista do Museu de Arqueologia e Etnologia. Suplemento. https://doi.org/10.11606/ISSN.2594-5939.REVMAESUPL.1997.113441

[68] Shepard, A.O. (1956). Ceramics for the Archaeologist. Ceramics for the Archaeologist.

[69] Panagopoulou, A. et al. (2018), Technological Examination Of Iznik Ceramics By Sem-

Edx, Raman, Xrd, Plm: A Case Study Scientific Culture, Scientific Culture
https://doi.org/10.5281/zenodo.1409802

[70]   Dabanlı, Ö. et al. (2021). Composition and Phase Analysis on Glazed Tiles of Southeast
       Anatolia: Production Process Identification. Mediterranean Archaeology and
       Archaeometry. https://doi.org/10.5281/ZENODO.5545709

[71]   Shoval, S. et al. (2006). The ceramic technology used in the manufacture of Iron Age
       pottery from Galilee. Geological Society Special Publication.
       https://doi.org/10.1144/GSL.SP.2006.257.01.08

[72]   Fabbri, B. et al. (2014). The presence of calcite in archeological ceramics. Journal of the
       European Ceramic Society. https://doi.org/10.1016/J.JEURCERAMSOC.2014.01.007

[73]   Leemann, A. (2017). Raman microscopy of alkali-silica reaction (ASR) products formed
       in concrete. Cement and Concrete Research.
       https://doi.org/10.1016/j.cemconres.2017.08.014

[74]   Spadavecchia, S. et al. (2024). Evaluation of the effectiveness of coatings for the
       protection of outdoor terracotta artworks through artificial ageing tests. Journal of Cultural
       Heritage.

[75]   Li, M. et al. (2022). In situ simultaneous quantitative analysis multi-elements of
       archaeological ceramics via laser-induced breakdown spectroscopy combined with
       machine learning strategy. Microchemical Journal.

[76]   Drebushchak, V.A. et al. (2006). The investigation of ancient pottery: Application of
       thermal analysis. Journal of Thermal Analysis and Calorimetry.
       https://doi.org/10.1007/S10973-005-0942-9

Chapter 3

Structural and Technological Classification of Ceramic Products

Ilker Ozkan[1*], Hussein Elaibi[2], Farah Mutlag[2,3], Cumhur Eren Isık[4], Selcuk Erdogan[2], Irem Turk[2], Iskender Isık[5], Fatih Sen[2*]

[1]Dokuz Eylül University, Torbalı Vocational School, Industrial Glass and Ceramics Department, 35000 Izmir, Türkiye

[2]Sen Research Group, Department of Biochemistry, Kutahya Dumlupinar University, Kutahya 43000, Türkiye

[3]Ministry of Education, Karbala Education Directorate, Karbala, 56001, Iraq

[4]Department of Handicrafts, Kutahya Fine Arts Vocational School, Kutahya Dumlupınar University, Evliya Çelebi Campus, 43000 Kutahya, Türkiye

[5]Department of Materials Science & Engineering, Faculty of Engineering, Kutahya Dumlupınar University, Evliya Çelebi Campus, 43100 Kutahya, Türkiye

ilker.ozkan@deu.edu.tr, fatihsen1980@gmail.com

Abstract

This chapter examines ceramics from a structural and technological point of view, with a focus on both traditional and advanced ceramics. Traditional ceramics such as terracotta, brickware, earthenware, stoneware, porcelain, vitreous china, refractories, and glass are studied in terms of raw materials, manufacturing procedures, phase development during firing, and the resulting physical and mechanical qualities. The chapter also aims to give detailed data on the compositional elements that influence the physical properties, such as color, porosity, and strength, in addition to their technical progress and current industrial applications. On the other hand, advanced ceramics are examined according to their synthesis, processes, structural features, and specialized functions. Advanced ceramics are studied in two groups: non-oxide ceramics and oxide ceramics. The data on their mechanical, chemical, and thermal properties are given. The chapter emphasizes the importance of compositional control, processing techniques, and microstructural engineering in determining ceramic qualities to modern technological needs.

Keywords

Ceramic Raw Materials, Advanced Ceramic Synthesis, Terracotta, Porcelain

1. Introduction

Ceramics are one of the oldest human-made materials. Ceramic technology is one of the most ancient technologies, dating back over 24.000 years, and at the same time is the most modern and developing technology. The application of ceramics is constantly increasing, and this situation gives ceramics strategic importance [1–3]. Ceramics have been used mostly in construction, dinnerware, and sanitary items for decades. Since the early 1990s, ceramic technology has allowed ceramics to be used in several industries, from traditional to cutting-edge. Aerospace, nuclear, electrical, and biological applications are among these modern uses. Traditional and sophisticated ceramics exist. This justifies this classification. [1,4,5]. Traditional ceramics are made from clay or silica and are inexpensive to make. In most cases, clay and silica dominate. Traditional ceramics are sometimes associated with poor technology, even though advanced manufacturing methods are used. Competition among producers boosts processing productivity and cost-effectiveness. Many people in developing nations still use traditional pottery [6–8]. Advanced ceramics are produced from high-purity ceramic powders, with properties tailored through precise control of processing parameters. They are more costly than traditional ceramics and are also known as "special," "technical," or "engineering" ceramics. These materials exhibit superior mechanical strength, exceptional resistance to corrosion and oxidation, and distinctive electrical, optical, or magnetic characteristics. While clay-based ceramics have been in use for over 25,000 years, the development of advanced ceramics has taken place primarily within the past century [4–6,8]. Terracotta, bricks, tiles, earthenware, stoneware, porcelain, vitreous china, refractories, and glass are traditional ceramics. Laser host materials, piezoelectric ceramics, and DRAM ceramics are produced in limited quantities at higher prices. Oxide and non-oxide advanced ceramics exist [9,10].

Table 1 summarizes the historical progression from primitive clay-silica composites to high-purity engineered ceramics, highlighting the transition from production focused on structure, aesthetics, and cost to an emphasis on functional design. This changeover has been propelled by enhancements in powder purity, phase regulation, and contemporary sintering methodologies.

Table 1. Historical timeline, classification, and examples of ceramics.

Period / Era	Key Developments	Main Categories	Examples
24,000 Years	Earliest known human-made ceramics	Pre-classification (Prehistoric Ceramics)	Pottery for cooking and storage
Ancient Civilizations	Ceramics used in daily life and architecture	Traditional Ceramics	Terracotta, bricks, tiles, earthenware
20th Century (Industrial)	Industrial-scale production, improved processing methods	Traditional Ceramics	Stoneware, porcelain, vitreous china, refractories, glass products
Modern Era (Since 1990s)	Expansion into high-tech applications, advanced materials	Advanced Ceramics	Laser host materials, piezoelectric ceramics, DRAM ceramics, oxide and non-oxide ceramics

2. Traditional Ceramics

2.1 Terracotta products

Terracotta includes pottery, roof tiles, bricks, flues, drainage pipes, and floor tiles. These goods were previously manufactured by modeling, drying, and firing common clays. Modern compositions are more complicated, containing additions or agents to improve manufacturing or final properties. Water is added to form plastic paste. Shaping requires a combination of proper rheology. Ventilated cells or tunnel dryers dry shaped items. Product drying is followed by firing at 900–1160 °C [4,6,11]. Terracotta is porous and mechanically resistant. They are produced and sold either glazed or unglazed. They are valued for their esthetic qualities, long-term stability, and hydrothermal and acoustic properties. They represent a highly automated industrial sector that is constantly evolving technologically, as in Figure 1 [12–16].

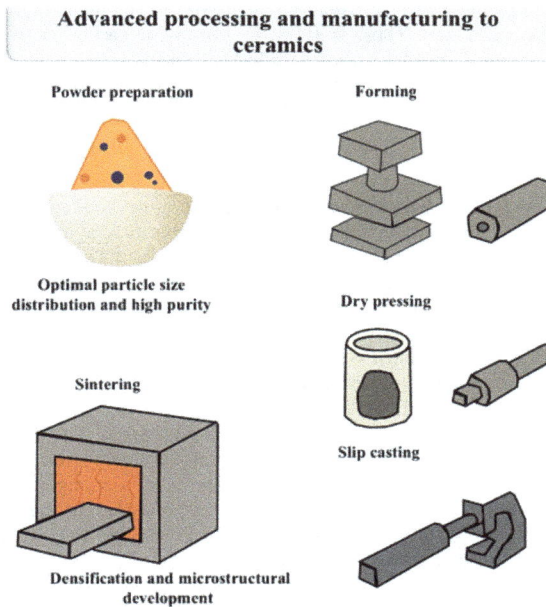

Advanced processing and manufacturing to ceramics

Powder preparation

Forming

Optimal particle size distribution and high purity

Dry pressing

Sintering

Slip casting

Densification and microstructural development

Figure 1. Advanced ceramic processing and manufacturing techniques.

A wide spectrum of colors can be found in terracotta items, from a yellowish white to a dark brown. To decide the color of the product, either the bonding of iron ions with inhibitors like calcium ions or the addition of coloring agents like titanium and manganese oxides is responsible for the color. It is possible to employ a ternary system, which is defined by the primary oxides Al_2O_3, SiO_2, and CaO, to explain the crystalline phases that are created during the firing process of a manufactured terracotta object. Three phases are generated the most frequently: wollastonite

(CaO.SiO$_2$), galenite (2CaO.Al$_2$O$_3$.SiO$_2$), and anorthite (CaO.Al$_2$O$_3$.2SiO$_2$) [17–19]. High temperature promotes anorthite at the expense of the other two phases. The Fe^{3+} ions in anorthite make it yellow. Hematite (Fe$_2$O$_3$) is a brownish-red mineral. Fe^{2+} ions give compounds bluish or greenish tones. These three effects combine to color the shard. When an oxidizing medium contains a lot of iron, it produces brownish-red hematite. A high CaO concentration in the original mixture promotes anorthite formation, which yellows the coloration [20,21]. Based on these considerations and experimental findings, the following guidelines can be established: if the Al$_2$O$_3$/Fe$_2$O$_3$ mass ratio is below 3, the resulting product is red; when the ratio is between 3 and 5, the product appears pink; and if the Fe$_2$O$_3$/CaO mass ratio is below 0.5, applying an appropriate heat treatment—high temperature in an oxygen-rich atmosphere produces a yellow product [22,23]. When the CaO/Al$_2$O$_3$ mass ratio is close to one, the color of the shard is particularly dependent on all the parameters that are likely to affect anorthite formation. It is also significantly influenced by the presence of other impurities [24,25]. Excessive oxidation in the firing atmosphere is detrimental to the formation of anorthite. The temperature of the final stage of firing can also be significant. Thus, at 1000 °C, iron oxide can be in the form of hematite (pink coloring), dissolved in anorthite (yellow coloring), and partially reduced at 1100 °C for a given composition (coloring turning to green). The color of the tile is thus determined by the raw material composition and the firing conditions (temperature, atmosphere, and setting load of the kiln) [26,27].

2.2 Brickware

Bricks, roof tiles, and other building elements are the most common types of brickware. Lightweight alternatives are now replacing traditional bricks (of cavity or porous types). The replacement saves money on manufacturing and reduces the weight of building structures. Considerable progress has been made in the production of large building elements (floor slabs, beams, panels) that are used in modern construction, in addition to the already common prefabricated panels and elements of other materials. Concretes, particularly lightweight concretes [28,29]. Brickmaking is distinctive in that it involves the production of large quantities of relatively inexpensive products, so raw material transportation costs play a significant role. As a result, the plants are typically built directly on the sites of the primary raw materials. The degree of mechanization and automation is determined by the size of the plant [30]. Mechanical strength and frost resistance are the two most important properties of brickware. Both improve as the firing temperature rises (densification degree). Compression strength ranges from 6 to 60 MPa. Frost resistance is determined by the distribution of pore sizes, as the ware is damaged by repeated freezing and thawing of water absorbed in the pores. The rate at which the pores fill with water can be used to assess frost resistance. Thermal conductivity and thermal expansion are two other important properties [31]. The bulk density of full bricks is approximately 1.8 g. cm^{-3}. The bulk density of bricks made lighter by combustible additions is reduced to about 1.0, with proportionally decreasing strength and thermal conductivity.

Vitrified bricks or klinkers are special bricks with a high degree of densification through sintering; their water absorption is less than 9% and their compressive strength is around 60 MPa [4,32]. The main raw material is low-grade colored-firing clay. Non-plastic materials such as sand, slag, and fly ash are used. The proportions added are determined by the required forming properties and firing behavior. The clay must not contain soluble salts that cause efflorescence on the products (e.g., MgSO$_4$, Na$_2$SO$_4$). Hydration of the salts causes expansion, which compromises the ware's integrity. The clay should also be free of pyrite, which can cause local defects due to excessive fusion. It may also form expanding gaseous blisters and cause sulphate efflorescence because of oxidation during firing. It is especially dangerous when present in large particles, such as limestone. The latter is unable to completely combine with the other body components during firing and remains in the ware as free lime, which hydrates and expands when exposed to moisture. This harmful effect of $CaCO_3$ can be suppressed by fine grinding [4,33]. The raw material is transformed into a plastic mass in the plants that contain around 25 % water or into a pressing mix that contains between 8 and 12% water; the shaping is accomplished using pug extruders or by pressing. Due to the size of the products, the latter "dry" approach offers the advantage of speeding up the drying process. On the other hand, the initial raw material can be dried quickly, easily, and economically [34]. Dryers that have a chamber or tunnel configuration are used to dry shaped material. Nowadays, firing occurs in tunnel kilns at 900-1000 °C. The products sinter under the influence of the melt that forms from around 850 °C and upwards. Several new crystalline phases, especially those from the SiO_2-Al_2O_3-CaO system, may form during firing [35,36].

2.3 Earthenware

White non-vitreous ceramics with medium to high porosity that are glazed or unglazed are referred to as earthenware. The glazing allows for the masking of the body's appearance and the reduction of the high permeability caused by the presence of open porosity varying from 5to 20%. Although available in the form of decorative items and pottery, earthenware materials are utilized most frequently as wall tiles [11,37].

The most popular variety is triaxial earthenware, which is made of clay, quartz, and feldspar. The mix composition varies across a considerable range, and some varieties also include $CaCO_3$. The following ingredients make up hard feldspar earthenware: 50–55 % of the composition is clay, 35–45 % quartz, and 6–12 % feldspar. The lower firing temperature makes it different from porcelain because it prevents complete densification during the sintering process. Because of this, earthenware is opaque and has relatively low strength. In open pores, water absorption ranges from 8 % to 22 % [38].

Products made of earthenware are shaped via atomized powder pressing, plastic paste jiggering, and slip casting. The raw product is fried into biscuits after drying. The resulting porous biscuit is subsequently glazed, and a glaze firing is then performed at a temperature that is occasionally lower than or equal to that of biscuit firing. Earthenware can be fired at temperatures between 1150 °C and 1280 °C. The glaze is often of the borate-lead variety and is applied to a fired body that is then fired a second time at a lower temperature. When compared to firing porcelain, the method is less demanding; since the ware is fired below the temperature needed for complete sintering and with a relatively low quantity of melt created, there is no risk of deformation. Some decorations placed on the glaze must be fixed with a third firing at a lower temperature, especially those with gold or platinum content and those referred to as "low fire" decorations [39–41].

Earthenware is used in the manufacture of wall tiles, domestic and decorative ware, and sometimes-sanitary ware. Since the body is porous, the glaze also has the additional benefit of making the surface water-resistant. Earthenware that has a glaze on it is also known as faience, which comes from the Italian town of Faenza, which was a major producer of "white majolica" in the 16th century. Majolica is a type of double-fired ceramic having a porous body and an impermeable lead-tin oxide glaze. Before firing, painted colored decoration is applied to the ceramic, and it is then heated to between 1000 °C and 1100 °C. The term majolica was originally used for the lustred ware of Spanish-Moorish porters imported into Italy through the Majorca harbor [4,42].

For filtration needs, non-glazed earthenware is employed (diaphragms, bacteriological filters). The term "earthenware" is used to refer to a wide variety of high-porosity fine ceramics in a broader sense. For instance, the clay used to make wall tiles contains around 9% CaO and reacts when fired to form anorthite, which is composed of $CaO.Al_2O_3.SiO_2$. Wollastonite, $CaSiO_3$, can be used in addition to, or instead of, $CaCO_3$. Due to their modest shrinkage, these materials benefit from low firing temperatures (about 1100°C) and extremely tight manufacturing tolerances [6,43].

Sanitary products and some household items are made from materials with low water absorption rates, such as 0.5% or less. They share a similar composition to porcelain and earthenware, but in addition to feldspar or feldspar pegmatite, the mixture frequently includes additional ingredients acting as fluxes (limestone, dolomite, talc). The firing temperatures range from 1250 to 1300 °C, and the technique is substantially the same. These materials lack the translucency of porcelain. They have white or colored glazes applied to their surface. When the glaze is applied to greenware after drying and the glaze and body are fired together, significant cost savings and manufacturing speed are achieved [44,45].

Many older products contain common earthenware. Their production is very limited nowadays. They mostly consist of earthenware and glazed pottery [46,47].

Ceramics with a fusible clay paste glaze are quite like terra cotta products. They are made from typical clayey soils that are relatively fusible and naturally include a certain amount of sand, just like them. Their bodies are still porous despite sintering occurring between 900 °C and 1060 °C [48].

These products, which include dinnerware like jugs and pots as well as construction materials like enameled bricks and tiles, are typically covered with an engobe with pores that are smaller than those on the body are. The purpose of this engobe is to provide a uniform, smooth surface that can be utilized as a base for decorations and to conceal the body's coloration [49].

Certain decorative objects are traditionally made with an argilo-calcareous paste made by combining argillaceous marls, limestone, and sometimes sand. Dolomite ($CaMg(CO_3)_2$) and magnesium carbonate ($MgCO_3$) can also be used. The biscuits are sintered at temperatures ranging from 900 to 1060 °C. They are usually covered in a tin dioxide-opacified glaze, which explains the name stanniferous earthenware [50].

Fine earthenware has a white or very lightly colored body, a thin and regular texture, high mechanical strength, and a bright and durable glaze. They are commonly used as decorative items and crockery, where the quality of their enamel is highly valued [51].

Clays and kaolin make up the paste's clayey component. Kaolin promotes paste refractoriness, while polymeric clays increase raw material mechanical strength. The selected clays are extremely low in colorings, however very low levels of impurities like Fe_2O_3 and TiO_2 can mildly tint the body. Thus, the kaolin/clay ratio must be adjusted to balance biscuit whiteness and raw part resistance. Kaolin makes up 25–50% of clay basic materials. Calcareous earthenware can also contain grog, silica, and chalk. The biscuit is burned at 950–1150 °C in an oxidizing environment. A lot of remaining quartz shrinks the shard when cooled, strengthening the glaze's mechanical strength [52].

Feldspathic earthenware is fired at temperatures ranging from 1140 °C to 1230 °C. A feldspathic earthenware body is made up of kaolin (40–70 % of the mass), quartz (25–58 %), and feldspar (3–14 %). During firing, the feldspar component promotes the formation of a vitreous phase. Because of the significant amount of vitreous phase formed during firing, the biscuit has increased stability after cooling. The porosity value ranges between 10, and 15 %, which is lower than that found in other earthenware. The glaze is fired at temperatures ranging from 1000 to 1140 °C. The use of finer silica, which is more conducive to the formation of cristobalite, or the reduction of body porosity all contribute to the improvement of the body/glaze combination. Feldspathic earthenware is particularly suitable for tiling applications due to its extremely high stability, particularly scratch-resistant glaze, and low open porosity [53].

2.4 Stoneware

Stoneware is a dense ceramic that is typically yellow to brown in color, with water absorption ranging between 0 and 4 %, reaching 7 % in the case of tubes. They are made from a combination of vitrifying plastic clays and flux, which is sometimes supplemented with sand or grog. They are made by extrusion (pipes, bricks, etc.) or by pressing granulated powder (tiles, slabs, etc.). Drain and sewer pipes, floor tiles, chemical and kitchenware, and electric insulators are examples of stoneware products. The technological properties of the mixtures allow us to produce large objects. Stoneware is classified as coarse or fine based on its texture [54,55].

The firing temperature is a critical parameter that typically ranges between 1120 °C, and 1300 °C. In fact, sintering at insufficient temperatures results in the persistence of a significant open porosity, whereas firing at extremely high temperatures causes deformation of the pieces due to the excessively large quantity and low viscosity of the formed amorphous phase. Stoneware is well-known for its excellent mechanical properties as well as resistance to erosion and chemical agents [6,9].

The primary raw materials are stoneware clays, which sinter even at temperatures as low as 1200-1300°C but soften at much higher temperatures. As a result, they have a wide firing interval, allowing safe firing even for large ware [56].

The raw material composition is as follows: 40-50 % clay, 35-45 % quartz, and 7-20 % feldspar. Crushed stoneware shards provide suitable grog, and additional components such as talc, magnesite, and so on are added specifically to technical stoneware [4,57].

The surface of stoneware is usually covered with salt glaze, which forms spontaneously because of sodium chloride being introduced into the kiln at the end of firing. In the presence of water vapor, sodium chloride vapor reacts with SiO_2 and the other components to form a smooth silicate layer; the other option is feldspar glazes [6].

Natural stoneware is made from natural vitrifying clays, which can form a significant 1 amount of vitreous phase at high temperatures. They can be improved by incorporating kaolinitic refractory clay. Fe_2O_3 can account for up to 3 % of the mass of these raw materials' composition [58].

Mullite forms needle-like around 1000–1100 °C and up to 1200 °C, regardless of the fire environment. The vitreous phase dissolves the finest quartz grains after burning. Oxidized bodies change ivory to dark brown. Iron oxide and other contaminants color raw materials. Natural stoneware is pale yellow with titanium dioxide and dark with manganese oxide. To minimize blisters from clay sulphates, natural stoneware is baked in a reducing environment. Fe^{3+} ions decrease above 570°C. Active ferrous oxide fluxes produce gray bodies. In addition to alkaline derivatives, high iron levels can soften and distort burning items. Clays are limited and fire is arduous, thus stoneware is produced traditionally [6].

Fine-grained stoneware differs from natural stoneware in that the flux grains are no longer contained in the clay but are added as feldspars. Fine-grained stoneware is made from clay that is very poor in coloring, kaolin, ball clay, and a mixture of orthoclase and albite. Coloring agents are sometimes used to achieve a specific color in the mass of the product. Fine-grained stoneware is used as crockery, wall or floor tiles, antacid tiles, and sanitary pipes [59].

Dihydroxylation of clays occurs after 450 °C during firing. Shortly before 1000 °C, orthoclase begins to react with silica, resulting in liquid. The formation of mullite begins between 1000 and 1100 °C. At this temperature, certain micaceous phases found in clays can begin to react with metakaolin decomposition products. Albite's interaction with silica begins at 1140 °C. Because the maximum firing temperature is between 1250 and 1280 °C, amorphous silica derived from kaolinite that is not dissolved in liquid can be transformed into cristobalite. The degree of crystallization of SiO_2 in the final product is determined by the stoneware thermal past, the nature of the flux, and the mineralizing impurities [4, 59].

Porcelain stoneware exhibits an open porosity of less than 0.5%, resulting in outstanding mechanical strength and excellent resistance to frost and corrosive agents. Its use as a flooring material has grown rapidly, with global production rising from a few million square meters in the 1980s to over 150 million by 1997. This growth is attributed to advancements in grinding (wet processing), forming (high-capacity presses), sintering (rapid mono-layer sintering in roller-hearth kilns), and decoration techniques (polishing and simultaneous pressing of multiple enamel powder layers). These innovations have lowered manufacturing costs while enhancing the aesthetic quality of the final product [6].

Stoneware tiles are typically made from a combination of plastic clays, kaolin, feldspathic sand, sodium or potassium feldspar, and trace amounts of talc, dolomite, and/or chlorite. The body's overall chemical composition is less pure than that of fine-grained stoneware tiles, but it is also higher in Al_2O_3 [6].

The heat treatment's maximum temperature typically ranges between 1120 and 1200 °C. The phase evolution during this firing is very similar to that described for fine-grained stoneware. Mullite, amorphous phase, and quartz dominate the porcelain stoneware bodies. These bodies have no open porosity and a closed porosity of 7 to 13 %. This property gives these materials a high Young's modulus and rupture modules (around 75 GPa and 85 MPa, respectively), making them suitable for use as floor tiles. These moduli increase in proportion to the Al_2O_3 content, the amount of formed mullite phase, and the body density (reduction in closed porosity).

Grogged stoneware is made up of vitrifying clay, which can be rich in kaolinite or quartz, a small amount of flux, and a large amount of grog and/or ground shard (40 to 60 % of the mass). Slip casting in porous plaster molds is commonly used to shape it. The use of large-sized grog grains (up to 0.8 mm) increases the rigid skeleton's permeability and drying speed. It reduces the capillary forces that cause cracks to form [6].

Materials Research Foundations 184 (2025)   https://doi.org/21741/9781644903834

Englobe and glaze dry items before firing. Engobe and glaze are dipped or pulverized and applied. To achieve sufficient thickness without excessive rewetting of the dry portion, layers must be applied. After drying, it burns once. The stiff grain lattice does not shrink sufficiently to eliminate porosity during sintering. Open-poor remains 8–15% after fire between 1250 and 1280 °C. An opaque engobe hides body color, smooths surface flaws, and fixes the glaze. Grogged stoneware, which dries faster than most goods, is perfect for bulky objects. Sanitary plumbing uses them in sinks, shower basins, etc [6].

2.5 Porcelain

Porcelain is a non-porous, white ceramic material that is partially translucent. Porcelain requires extremely pure raw materials for production, and as a result, porcelain bodies are white and translucent. High-quality kaolin is required to produce porcelain. Porcelain manufacturing has thus developed in countries with suitable kaolin deposits [4,6].

Domestic porcelain is desirable for its whiteness and translucency. The amount of glassy phase, the fineness of the solid phases, and the differences in refractive indexes determine the translucency of porcelain. Porcelain glass phase, quartz, and mullite have refractive indexes of about 1.5, 1.54, and 1.64, respectively. The values are similar, which contributes to porcelain's characteristic translucency. The only phase with a significantly different refractive index is air contained within pores ($n = 1$). This is why translucency is strongly dependent on firing temperature, increasing to a maximum and then decreasing. The decrease is explained by higher porosity caused by gas separation from the oversaturated melt. Even with very low porosity, translucency is significantly reduced. It has been reported that with corundum ceramics, a porosity of 0.3 vol.% is sufficient to reduce translucency to 10% of that of non-porous dense material. In the case of porcelain, where translucency is a highly sensitive indicator of porosity, a similar relationship can be expected.

Another important property, particularly in the case of technical ware, is mechanical strength. The strength of standard porcelain is roughly equivalent to that of glass and is thus lower than that of typical ceramic crystalline phases. This implies that the glassy phase is an important factor in the strength of porcelain [6,9].

The firing temperature has a significant impact on the strength of porcelain. Peak strength is typically achieved below the temperature of complete sintering. The formation of secondary pores and the increased content of the glassy phase can explain the decreased strength caused by exceeding this temperature. When the glassy phase content exceeds a certain threshold, the effect of the crystalline phases present ceases to be significant. The relationship between strength and porosity is exponential, with a small number of pores causing a large decrease in strength [4,6].

The dependence of measured strength values on specimen size, loading rate, surface treatment, and so on is like that of other materials, such as glass. Despite its thinness in comparison to the body, the glaze has a strong effect. Because fracture begins due to tensile stressing at the surface, compressive pre-stressing in the surface layer is extremely effective. Glazing increases strength by 20-50%. Hard porcelain has quite favorable electrical properties, which can be improved further by lowering the alkali content and introducing divalent oxides, talc, etc [4,6].

Among porcelain's other characteristics, it should be noted that it has a moderately low expansion coefficient, adequate thermal shock resistance, and good chemical resistance. In laboratory porcelain, both characteristics are used [4].

They have no open porosity (0.5%) but are likely to have some large, closed pores. The porcelain body fracture is brilliant and has a vitreous appearance. The surface of the pieces is remarkably smooth and brilliant after glazing.

The classic porcelain body composition (hard porcelain) is 50 % kaolin, 25 % quartz, and 25 % feldspar. After firing, this mixture has the following theoretical composition: 69 % silica, 26 % aluminum oxide, and 5 % alkalis (calculated as K_2O from potassium feldspar). The raw material mix's component ratio varies depending on the desired properties of the ware. The ratio of feldspar to clay component, which is usually kaolin, but some may be replaced by a suitable clay, which imparts more favorable forming properties to the mix, plays a significant role. The amount of eutectic melt increases as the feldspar content increases, lowering the firing temperature. The product becomes highly translucent, but its mechanical and electrical properties deteriorate [4,6].

When porcelain is fired, a liquid phase surrounds and dissolves the finest grains. Viscosity is sufficiently high during this stage, known as "pasty fusion," for the deformation of the pieces to remain within acceptable limits. The formation of a large amount of vitreous phase results from the solidification of the liquid phase during cooling.

Manufacturing processes are constantly evolving. Thus, pressure casting and shaping by isostatic pressing gradually replace jiggering and casting in plaster molds when the geometry of the parts allows it. Fast firing techniques are becoming more popular for glaze and decoration firing. They improve part quality by reducing the possibility of deformation.

Hard porcelains are made from almost entirely kaolin, quartz, and feldspar mixtures. Hard porcelain is used in chemical and electro-technical applications, as well as tableware, among other things. To promote the formation of the viscous liquid phase, a small amount of chalk (about 2 % of the total mass) can be added. This mixture is very similar to that used in the production of fine earthenware. It is distinguished only by the almost exclusive use of kaolin as clay and the proportions of the various components [4,6].

Sintering occurs at 1350–1430 °C. Use of a reducing environment to reduce Fe^{3+} ions to Fe^{2+} ensures a white body (bluish reflections possible). A double-firing technique begins with a bisque firing in an oxidizing environment at 900–1050 °C. The product features glazing-ready stiffness and open porosity. New passage kilns with numerous heating zones and atmospheres can prepare specific hard-paste porcelain in one fire. An oxidizing environment crosses the heating zone up to 1050 °C to prevent carbon monoxide from depositing carbon (Boudouard equilibrium) in the remaining porous paste. Only in the sintering zone does the reducing atmosphere circulate.

Soft porcelains differ from the preceding porcelains in that they are more translucent and have a lower sintering temperature. This category includes Chinese porcelains and porcelains for dental implants. English porcelains, also known as bone china, belong to a distinct class.

Fine bone china porcelains, the most expensive, are prized for their beauty. They were named from the bone ash added to basic materials. The average fine bone china porcelain body has 37–50% bone ash, 22–32% potassium feldspar, 22–41% kaolin, and 0–4% quartz. Bone ash contains hydroxyapatite, $Ca_5(PO_4)_3OH$, which forms a low-viscosity liquid phase that rapidly increases at the solidus temperature. At over 1200 °C, this liquid phase dissolves free quartz. Only calcium phosphate, $Ca_3(PO_4)_2$, and liquid phase are present at the biscuit's maximum firing temperature

https://doi.org/21741/9781644903834

of 1250–1280 °C. Sintering shrinkage is more susceptible to heat treatment than hard porcelains and vitreous china because it depends on liquid phase amount and viscosity. Firing temperature and sintering duration must be adjusted to control piece size and deformation. Bone china porcelain is mostly calcium phosphate (35–45%), vitreous phase (27–30%), and anorthite (25–30%) [4].

Because of the lack of porosity in the body, the fixing of glaze presents challenges. It is fired at high temperatures ranging between 1120 and 1160 °C. This second fire must be done in a strictly oxidizing atmosphere to avoid the appearance of efflorescence caused by the decomposition of the glaze. The resulting brilliance of the enamel is highly dependent on the lead oxide content [4].

Beautiful quality bone china is the most expensive. For adding bone ash to simple materials, they were termed. Normal fine bone china porcelain comprises 37–50% bone ash, 22–32% potassium feldspar, 22–41% kaolin, and 0–4% quartz. At solidus temperature, hydroxyapatite, $Ca_5(PO_4)_3OH$, in bone ash rapidly increases in viscosity. Over 1200 °C, this liquid melts quartz. Biscuits contain solely calcium phosphate, $Ca_3(PO_4)_2$, and liquid phase at 1250–1280 °C. The liquid phase quantity and viscosity determine sintering shrinkage, making it more heat-sensitive than hard porcelains and vitreous china. Set firing temperature and sintering time to control item size and distortion. Bone china porcelain is mostly calcium phosphate (35–45%), vitreous phase (27–30%), and anorthite (25–30%).

Extra-aluminous porcelain is also available, with Al_2O_3 content ranging from 50 to 95% of the mass. To assist in their formation, 3 to 5% of very pure plastic clays and/or organic binders are added to the paste. When glaze is required, the raw parts are usually covered by dipping them in the glaze suspension. The body exhibits very little vitreous phase after sintering at temperatures ranging from 1430 to 1600°C.

These porcelains are used for their electrical properties (insulators, disconnecting switches, spark plugs, high voltage dielectrics) and hardness (cutting tools, wire guides, grinding balls). Because of low tolerances (machining after firing) or the large size of the piece, the dimensions of these technical ceramics can be severely constrained. As a result, the production conditions for the 4 m high and 1 m diameter aluminous porcelain insulators used for very high voltage electricity transmission had to be mastered [4,6].

2.6 Vitreous China

The term "vitreous china" refers to dense products made from pastes like those used to make feldspathic earthenware. The feldspar content of these pastes is increased to generate enough liquid phase to eliminate open porosity (0.5%) during firing. They are primarily used to produce sanitary articles and extremely durable crockery (wash basins, community crockery). Vitreous materials fall somewhere between white paste stoneware and porcelain.

These goods are made by jiggering, casting, or isostatic pressing. Understanding raw materials and shaping enables the fabrication of raw components with the mechanical strength to survive glazing. Sanitary goods are vitrified and glazed in one treatment at 1200–1280 °C in an oxidizing environment. Crockery is usually double-fired. The first fire was 900–950 °C. The glaze forms and open porosity disappear during the second firing at 1200–1250 °C. The opacified glaze can hold many decorations [4,6].

2.7 Refractories

Refractories are heat-resistant materials that can tolerate varying degrees of mechanical and thermal stress and strain, corrosion/erosion from solids, liquids, and gases, gas diffusion, and mechanical abrasion at various temperatures. The chemical compositions, production processes, physical forms/shapes, and application temperatures of refractories can all be used to classify them [60].

Chemical makeup classifies them as acidic, basic, or neutral. Slag, fume, and gas-resistant acidic refractories are resistant to high temperatures. These refractories are only used in acidic slag and air. SiO_2 and fireclay are acid refractories. Basic refractories, such as magnesite and dolomite, are damaged by acidic components yet remain stable in alkaline slags, as well as in the presence of dust, fumes, and high temperatures. Neutral refractories are employed in acidic or basic slag and atmospheres because they are chemically stable. Neutral refractories contain alumina and chromium oxide [1,60].

Different methods or processes are used to manufacture refractories. The manufacturing methods of refractors differ to achieve specific shapes, sizes, and properties. This classification includes pressed and fired refractories, fused cast refractories, and hand-molded refractories.

Classification is also important in terms of the physical form of the refractory. Refractories are classified into two types based on their physical shapes: shaped refractories and unshaped refractories. Shaped refractories, like bricks, have a fixed shape and size during production. Unshaped refractories, such as castables, are well-mixed granular masses with no specific shape or dimension during manufacturing [60].

Another consideration in refractory classification is application temperature (or heat duty). This classification denotes the maximum application temperature or the maximum temperature that a refractory can withstand. Refractory types are classified according to heat duty as super heat duty, high heat duty, medium heat duty, and low heat duty [60].

2.8 Glass

Glass is an amorphous solid that is typically formed by solidifying a melt without crystallization. Glass lacks the regular arrangement of atoms in a periodic lattice that characterizes crystals. There are several inorganic and organic substances that form glass when cooled from a liquid state at a rate that does not allow for the formation of a regular structural lattice [4,61].

For centuries, glass has been employed in structural applications such as windows, in packaging such as bottles, and household ware, including drinking glasses and goblets. The advent of radio, television, and advanced electronics in the twentieth century led to numerous new uses for glass, driving the development of compositions with enhanced electrical properties. Glasses are amorphous materials and are produced in a wide range of compositions [4,61].

Commercial glasses are classified into several categories. Vitreous silica, the least complex commercial glass in terms of composition, is used for optical fibers, ultraviolet and infrared transmitting optics, and high-temperature applications. Most glass produced, by weight, is either flat glass used for windows, mirrors, and automotive applications, container glass, or incandescent lamp glass. These glasses, which have very similar compositions, are known as soda-lime-silica glass [4,61].

Different glass varieties are made for certain uses. Cookware, chemical laboratory equipment, flat-panel displays like televisions, handheld gadgets, and fluorescent lamps employ borosilicate glasses for their thermal shock resistance and chemical longevity. Thermal and acoustic insulation and temperature-resistant fiberglass cloth use glass fibers, which contain minor amounts of boric oxide but are not borosilicate glasses. Traditional televisions and cathode ray tubes use lead-based glasses. Many more glass compositions are made in small quantities for optical, electrical, biological, and other uses [4,61].

The density, refractive index, thermal expansion coefficient, glass transformation temperature, strength, elastic modulus, and chemical durability are the most important properties of commercial glasses [4,61].

Table 2 encapsulates Section 2, Traditional Ceramics, emphasizing the primary categories, raw materials, processing techniques, firing conditions, essential qualities, aesthetic elements, and uses. It synthesizes intricate technical details into a comparative scientific format for enhanced reference and analysis.

Table 2. Summary of traditional ceramics categories, composition, processing, and applications.

Section & Category	Main Products	Raw Materials & Composition	Processing & Firing Conditions	Key Properties	Color & Appearance Factors	Applications
Terracotta	Pottery, roof tiles, bricks, flues, drainage pipes, floor tiles	Common clays + additives (plastic paste with water)	Shaping → Drying (ventilated/tunnel) → Firing 900–1160 °C	Porous, mechanically resistant, hydrothermal & acoustic stability	Fe^{3+} in anorthite → yellow; hematite → red-brown; Fe^{2+} compounds → blue-green	Decorative, architectural, functional construction
Brickware	Bricks, roof tiles, building elements, floor slabs, panels	Low-grade colored clays + sand, slag, fly ash; no soluble salts or large $CaCO_3$ particles	Extrusion or pressing → Drying → Tunnel kiln firing 900–1000 °C	Strength 6–60 MPa; frost resistance; density 1.0–1.8 g/cm^3	Depends on pore size distribution & firing temperature	Structural construction, flooring, sewer/drainage
Earthenware	Wall tiles, decorative ware, pottery, sanitary ware	Clay, quartz, feldspar (triaxial mix), sometimes $CaCO_3$	Powder pressing/jiggering/slip casting → Biscuit firing 1150–1280 °C → Glaze firing	Medium-high porosity (8–22%); opaque; low strength	Glazed (borate-lead) for water resistance & aesthetics	Tiles, domestic ware, decorative pottery
Stoneware	Pipes, bricks, tiles, slabs, sanitary ware	Plastic clays, quartz, feldspar, sand/grog; technical stoneware with talc/magnesite	Extrusion/pressing → Firing 1120–1300 °C	Dense, high strength, low water absorption (0–4%)	Salt/feldspar glazes; iron & impurity content affect color	Sewer pipes, flooring, chemical ware, insulators

Porcelain	Tableware, technical ceramics, bone china, aluminous porcelain	Kaolin, quartz, feldspar; bone china with bone ash; aluminous with high Al_2O_3	Forming → Biscuit firing 900–1050 °C → Glaze firing 1350–1430 °C	White, translucent, non-porous (<0.5%), high mechanical strength	Color/translucency depends on glassy phase & porosity	Tableware, lab ware, electrical insulators, dental
Vitreous China	Sanitary ware, durable crockery	Feldspathic earthenware-like body with higher feldspar	Jiggering/casting/isostatic pressing → Single or double firing (1200–1280 °C)	Dense, vitrified, low porosity (0.5%)	Opacified glaze with decorations	Wash basins, toilets, community crockery
Refractories	Bricks, castables, shapes for high-temp use	Acidic (SiO_2, fireclay), basic (MgO, dolomite), neutral (Al_2O_3, Cr_2O_3)	Pressing, firing, fused casting, hand molding	Withstand high temp & corrosion; classified by heat duty	Depends on slag/atmosphere compatibility	Furnaces, kilns, reactors
Glass	Windows, bottles, tableware, optical fibers	Soda-lime-silica, borosilicate, vitreous silica, lead glass	Melting → Forming → Cooling without crystallization	Amorphous, transparent, durable, variable refractive index	Composition controls thermal, optical properties	Construction, electronics, optics, packaging

3. Advanced Ceramics

3.1 Non-oxide ceramics

Non-oxide ceramics are classified into three main categories: Nitrides, carbides, and borides. These materials have low density, high hardness, high melting points, and thermal and mechanical properties that make them suitable for advanced engineering applications. However, some types may suffer from brittleness or performance degradation when exposed to ultra-high temperatures [62–64].

Aluminum nitride (AlN) is an advanced ceramic material with high thermal conductivity and outstanding thermal shock resistance, making it an ideal choice for heat management applications in high-density electronic systems, where performance requirements dictate thermal and mechanical stability in harsh operating conditions. Boron nitride (BN) exists in two main crystalline forms: The hexagonal form (h-BN), which acts as a solid lubricant that is resistant to oxidation and chemically stable at high temperatures, and the cubic form (c-BN), which is the second hardest known material after diamond and is widely used in high-performance cutting tools due to its hardness and high wear resistance. In contrast, silicon nitride (Si_3N_4) has a unique combination of high thermal shock resistance, mechanical hardness, and structural stability, making it an ideal material for mechanical, turbine, and engine components subjected to high temperatures and repeated thermal stresses, while maintaining a long service life and consistent performance [62,64–66].

3.2 Oxide ceramics

Alumina occurs naturally in minerals such as corundum, diaspore, and gibbsite, and is industrially extracted from bauxite and laterite ores via the Bayer process, which involves grinding, digestion with sodium hydroxide, separation, and calcination. Alumina can be used in single-crystal form (sapphire and ruby) or as polycrystalline Al_2O_3. Colorless synthetic sapphire is used as a high-strength optical component in aircraft cabin windows and laser assemblies, whereas polycrystalline alumina is widely employed in refractory applications, electrical insulators, wear-resistant mechanical parts, and abrasive materials. Alumina is also biocompatible, enabling its use in medical implants such as artificial eye lenses and orthopedic components. Furthermore, it is a key material in ceramic armor due to its combination of high mechanical strength, low density, and availability [67,68].

Beryllium oxide (BeO), or beryllia, combines exceptionally high thermal conductivity comparable to that of pure metals with excellent electrical insulation, making it highly suitable for applications that require efficient heat dissipation without compromising electrical isolation. It possesses high resistance to wetting and corrosion by many metals and chemicals, mechanical properties comparable to those of high-purity alumina, and valuable nuclear characteristics, including a very low thermal neutron absorption cross-section and the ability to moderate fast neutrons. These properties enable its use in microwave tube components, high-power transistor substrates, crucibles for melting uranium, thorium, and beryllium, as well as various nuclear reactor components. BeO surfaces can be metallized using thick- and thin-film techniques for hermetic electronic packaging; however, inhalation of submicron beryllia particles presents serious health hazards to a small percentage of the population, which significantly restricts its industrial applications despite its outstanding technical performance.[69,70].

Zirconium oxide (ZrO_2), or zirconia, is a high-performance ceramic distinguished by its high hardness, density, thermal insulation capability, chemical inertness, thermal expansion coefficient comparable to steel, and relatively high fracture toughness compared to many other ceramics. The tetragonal-to-monoclinic phase transformation during cooling causes a volume increase and structural cracking, limiting the high-temperature applications of pure zirconia. To overcome this, stabilization is achieved by adding oxides such as CaO, MgO, or Y_2O_3 in controlled amounts. Full stabilization produces a cubic structure that remains stable over a broad temperature range, whereas partial stabilization yields tetragonal zirconia polycrystal (TZP), which exhibits superior strength and fracture toughness through transformation toughening—a mechanism in which tetragonal grains transform to monoclinic at crack tips, absorbing fracture energy and inhibiting crack propagation. These enhanced properties make stabilized zirconia suitable for applications such as ceramic knives, dental crowns, pump components, bearings, and thermal insulation elements in turbine systems [71–73].

Table 3 encapsulates Section 3 on Advanced Ceramics, categorizing materials into non-oxide (nitrides, carbides, borides) and oxide ceramics, while elucidating their properties and uses. It emphasizes the correlation among material composition, performance, and appropriateness for extreme or advanced technological settings.

Table 3. Summary of advanced ceramics: non-oxide and oxide categories, properties, and applications.

Category	Subcategory Material	Key Properties	Main Applications
Non-oxide Ceramics	Nitrides	High hardness, thermal stability, good wear and corrosion resistance; some with high thermal conductivity (e.g., AlN)	Crucibles, cutting tools, electronic substrates, armor
	Aluminum Nitride (AlN)	High thermal conductivity, low expansion, high electrical resistivity, corrosion resistance	Electronics cooling, substrates, armor, molten metal handling
	Boron Nitride (BN)	Forms: hexagonal (lubricant, insulator), cubic (hardness ~diamond, high thermal conductivity)	Grinding tools, heat sinks, refractories, crucibles
	Silicon Nitride (Si3N4)	High strength, low expansion, excellent thermal shock resistance	Automotive parts, turbines, bearings, armor
Non-oxide Ceramics	Carbides	Extremely hard, high wear resistance, high temperature stability	Cutting/drilling tools, armor, wear parts
	Tungsten Carbide (WC)	Very hard, wear-resistant, often in metal-matrix composites	Cutting/drilling tools, dies, and industrial wear parts
	Boron Carbide (B4C)	Very hard, lightweight, neutron absorber	Armor, nuclear applications, abrasives
	Silicon Carbide (SiC)	High strength/stiffness, corrosion resistance, low density	Structural components, abrasives, high-temp parts
Non-oxide Ceramics	Borides	High melting point, good electrical conductivity, wear resistance	Cutting tools, armor, electrodes
	Titanium Diboride (TiB2)	High hardness, thermal/electrical conductivity	Armor, cutting tools, wear coatings
	Zirconium Diboride (ZrB2)	Ultra-high temp stability, corrosion resistance	Hypersonic vehicle components, electrodes
Oxide Ceramics	Aluminum Oxide (Al2O3)	High hardness, wear resistance, chemical stability, biocompatibility	Armor, refractories, medical implants, electronics
	Beryllium Oxide (BeO)	High thermal conductivity, good dielectric, nuclear properties	Electronics, nuclear reactor parts
	Zirconia (ZrO2)	High fracture toughness, wear resistance, transformation toughening	Cutting tools, dental implants, oxygen sensors

4. Advanced processing and manufacturing techniques of ceramics

Advanced ceramic manufacturing depends on meticulous powder preparation to attain optimal particle size distribution and excellent purity, hence providing exceptional final qualities. Forming processes encompass traditional dry pressing, slip casting, and extrusion, as well as sophisticated technologies like injection molding and additive manufacturing (3D printing), facilitating intricate geometries with minimal imperfections. Sintering techniques, such as conventional firing, hot pressing, hot isostatic pressing (HIP), microwave sintering, and spark plasma sintering (SPS), are utilized to improve densification and regulate microstructure. Surface treatments and coatings, including physical and chemical vapor deposition (PVD, CVD); enhance wear resistance, corrosion protection, and functional efficacy. Current advancements aim to decrease energy consumption, reduce shrinkage and cracking, and provide eco-friendly manufacturing methods, enhancing the competitiveness of advanced ceramics in high-performance applications [74–76].

5. Smart Ceramics and Applications

Smart ceramics are a class of advanced ceramic materials designed to exhibit responsive and adaptive properties under external stimuli such as mechanical stress, electric or magnetic fields, temperature variations, or chemical environments. They combine the traditional advantages of ceramics, such as high hardness, thermal stability, and chemical inertness, with functional capabilities that enable sensing, actuation, and energy conversion [77–79].

5.1 Types of Smart Ceramics

Smart ceramics can be classified according to their functional mechanisms into several main types. Piezoelectric ceramics, such as lead zirconate titanate (PZT), are capable of converting mechanical stress into an electrical signal and vice versa, which makes them essential in sensors, actuators, ultrasound transducers, and precision positioning systems. Electro-optic ceramics change their optical properties when subjected to an electric field, enabling their use in laser modulation, optical switching, and advanced telecommunication systems. Magnetoelectric ceramics exhibit coupling between magnetic and electric fields, opening the way for multifunctional devices that integrate sensing and energy harvesting capabilities. Finally, thermoelectric ceramics transform temperature gradients into electrical voltage, offering promising applications in waste-heat recovery systems and renewable energy technologies [80–82].

5.2 Processing and Manufacturing Techniques

The fabrication of smart ceramics demands precise control over microstructure and composition to achieve the desired functional responses. Advanced processing approaches include additive manufacturing (3D printing), which enables the production of complex geometries with integrated functionalities; Spark Plasma Sintering (SPS), a rapid densification technique that minimizes grain growth while preserving microstructural uniformity; and Chemical and Physical Vapor Deposition (CVD, PVD), which are employed to deposit thin-film coatings that enhance surface properties, improve wear and corrosion resistance, or introduce additional functional layers[80–82].

5.3 Applications

Smart ceramics are increasingly used across diverse advanced technology sectors due to their unique functional capabilities. In biomedical engineering, they enable the development of implantable biosensors, drug delivery systems, and prosthetic feedback devices that enhance patient care and rehabilitation. Within energy systems, thermoelectric generators, piezoelectric energy harvesters, and self-powered sensor networks leverage their ability to convert mechanical or thermal energy into electricity. In aerospace and defense, smart ceramics are employed in adaptive armor, vibration-damping structures, and high-performance navigation systems capable of operating under extreme conditions. In robotics and automation, they serve as key components in micro-actuators and precision control systems, supporting the miniaturization and efficiency of next-generation robotic platforms [65,83–85].

6. Conclusion

This chapter has detailed the structural and technological classification of ceramic materials, providing an extensive overview of both traditional and advanced ceramics. Traditional ceramics, which have been in use for centuries, are still evolving due to advances in raw material processing, compositional optimization, and firing techniques. Terracotta, brickware, and porcelain continue to be used in architectural, residential, and sanitary applications where porosity, mechanical strength, and aesthetic quality are critical performance indicators. When determining final qualities like color and phase development, the interaction of firing conditions and chemical composition is particularly important. On the other hand, advanced ceramics are the frontier of material science and are distinguished by their remarkable mechanical strength, corrosion resistance, thermal stability, and functional diversity. Materials like silicon nitride, aluminum nitride, and zirconium diboride demonstrate the complexity and engineering precision required to meet the needs of industry. Sintering additives, microstructure control, and processing methods are highlighted as critical components in realizing the full potential of these ceramics.

Ceramic materials have great potential in industry. Some suggestions can be offered to expand the influence of ceramics across both traditional applications and emerging high-technology fields. The industry can integrate advanced characterization techniques in the production processes to prevent failure. Research should be carried out on eco-friendly and cost-effective raw material alternatives in addition to recycling studies. Also, Artificial Intelligence (AI) assisted with new smart production and characterization techniques should be studied. In conclusion, this chapter's structural and technological classification framework offers a fundamental comprehension of the wide range of ceramic materials. Furthermore, it provides a basis for future innovation that aims to satisfy the evolving demands of today's society and industry.

References

[1]   R. Sarkar, Refractory technology: fundamentals and applications, CRC Press, 2023.

[2]   A. Bin Rashid, M. Haque, S.M.M. Islam, K.M.R. Uddin Labib, P. Chowdhury, Breaking boundaries with ceramic matrix composites: A comprehensive overview of materials, manufacturing techniques, transformative applications, recent advancements, and future prospects, Adv. Mater. Sci. Eng. 2024 (2024) 2112358.

[3]   J. Li, X. Wang, S. Guo, D. Zhang, J. Qi, Y. Wang, Recent progress in the fabrication strategies and toughening mechanism of flexible ceramics and their applications, J. Mater. Chem. C (2024).

[4]   J. Hlavac, The Technology of Glass and Ceramics: An Introduction, Elsevier Sci. Pub. Co. (1983).

[5]   W.R. Matizamhuka, Advanced ceramics-the new frontier in modern-day technology: Part I, J. South. African Inst. Min. Metall. 118 (2018) 757–764.

[6]   J.P. Bonnet, J.M. Gaillard, Ceramic Materials: Processes, Properties, and Applications - Google Kitaplar, (2007).

[7]   E. Rambaldi, Pathway towards a high recycling content in traditional ceramics, Ceramics 4 (2021) 486–501.

[8]   A. Abdullayev, M.F. Bekheet, D.A.H. Hanaor, A. Gurlo, Materials and Applications for Low-Cost Ceramic Membranes., Membranes (Basel). 9 (2019). https://doi.org/10.3390/membranes9090105.

[9]   C.B. Carter, M.G. Norton, Ceramic materials: Science and engineering, Ceram. Mater.

Sci. Eng. (2013) 1–766. https://doi.org/10.1007/978-1-4614-3523-5/COVER.

[10] V. Rastogi, S. Chaurasia, Advances in and Future Perspectives on High-Power Ceramic Lasers, in: Photonics, MDPI, 2024: p. 942.

[11] B.T. Touolak, F.T. Nya, More value of Maroua clay in the formulation of ceramic products (terracotta, earthenware, stoneware, porcelain), Adv. Mater. Phys. Chem. 4 (2014) 284–299.

[12] S. Devarajan, K. Ambika, Analyzing the Acoustic Performance of Unglazed Terracotta in an Indoor Office Environment, J. Salut. Archit. 3 (2024) 64–79.

[13] A. Almusaed, I. Yitmen, A. Almssad, Contemporary innovations and sustainable practices in the application of clay materials within architectural design and construction methodologies, in: Dev. Clay Sci. Constr. Tech., IntechOpen, 2024.

[14] X. Yin, C. Guo, B. Sun, H. Chen, H. Wang, A. Li, The state of the Art in digital construction of clay buildings: reviews of existing practices and recommendations for future development, Buildings 13 (2023) 2381.

[15] A. Rayhan, D. Gross, P. Hemayetpur, Optimizing Efficiency and Viability in Automatic Clay Brick Production: A Case Study, (n.d.).

[16] A. Taher, S. Aşut, W. van der Spoel, An integrated workflow for designing and fabricating multi-functional building components through additive manufacturing with clay, Buildings 13 (2023) 2676.

[17] G. Thalmaier, N. Cobîrzan, A.-A. Balog, H. Constantinescu, A. Ceclan, M. Voinea, T.F. Marinca, Assessment of limestone waste addition for fired clay bricks, Materials (Basel). 15 (2022) 4263.

[18] A. Idoum, R. Tiskatine, S. Mounir, L. Bammou, A. Aharoune, Characterization and Suitability Assessment of Ameskroud Clay Deposits for Ceramic Applications, Adv. Mater. Sci. Eng. 2024 (2024) 5531545.

[19] J. Shi, Q. Chun, Z. Mi, S. Feng, C. Liu, Z. Liu, D. Wang, Y. Zhang, Comparative study on material properties of ancient fired clay bricks of China, Case Stud. Constr. Mater. 19 (2023) e02463. https://doi.org/https://doi.org/10.1016/j.cscm.2023.e02463.

[20] A. Pinto, J. Groenen, B. Zhao, T. Zhu, P. Sciau, Chromogenic mechanisms in blue-and-white porcelains, J. Eur. Ceram. Soc. 40 (2020) 6181–6187.

[21] J.E. Shigley, Analytical Techniques in Gemology: A Historical Overview, GEMS Gemol. 60 (2024).

[22] M. Pyzalski, T. Brylewski, A. Sujak, K. Durczak, Changes in the Phase Composition of Calcium Aluminoferrites Based on the Synthesis Condition and Al_2O_3/Fe_2O_3 Molar Ratio., Mater. (Basel, Switzerland) 16 (2023). https://doi.org/10.3390/ma16124234.

[23] L. Bloomfield, Colour in glazes, Bloomsbury Publishing, 2022.

[24] Y. Xu, P. Song, W. Cao, H. Li, J. Liang, Effect of $Al2O3$–$SiO2$ Addition on Gehlenite Growth and the Mechanical Performance of Steel Slag, Crystals 11 (2021) 936.

[25] T. Chen, T. Peng, H. Sun, X. Li, Effect of $SiO2/Al2O3$ ratio on sintering behavior, crystallization behavior and properties of diopside-anorthite glass-ceramics, J. Asian Ceram. Soc. 11 (2023) 316–329.

[26] K. Wiśniewska, W. Pichór, E. Kłosek-Wawrzyn, Influence of Firing Temperature on

Phase Composition and Color Properties of Ceramic Tile Bodies., Mater. (Basel, Switzerland) 14 (2021). https://doi.org/10.3390/ma14216380.

[27] K.S. Park, R. Milke, I. Efthimiopoulos, R.-R. Pausewein, S. Reinhold, Pyrometamorphic process of ceramic composite materials in pottery production in the Bronze/Iron Age of the Northern Caucasus (Russia), Sci. Rep. 9 (2019) 10725.

[28] D. Behera, K.-Y. Liu, F. Rachman, A.M. Worku, Innovations and Applications in Lightweight Concrete: Review of Current Practices and Future Directions, Buildings 15 (2025) 2113.

[29] M. Dwarampudi, B. Venkateshwari, Performance of light weight concrete with different aggregates—a comprehensive review, Discov. Civ. Eng. 1 (2024) 46.

[30] M.A.H. Emu, Business analysis of brick manufacturing company: Messrs. Munshi Bricks, (2025).

[31] J.-B. Liu, Z.-J. Zhang, Characteristics and weathering mechanisms of the traditional Chinese blue brick from the ancient city of Ping Yao., R. Soc. Open Sci. 7 (2020) 200058. https://doi.org/10.1098/rsos.200058.

[32] I. Netinger Grubeša, M. Vračević, V. Ducman, B. Marković, I. Szenti, Á. Kukovecz, Influence of the size and type of pores on brick resistance to freeze-thaw cycles, Materials (Basel). 13 (2020) 3717.

[33] A.B. Poole, R. Hughes, Materials for construction in deserts, (2012).

[34] D. Johnson, Brickmaking: History and Heritage, Amberley Publishing Limited, 2021.

[35] F.L. Olsen, The kiln book, Dorrance Publishing, 2023.

[36] V. Du Phan, X.H. Nguyen, V.N. Dinh, T.S. Dang, V.C. Le, S.P. Ho, H.C. Ta, D.T. Duong, T.A. Mai, Development of an adaptive fuzzy-neural controller for temperature control in a brick tunnel kiln, Electronics 13 (2024) 342.

[37] M. Gasparian, Ceramics in Architecture: Enabling the design, manufacturing and integration of decorative functional architectural ceramics through design and ceramic practices, (2021).

[38] Y. Sawadogo, M. Sawadogo, N. Sory, M. Ouedraogo, K. Dao, M. Seynou, P. Blanchart, L. Zerbo, Porcelain: raw materials, technological properties and applications--a review., J. La Société Ouest-Africaine Chim. 53 (2024).

[39] J.A. Burrison, Beautiful Clay: A Traditional Craft as Art, Indiana University Press, 2025.

[40] N.P. Sanchez, Staining and Aging Dependent Changes on Color and Translucency of 3D-Printed Resin-Modified Ceramics, (2023).

[41] P.J. Knobloch, M.D. Glascock, B.L. MacDonald, Picking up the pieces: Instrumental Neutron Activation Analysis (INAA) of early intermediate period and middle horizon pottery from Ayacucho, Peru, Ñawpa Pacha 43 (2023) 55–98.

[42] N. Chaukura, W. Moyo, T.A. Kajau, A.A. Muleja, B.B. Mamba, T.T.I. Nkambule, Low-cost ceramic filtration for point-of-use water treatment in low-income countries, Water Secur. 20 (2023) 100145.

[43] H. Alves de Oliveira, C. Pereira dos Santos, Limestone Clays for Ceramic Industry, in: Clay Sci. Technol., IntechOpen, 2021. https://doi.org/10.5772/intechopen.92506.

[44] S. Conte, C. Molinari, M. Ardit, G. Cruciani, M. Dondi, C. Zanelli, Porcelain versus Porcelain Stoneware: So Close, So Different. Sintering Kinetics, Phase Evolution, and Vitrification Paths., Mater. (Basel, Switzerland) 16 (2022). https://doi.org/10.3390/ma16010171.

[45] A. Bernasconi, V. Diella, A. Pagani, A. Pavese, F. Francescon, K. Young, J. Stuart, L. Tunnicliffe, The role of firing temperature, firing time and quartz grain size on phase-formation, thermal dilatation and water absorption in sanitary-ware vitreous bodies, J. Eur. Ceram. Soc. 31 (2011) 1353–1360.

[46] M.S. Tite, Ceramic production, provenance and use—a review, Archaeometry 50 (2008) 216–231.

[47] P. Colomban, Glass, Pottery and enamelled objects: Identification of their technology and origin, Conserv. Sci. Herit. Mater. 2nd Ed. P. Garside E. Richardson Eds, RSC (2020).

[48] E. Gliozzo, Ceramic technology. How to reconstruct the firing process, Archaeol. Anthropol. Sci. 12 (2020) 260. https://doi.org/10.1007/s12520-020-01133-y.

[49] L.A. Dobrzański, L.B. Dobrzański, A.D. Dobrzańska-Danikiewicz, M. Kraszewska, Manufacturing powders of metals, their alloys and ceramics and the importance of conventional and additive technologies for products manufacturing in Industry 4.0 stage, Arch. Mater. Sci. Eng. 102 (2020).

[50] J. Coll Conesa, Valencian Tin-Glazed Earthenware and Technological Change: A Mediterranean Industry, Hist. Archaeol. 57 (2023) 1212–1230.

[51] P. Colomban, Glass, Ceramics and Enamelled Objects, Conserv. Sci. Herit. Mater. (2023) 200–247. https://doi.org/10.1039/bk9781788010931-00200.

[52] G. St George, How Clays Work: Science and Applications of Clays and Clay-Like Minerals in Health and Beauty, Galina St George, 2021.

[53] A.I.M. Ismail, M.S. Elmaghraby, B.N.A. Shalaby, Flux ceramic tiles based on Egyptian trachyte, Bull. Natl. Res. Cent. 46 (2022) 226.

[54] E. County, H. Handbook, Ceramics Greenware & Handbuilding Checklist, (n.d.).

[55] S. Pryke, L. Bloomfield, Design and Create Contemporary Tableware: Making Pottery You Can Use, Bloomsbury Publishing, 2023.

[56] M.F. Zawrah, M.A. Taha, R.A. Youness, Advanced ceramics: stages of development, in: Adv. Ceram., Springer, 2023: pp. 1–46.

[57] P.S. Quinn, Thin section petrography, geochemistry and scanning electron microscopy of archaeological ceramics, (2022).

[58] S. Conte, C. Molinari, M. Ardit, D. Giordano, M. Dondi, C. Zanelli, Vitrification paths in porcelain Stoneware: Dependence on bulk chemical composition and effect on sintering behaviour, Ceram. Int. (2025).

[59] A.E. Dodd, Dictionary of ceramics, CRC Press, 2024.

[60] S. Banerjee, Properties of refractories , (2004).

[61] E.J. Shelby, Introduction to Glass Science and Technology, 3rd Edition , (n.d.).

[62] R.B.. Heimann, Classic and advanced ceramics : from fundamentals to applications, (2010) 553.

[63] B. Basu, K. Balani, Advanced Structural Ceramics, Adv. Struct. Ceram. (2011).
      https://doi.org/10.1002/9781118037300;CTYPE:STRING:BOOK.

[64] M. Srinivasan, W. Rafaniello, Non-Oxide Materials: Applications and Engineering,
      Carbide, Nitride Boride Mater. Synth. Process. (1997) 3–42. https://doi.org/10.1007/978-
      94-009-0071-4_1.

[65] R. Nagraik, A. Sharma, D. Kumar, S. Mukherjee, F. Sen, A.P. Kumar, Amalgamation of
      biosensors and nanotechnology in disease diagnosis: Mini-review, Sensors Int. 2 (2021)
      100089. https://doi.org/https://doi.org/10.1016/j.sintl.2021.100089.

[66] F. Mutlag, H. Elaibi, R. Mahious, E. Halvacı, F. Şen, Recent advances in enzymatic fuel
      cells, biocatalytic fuel cells, biofuel cells for clean and efficient energy harvesting, Int. J.
      Boron Sci. Nanotechnol. (2025) 1–23.

[67] A.M. Abyzov, Aluminum oxide and alumina ceramics (review). Part 1. Properties of
      Al2O3 and commercial production of dispersed Al2O3, Refract. Ind. Ceram. 60 (2019)
      24–32.

[68] A. Hojjati-Najafabadi, A. Aygun, R.N.E. Tiri, F. Gulbagca, M.I. Lounissaa, P. Feng, F.
      Karimi, F. Sen, Bacillus thuringiensis Based Ruthenium/Nickel Co-Doped Zinc as a
      Green Nanocatalyst: Enhanced Photocatalytic Activity, Mechanism, and Efficient H2
      Production from Sodium Borohydride Methanolysis, Ind. Eng. Chem. Res. 62 (2023)
      4655–4664. https://doi.org/10.1021/acs.iecr.2c03833.

[69] C.A. Harper, Handbook of ceramics, glasses, and diamonds , (2001).

[70] F. Mutlag, H.K. Elaibi, HasanAA (2023) The Effect of Vitamin D and Parathyroid on
      Hepatitis C Patients and Non-Hepatitis C Patients on Chronic Hemodialysis Patients, J
      Appl Microbiol Biochem 7 (2023) 187.

[71] S. Eris, Z. Daşdelen, F. Sen, Investigation of electrocatalytic activity and stability of
      Pt@f-VC catalyst prepared by in-situ synthesis for Methanol electrooxidation, Int. J.
      Hydrogen Energy 43 (2018) 385–390.
      https://doi.org/https://doi.org/10.1016/j.ijhydene.2017.11.063.

[72] F.F. MUTLA, H.K. ELAIBI, The Change Complete Blood Count And Other
      Inflammatory Markers Before And After Sinopharm Coronavirus Vaccine, (n.d.).

[73] U.M.B. Al-Naib, Introductory Chapter: Introduction to Zirconia Ceramic–A Versatile and
      Durable Material with a Wide Range of Applications, in: Zirconia-New Adv. Struct. Fabr.
      Appl., IntechOpen, 2023.

[74] A. Ullah, M. Shah, Z. Ali, K. Asami, A. Ur Rehman, C. Emmelmann, Additive
      manufacturing of ceramics via the laser powder bed fusion process, Int. J. Appl. Ceram.
      Technol. 22 (2025) e15087.

[75] K.S. Randhawa, A state-of-the-art review on advanced ceramic materials: fabrication,
      characteristics, applications, and wettability, Pigment Resin Technol. 53 (2024) 768–785.

[76] K.P. Misra, R.D.K. Misra, Advanced ceramics, in: Ceram. Sci. Eng., Elsevier, 2022: pp.
      21–43.

[77] A.K. Mishra, Smart ceramics: preparation, properties, and applications, CRC Press, 2018.

[78] H.K. Bowen, Advanced ceramics, Sci. Am. 255 (1986) 168–177.

[79] R.E. Newnham, Smart Ceramics, Ceram. Soc. (1995) 7–30.

[80] R.E. Newnham, G.R. Ruschau, Smart electroceramics, J. Am. Ceram. Soc. 74 (1991) 463–480.

[81] R. Roy, D. Bhakta, S. Roy, Applications and types of smart materials, in: Model. Charact. Process. Smart Mater., IGI Global, 2023: pp. 266–296.

[82] Y. Yıldız, İ. Esirden, E. Erken, E. Demir, M. Kaya, F. Şen, Microwave (Mw)-assisted synthesis of 5-substituted 1H-tetrazoles via [3+ 2] cycloaddition catalyzed by Mw-Pd/Co nanoparticles decorated on multi-walled carbon nanotubes, ChemistrySelect 1 (2016) 1695–1701.

[83] J.A.S. Khdir, D. Aziz, I. Qader, B.I.M. Meena, B.M. İbrahim, Nano-Catalytic Synthesis of 5-Substituted 1H Tetrazole Derivatives and Biological Applications, J. Turkish Chem. Soc. Sect. A Chem. 11 (n.d.) 1495–1514.

[84] S. Kothari, C. Ameta, K.L. Ameta, B.K. Sharma, R. Ameta, R. Ameta, Microwave-Assisted Organic Synthesis: A Need of the Day, in: Green Chem. 2nd Ed., Apple Academic Press, 2023: pp. 399–452.

[85] F. Gol, E. Kacar, Z.G. Saritas, S. Cibuk, C. Ture, M. Arslan, F. Sen, The use of boron based materials on efficiency of environmentally friendly porous ceramics, Environ. Res. 216 (2023) 114454. https://doi.org/10.1016/j.envres.2022.114454

A Journey from Raw Materials to Ceramics

Materials Research Foundations 184 (2025)

Materials Research Forum LLC

https://doi.org/21741/9781644903834

Chapter 4

Industrial Ceramic Clay Preparation Methods

Nihal Derin Coskun [1*], Nazım Kunduracı[2], Selcuk Erdogan[3], Cumhur Eren Isık[4], Iskender Isık[5], Fatih Sen[3*]

[1]Ceramic and Glass Department, Ordu University, 52200 Ordu, Türkiye

[2]Bulent Ecevit University, Department of Metallurgical and Materials Engineering, 67000 Zonguldak, Türkiye

[3]Sen Research Group, Department of Biochemistry, Kutahya Dumlupinar University, Kutahya 43000, Türkiye

[4]Department of Handicrafts, Kutahya Fine Arts Vocational School, Kutahya Dumlupınar University, Evliya Çelebi Campus, 43000 Kutahya, Türkiye

[5]Department of Materials Science & Engineering, Faculty of Engineering, Kutahya Dumlupınar University, Evliya Çelebi Campus, 43100 Kutahya, Türkiye

nihalderincoskun@odu.edu.tr, fatihsen1980@gmail.com

Abstract

The preparation of industrial ceramic clay is a key factor in determining final product quality, directly influencing plasticity, forming ease, and defect reduction during drying and firing. This work aims to outline the fundamentals and main stages of clay preparation, emphasizing the impact of raw material selection, particle size distribution, and plasticity control, while reviewing modern quality control practices and sustainability measures. The methodology includes selecting raw materials based on purity and optimal SiO_2/Al_2O_3 ratios, followed by crushing, grinding, and homogenization. Both wet and dry processing routes are employed, with additional treatments such as magnetic separation for iron removal and property adjustments for moisture, viscosity, and density. Drying and pre-firing procedures are optimized to minimize deformation and cracking. Results show that wet processing achieves high homogeneity and ideal plasticity, while dry processing offers water savings and suitability for specific industrial uses. Accurate control of plasticity, moisture content, and particle size distribution significantly improves mechanical strength and reduces defects. It is concluded that enhancing industrial ceramic clay quality requires integrating proper raw material selection with advanced preparation and property control techniques. Adopting sustainable practices such as waste recycling and energy optimization further improves economic efficiency and reduces environmental impact in ceramic manufacturing.

Keywords

Wet Processing, Dry Processing, Particle Size Distribution, Quality Control, Sustainability, Raw Material Selection

1. Introduction

The preparation of industrial ceramic clay is a critical factor in ensuring the quality of final products, as it directly affects plasticity, ease of forming, and the reduction of defects during drying and firing. The process begins with the selection of suitable raw materials based on purity and chemical composition, particularly the optimal ratio between silicon dioxide (SiO_2) and aluminum oxide (Al_2O_3), while also considering rheological properties and pH to ensure balanced performance during processing [1–3].

The preparation route extends from the quarry to the stage of forming readiness, passing through crushing and milling to achieve optimal particle size distribution, followed by mixing and homogenization to produce a consistent clay body. Deflocculation is then applied in wet processes to improve flowability. In certain applications, advanced techniques such as magnetic separation are used to remove iron impurities, and calcination is employed to adjust the crystalline structure [4–6].

The preparation process concludes with quality control testing, including Atterberg limits (plastic and liquid limits), moisture content measurement, viscosity assessment, and verification of particle size distribution. In line with modern trends, the industry focuses on sustainability by optimizing energy consumption, recycling clay waste, and minimizing raw material losses, thereby improving economic efficiency and reducing environmental impact [7–9].

This chapter aims to present the fundamentals and stages of industrial ceramic clay preparation, highlighting the impact of raw material selection, particle size distribution, and plasticity on final product quality, while reviewing the latest technical practices for quality control and promoting sustainability in the industry.

2. Types of Industrial Ceramic Clays

Industrial clay can be classified into earthenware, stoneware, porcelain, and specialty clays. Earthenware has high plasticity but low firing temperatures; stoneware is fairly plastic and fired between 1100°C and 1300°C; porcelain has low plasticity; and specialty clays are prepared to meet requirements and are not of interest here.

2.1 Earthenware

Industrial ceramic clays, the foundation of the ceramic industry, are beginning raw materials used for ceramics and pottery. Based on their chemical composition, sensitivity towards temperature, occurrence, and firing shrinkage, they can be divided into stoneware, earthenware, specialty clay, and porcelain. The nature of clay affects the preparation method during mixing, shaping, and firing. Earthenware is a kind of natural clay that occurs between sedimentary beds. It consists mainly of aluminum silicate, while iron oxide is common [10–14].

Its color ranges from white through yellow to brown and red. The composition is variable in terms of chemical properties, grain size, and plasticity. The physicochemical properties depend on the degree of weathering and temperature. Raw earth clays are characterized by their

63

plasticity, shrinkage, and firing temperature. A high amount of quartz produces poor firing due to excessive sintering. When fired to a temperature between 1000°C and 1150°C, earthenware plates, bricks, or ornaments (< 0.7 mm) have the most widespread industrial application [15–17].

2.2 Stoneware

Stoneware clays set themselves apart by their maturation temperatures, which span from 1100°C to 1300°C. They can achieve either a vitreous or a semi-vitreous state. The 'vitreous' property denotes the development of a glassy phase that imparts low porosity and the capacity to be translucent and impermeable [18–20].

2.3 Porcelain

Porcelain clays are defined by their fusion of white-firing clays and feldspar. Typically, ball clay and kaolinite are the primary components with minor amounts of quartz, mica, or other accessory minerals. The addition of feldspar widens the plastic range, increases drying strength, and lowers the firing temperature. Unlike earthenware and stoneware clays, porcelains are intensely fired yet display fewer ceramic characteristics; their carefully balanced flux-to-silica ratio promotes densification without excessive mullite formation. Industrial porcelain bodies thus require particular preparation methods, tailored to achieving high firing temperatures with precise thermal behavior [21–24].

2.4 Specialty Clays

Specialty clays have myriad applications across ceramics, agriculture, pharmaceuticals, power generation, and fine chemicals. The associated preparation techniques, therefore, vary broadly and are generally dictated by ensemble properties. Reactive alumina (eta, gamma, and theta phases), fumed silicas, titania, and various pigments are added according to the targeted phase compositions that seek to optimise the firing temperature, colour, translucency, or density [25].

Microporous and nanoscale clays such as kaolin, montmorillonite, and illite are particularly amenable to chemical modification during preparation, as the ion exchange processes extend to the inter-layer galleries. Alkylammonium salts, quaternary ammonium, and alkali metal cations are, for example, intercalated or adsorbed to generate improved mechanical strength and electrical characteristics. The incorporation of alkali ions within the structure of ceramics is also often undertaken to form aluminosilicate absorbents [26–28].

Overall, the use of specialty clays continues to grow, and innovative hybrid processing routes are also emerging to enhance the powder and powder mix characteristics in order to maximise density and other bulk qualities following sintering.

Table 1 illustrates. Section 2 Summary Classification of industrial ceramic clays and their main properties, comparing the four main types of pottery, porcelain, and specialty clays based on their chemical composition, mineral properties, firing temperature, flexibility, and typical applications. The table also highlights the most important technical observations that affect preparation methods, performance, and suitability for various industrial uses.

Table 1. Classification and key characteristics of industrial ceramic clays.

Type	Typical Chemical Composition (wt%)	Dominant Mineral Phases	Firing Temperature (°C)	Plasticity	Common Applications	Technical Notes
Earthenware	SiO_2: 50–65% Al_2O_3: 15–25% Fe_2O_3: 3–8% Na_2O/K_2O: 2–5%	Kaolinite, Illite, Quartz, Feldspar	950–1150	Medium to High	Bricks, roof tiles, pipes, low-temperature tiles, decorative ware	High quartz content reduces plasticity and increases cracking during firing.
Stoneware	SiO_2: 60–70% Al_2O_3: 20–30% Fe_2O_3: 1–3% Na_2O/K_2O: 2–5%	Kaolinite, Mica, Quartz, Feldspar	1100–1300	Medium to High	Tableware, sanitary ware, kitchenware, water-resistant pottery	It can reach a near-vitreous state, giving high water resistance.
Porcelain	SiO_2: 65–75% Al_2O_3: 20–28% Fe_2O_3: <1% Na_2O/K_2O : 3–6%	Kaolinite, Quartz, Feldspar	1200–1400	Low	Sanitary porcelain, electrical insulators, laboratory ware, fine china	Requires precise preparation to ensure homogeneity; exhibits very low porosity after firing.
Specialty Clays	Variable depending on application; may include TiO_2, MgO, ZrO_2, rare oxides, nanomaterials	Kaolin, Montmorillonite, Illite, Added Oxides	1000–>1600 (depending on system)	Variable	Technical ceramics, electronics, wear- and heat-resistant materials, transparent ceramics, medical applications	Chemically or physically modified to achieve specific properties such as transparency or electrical conductivity.

3. Raw Material Sourcing

Sourcing of raw materials commonly employed in the preparation of industrial ceramic clays involves the procurement of red-burning, ball, china, bentonite, and kaolin clays from diverse deposits in the United States and abroad. These clays are supplemented with various commercially available additives such as silica, alumina hydrate, fluxes, and colorants to enhance specific properties of the final product [29–31]. Figure 1 illustrates raw material sourcing.

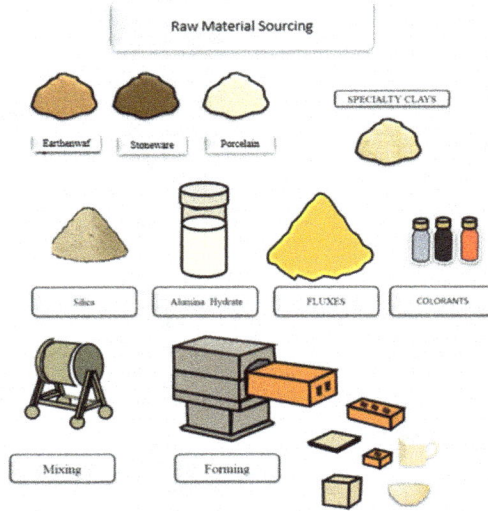

Figure 1. Raw material sourcing.

Raw materials suitable for the manufacture of ceramics in the form of bricks and tiles are typically derived from non-plastic residues and other industrial waste by-products. Natural clays from deposits in Southeast Asia offer low-cost and readily available solutions that impart excellent plasticity to ceramic bodies [32–34]

The ceramic industry for building materials seeks to utilize clays from proximal sources in order to optimize costs and logistics. Characterization and quality control of each clay source constitute critical steps in predicting the technical performance of local ceramic products and in the assessment of potential applications for construction materials such as bricks and tiles. The manufacture of traditional ceramics often involves a mixture of disparate raw materials followed by casting, extrusion, or compaction and subsequent firing at elevated temperatures [35–37].

3.1 Clay Types and Sources

Industrial ceramic clays originate from secondary (sedimentary) deposits. Because of their easily accessible characteristics, these clays become the raw materials for clay production. They are widely distributed worldwide; Southeast Asia is known for kilns producing various clays due to soil-material characteristics [38–41].

The clay types can significantly impact the physical and ceramic properties, so they must be carefully investigated before use for long-term stability. Several advantages of these clays are noteworthy. They are available in many parts of the world, especially where there are river banks or sedimentary deposits, which makes them widespread; they are friendly to the environment and have a good plastic forming ability, they have a good binding ability, and they present a consistent physico-chemical profile [42–44].

Depending on the characteristics and chemical composition, industrial ceramic clays can be classified as earthenware, stoneware, porcelain, and specialty clays. Other considerations during the sourcing of raw materials for ceramic clay include mineralogy, particle size, thermal behaviour, and plasticity. In addition, many of the materials or additives are mineral fluxes; therefore, they contain fluxing agents that cause the formation of a glassy phase during firing. Fluxing agents play a significant role in favouring the sintering of the material. Aluminum oxide acts as a refractory material, and calcium compounds help densify the material during burning. Several additives act as deflocculators or plasticity controllers; plasticity is an essential characteristic from the forming point of view, and a reasonable amount of control over plasticity is necessary. Additives also affect colour [11,24,45,46].

3.2 Additives and Fluxes

The ceramic industry requires a certain combination of minerals, together with special technical operations. Besides the clays, some additional materials can be added to improve the physicochemical properties of ceramic materials or to reduce the firing temperature. Those compounds are called additives and fluxes. Additives are usually very small mineral particles that increase the plasticity of the compound in the shaping state or the mechanical properties of the final product. Polymer addition increased flexural strength and modified the fracture mode and porosity of traditional ceramics [10,47,48].

Several organic and inorganic materials can be used; their classification should be done according to selected properties of the final ceramic product. Fluxes are high-calcium feldspars that produce a liquid phase at relatively low temperatures .Another flux used in the ceramic industry is calcium carbonate. It is frequently added to decrease the firing temperature. Table 2 illustrates. Section 3 Summary, Processing methods, analytical techniques, and equipment for industrial ceramic clay manufacturing [49–51].

Table 2. *Processing methods, analytical techniques, and equipment for industrial ceramic clay manufacturing.*

Category	Technique / Method	Scientific Description	Objectives	Advantages	Limitations	Common Equipment	Industrial Applications
Dry Processing	Screening & Sieving	Mechanical separation of particles based on mesh size (ASTM E11 Standard).	Remove oversized particles; unify particle size distribution.	Simple, low-cost, water-free.	Does not remove fine impurities, dust.	Vibratory Sieve Shaker, Rotary Screen	Clay preparation for tiles and red bricks.
	Dry Milling	Particle size reduction using Ball Mills or Roller Mills without water.	Improve homogeneity and reduce particle size.	High uniformity, easy storage.	High energy consumption; dust generation.	Planetary Ball Mill, Raymond Mill	Kaolin powder, electrical ceramics.

Wet Processing	Soaking & Slaking	The hydration of clay breaks down aggregates into fine particles.	Improve plasticity; remove soluble salts.	Excellent plasticity, reduced dust.	High water and time consumption.	Slaking Tank, Agitator Tank	Traditional pottery, handmade ceramics.
	Wet Ball Milling	Milling in an aqueous medium with Alumina Grinding Media to avoid contamination.	Achieve fine particle size; remove metallic impurities.	Superior uniformity, fine PSD.	Requires drying stage; higher energy use.	Ceramic Ball Mill, Attritor Mill	High-purity industrial ceramics.
Modern Mechanical Techniques	High-Speed Mixing	Intensive mixing using rotor–stator systems for uniform slurry preparation.	Reduce preparation time; ensure additive distribution.	High productivity, precise control.	High equipment cost.	High-Shear Mixer, Pug Mill	Porcelain tiles, tableware production.
	Continuous Processing	Automated feeding, mixing, and filtering in uninterrupted production lines.	Improve efficiency and product consistency.	Ideal for large-scale manufacturing.	Complex maintenance; less economical for small-scale.	Continuous Mixer, Belt Filter Press	Large-scale ceramic plants.
Control & Treatment Methods	Magnetic Separation	Removal of ferrous contaminants using high-intensity magnetic fields (Gauss > 10,000).	Improve purity; prevent firing blemishes.	Fast, efficient.	Ineffective for non-ferrous impurities.	High-Intensity Magnetic Separator	White ceramic products.
	Moisture Control	Adjusting water content via drying or humidification to target optimal plasticity.	Prevent cracks; improve shaping.	Better forming quality.	Requires continuous monitoring.	Infrared Moisture Analyzer, Rotary Dryer	Refractory bricks, technical ceramics.

4. Clay Preparation Techniques

The clay body suitable for the manufacture of ceramic products consists of raw materials that must be processed before mixing. This processing may include wet or dry methods such as grinding or crushing. The purpose of clay preparation is to form a homogeneous and uniform ceramic body that can be shaped in subsequent stages. Various additives can be added to the raw clay mixture either in the form of a slurry or dry powder before the final mixture is formed. Figure 2, illustrates Clay preparation techniques.

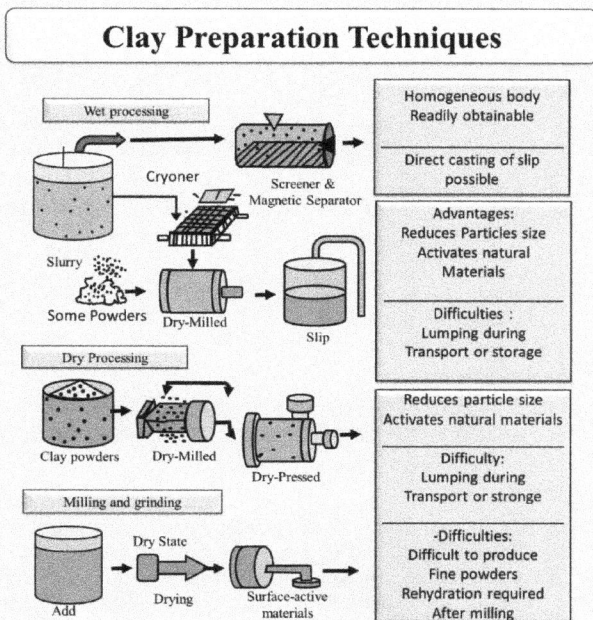

Figure 2. Clay preparation techniques.

4.1 Wet Processing

Wet processing is the traditional method of preparing clay on an industrial scale. In this method, the raw materials are ground in water until they reach the desired fineness, then screened and magnetically separated, followed by water classification to remove large particles, excess sand, and unwanted heavy metals [52,53].

The individual components are then filtered and screened to obtain pure solids that can be stored for later use. The components of the mixture are weighed as dry powders and then mixed with appropriate amounts of water to form a slip with the required plasticity [54,55].

The economic advantage of wet processing lies in the possibility of casting the slip directly to obtain semi-finished products in the desired shape, with reduced cutting and trimming operations. It is also easier to achieve the ideal plasticity with this method, which ensures a homogeneous body, especially when using raw materials with a low melting point [56,57].

4.2 Dry Processing

Dry processing is used when water is not available or impractical, or in some industrial processes that do not rely on traditional plastic clay bodies. An example is dry pressing, where dry clay powders are pressed with high force to form products [38,58].

Dry processing usually begins with mixing clay powders according to the ratios specified in the mixture sheet. In some cases, raw materials to be dry-ground need to be wet-processed first, then dried to obtain suitable powders [59,60].

There are also special ceramic powders that can be converted into synthetic clay simply by mixing them with water, such as heat-treated kaolin or ball clay powders, but these are often expensive [61–63].

The advantages of dry grinding are that it reduces particle size, homogenizes the material, and produces a powder with a particle distribution suitable for the intended use. This method may also activate the raw materials, improving chemical reactions during firing in an oxidizing atmosphere [38,52].

However, it should be noted that storing powders for long periods of time can cause them to agglomerate, and that conveyor systems must be designed to prevent agglomerates from sticking to the equipment.

4.3 Milling and Grinding

Grinding is necessary in wet processing to produce homogeneous clay bodies. In large factories, grinding stations allow large quantities to be processed quickly. Drying the solids before wet grinding helps to increase grinding efficiency, and hot air grinding can achieve drying and grinding in one step [64–66]. This technique is used in the manufacture of refractory bricks and traditional bricks, where it prevents the phenomenon of "lime blowing" that can damage the product.

Dry grinding increases the surface energy of the particles, making it more difficult to produce ultra-fine powders, but the use of planetary ball mills increases grinding efficiency and the degree of material activation. After grinding, clay minerals may lose their structural water, so they are left to reabsorb moisture, or additives such as PEG 600 are added to accelerate absorption, taking care not to overuse them so as not to weaken the mechanical properties [67–69].

Controlling micro-porosity and nano-porosity through grinding and chemical activation can also improve the mechanical properties in ceramic systems containing quartz, feldspar, calcite, dolomite, and illite clay [70–72]. Table 3 illustrates a comparison of clay preparation techniques in the ceramic industry.

Table 3. Comparison of clay preparation techniques in ceramic manufacturing.

Method	Main Process Steps	Advantages	Disadvantages / Limitations
Wet Processing	1. Grind raw materials in water to the required fineness 2. Screen and magnetically separate 3. Water classification to remove large particles, sand, and heavy metals 4. Filter and store solids 5. Weigh dry powders and mix with water to form a slip	- Produces a highly homogeneous body - Easy to achieve ideal plasticity - Slip can be cast directly into shapes, reducing cutting/trimming - Especially good for low-melting-point materials	- Requires large amounts of water - More equipment and energy for drying products - Slower if large volumes need water removal
Dry Processing	1. Mix clay powders according to the recipe 2. Dry-grind to the desired particle size 3. (Optional) Wet-process, then dry to produce powders. Press powders into shape	- No need for water in processing - Suitable for dry pressing methods - Good for certain synthetic clays and special industrial uses - Activates raw materials for better firing reactions	- Expensive if using special powders - Powders may agglomerate in storage - Requires careful design of conveyors to avoid sticking
Milling & Grinding	1. Grind to produce homogeneous clay bodies 2. In some cases, dry solids before wet grinding for efficiency 3. Use hot air grinding for drying & grinding in one step 4. Use planetary ball mills for higher efficiency	- Can improve micro- and nano-porosity control - Enhances mechanical properties - Prevents defects (e.g., "lime blowing" in bricks) - Can activate raw materials	- Ultra-fine grinding is harder due to increased surface energy - Overuse of additives like PEG 600 may weaken mechanical strength - Clay minerals may lose structural water and need rehydration

5. Control of clay properties after preparation in the ceramic industry

After the clay is prepared, whether by wet or dry processing, its physical and chemical properties must be controlled to ensure the quality of the final product. These properties include plasticity, moisture content, viscosity, density, and grain size, each of which is adjusted according to the desired molding method and type of finished product.

5.1 Plasticity

Plasticity is one of the most important properties of clay, as it determines how easily it can be molded and bent without cracking. Plasticity is controlled by adding water in specific proportions to improve molding, as more water makes the clay softer and less water makes it harder. Additives such as bentonite can be used to increase plasticity, or calcined kaolin to improve dry strength. In addition, adjusting the aging time plays an important role in improving the consistency and uniformity of the clay's properties. The importance of plasticity control lies in its direct effect on the ease of molding, whether pressing, extruding, or casting, and it also minimizes cracks that may appear during the drying phase [73–77].

5.2 Moisture content

Controlling the moisture content is a critical factor in the quality of the slurry before molding. In wet processing, the water content is adjusted to obtain a homogeneous consistency, while in dry processing, the residual moisture after drying is controlled. Air dryers or drying ovens at low temperatures (60-120°C) are used to ensure that excess moisture is removed without causing cracks. The importance of this step is that too much moisture causes deformations during molding, while too little moisture makes the clay brittle and prone to cracking. Optimal moisture content also ensures better homogeneity during the final firing stage [78–80].

5.3 Viscosity

The viscosity of ceramic clays directly affects the molding process, especially in screw casting techniques. Viscosity is measured using devices such as a rotational viscometer, where it is adjusted by adding diluents (such as sodium carbonate) to lower the viscosity, or thickeners (such as polymers) to increase it. Proper viscosity is necessary to ensure a uniform flow of clay during molding, preventing the formation of air bubbles or defects in the final product. Improper viscosity can cause issues in the extrusion or pressing process [81,82].

5.4 Density

Clay density is a key physical property influencing the material's cohesion, friability, and behavior during processing and firing. It is expressed as either bulk density or true density, measured using specialized instruments such as a pycnometer. Density can be modified by incorporating high-density fillers (e.g., quartz) or low-density porous additives (e.g., vermiculite). Optimal density ensures desired product characteristics such as durability, absorbency, and resistance to thermal shock, while improper density may cause warping or non-uniform shrinkage during firing [83–86].

5.5 Particle Size

The particle size distribution of clay significantly influences its mechanical properties and firing behavior. It is typically measured using sieves or advanced instruments such as laser particle analyzers. Particle size can be modified through additional grinding processes or by blending materials with different grain sizes. A homogeneous distribution of particles enhances the final product's strength and reduces porosity, while irregular distributions can lead to defects such as cracking or uneven shrinkage. Very fine particle sizes increase plasticity but may slow the drying process [82,87,88].

Table 4 illustrates a systematic summary of how to control the physical and chemical properties of clays after preparation. It relates control methods and measurement tools to the direct impact on the quality and durability of ceramic products.

Table 4. Scientific and technical control of clay properties after preparation in the ceramic industry.

Property	Scientific / Technical Notes	Control Methods	Measurement Standards & Equipment	Impact on Final Product	Industrial Examples
Plasticity	Defines clay's ability to undergo deformation without cracking; governed by particle size, shape, and water content.	Adjust water ratio; add **Bentonite** (↑ plasticity) or **Calcined Kaolin** (↑ dry strength); optimize **aging time** for improved particle bonding.	**Atterberg Limits** (ASTM D4318); manual forming tests.	Enhances moldability in **pressing, extrusion,** and **slip casting**; minimizes drying cracks.	Porcelain tiles, sanitaryware, fine tableware.
Moisture Content	Critical for molding consistency; affects drying rate and shrinkage behavior.	Optimize water content in wet processing; control residual moisture in dry processing; use **low-temp dryers (60–120 °C)**.	**Oven-Dry Method** (ASTM C566); moisture analyzers.	Prevents deformation (high moisture) or brittleness (low moisture); ensures uniform firing shrinkage.	Hollow bricks, roof tiles, ceramic pipes.
Viscosity	Determines flow behavior in slip casting and extrusion; linked to solid loading and particle interactions.	Add **Sodium Carbonate** (↓ viscosity) or **polymers** (↑ viscosity); control deflocculants.	**Rotational Viscometer** (ISO 3219); flow cone.	Ensures smooth flow in molds; prevents air entrapment and surface defects.	Slip-cast sanitary fixtures, ceramic art pieces.
Density	Bulk density affects mechanical strength; true density relates to mineral composition.	Increase density with **Quartz** fillers; reduce density with lightweight fillers like **Vermiculite**.	**Pycnometer** (ASTM C373); bulk density apparatus.	Improves durability, reduces porosity, and enhances thermal shock resistance.	Technical ceramics, refractory bricks, floor tiles.
Particle Size Distribution	Governs plasticity, shrinkage, and sintering rate; fine particles increase plasticity but slow drying.	Achieve desired PSD via grinding, milling, or blending coarse/fine materials.	**Sieve Analysis** (ASTM C136); **Laser Particle Size Analyzer**.	Improves strength, reduces porosity, prevents cracking, and uneven shrinkage.	Structural clay products, high-strength porcelain.

6. Drying and Pre-firing

Drying is an indispensable step in ceramic preparation and may also form part of the slip-casting process. Uncontrolled evaporation generates drying stresses and related defects such as cracking and distortion. Studies of drying in industrial kilns show overall drying rates similar to those found for the laboratory-scale samples, although the ability of infrared and microwave irradiation for drying, with very short-time processing, to reduce cracking. Figure 3, illustrates drying and Pre-firing ceramics.

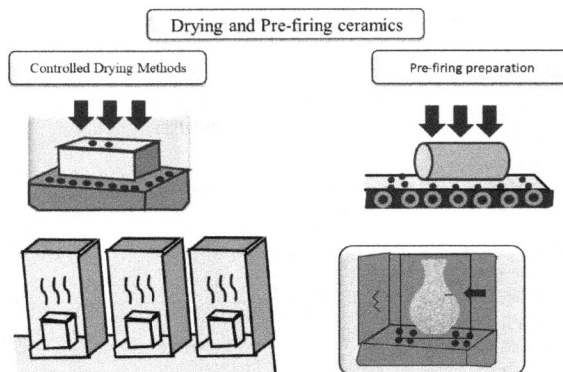

Figure 3. Drying and Pre-firing ceramics.

6.1 Controlled Drying Methods

The quality of clay drying is crucial for obtaining ceramic products without fissures or deformation. The aim is to maintain moisture and temperature gradients as low as possible during drying. Firing after drying causes clay shrinkage, providing mechanical resistance and stability in the final product. A carefully controlled drying stage ensures products with low deformation and fissures, potentially eliminating the need for recasting the drying bricks with a recast [89–91].

6.2 Pre-firing Considerations

Pre-firing preparation of industrial ceramic bodies ensures uniform heating and dimensional stability throughout the pre-firing (greenware) process. Key requirements include achieving consistency and suitable water content in the mixture, forming processes that maintain the dimensional stability of the greenware, and controlling drying conditions, typically under closed drying conditions to allow the clay to adjust and reduce internal stresses [92,93].

Porous clays and bodies that have a high capacity for water and salts are more vulnerable to thermal shock during pre-firing and require attention to several variables. Moreover, during firing, clays may lose moisture, gases, and organic materials, which can induce warping and cracks, underscoring the necessity of the earlier-mentioned preparation measures [94,95].

Industrial shapers of ceramic bodies incorporate gaging, control, or venting holes to ensure a uniform drying rate across the body. Although manual processes allow for these types of mechanical interventions, in industrial deformation processes where such precision is less feasible, the focus shifts toward achieving the requisite plasticity before shaping to mitigate firing issues. Table 5 illustrates the Drying and Pre-firing Stages, Requirements, Risks, and Best Practices in Ceramics.

Table 5. Drying and pre-firing stages, requirements, risks, and best practices in ceramics.

Step	Objective	Key Requirements	Challenges / Risks	Industrial Best Practices	Practical Examples
Drying	Prevent defects (cracking, distortion) by controlling evaporation.	Use lab-scale or industrial drying rates; apply infrared/microwave for fast drying and reduced cracking.	Uncontrolled drying causes cracks and distortion.	Match lab-scale drying rates, use fast-drying technologies.	Using an industrial infrared kiln to dry ceramic tiles in 30 minutes instead of 24 hours.
Controlled Drying Methods	Minimize moisture and temperature gradients to avoid fissures/deformation.	Maintain low gradients, control humidity, prevent warping; reduce the need for recasting.	Rapid drying without control leads to shrinkage-related defects.	Closed-environment drying, monitor humidity levels, gradual heating.	Placing products in a closed drying chamber with electronic control of humidity (RH = 60%) and temperature (40°C).
Pre-firing Considerations	Ensure uniform heating and dimensional stability in greenware.	Correct water content, stable forming process, closed drying to relieve internal stresses.	Porous/salt-rich clays prone to thermal shock; gas/organic loss during firing can cause warping.	Use gaging/control/vent holes, achieve correct plasticity before shaping.	Drilling vent holes in thick clay objects (e.g., pottery) before firing to allow steam to escape without cracking.

7. Conclusion

The preparation of industrial ceramic clay is a decisive step in ensuring product quality, where the correct choice of raw materials, control of particle size distribution, and adjustment of plasticity play a central role in achieving strength, dimensional stability, and defect-free surfaces. Wet processing remains advantageous for achieving high homogeneity and ideal workability, while dry processing offers water savings and benefits for certain specialized applications. Maintaining precise control over moisture content, viscosity, density, and particle size ensures consistent shaping, minimizes defects during drying and firing, and enhances mechanical performance. Advanced techniques, such as magnetic separation, fine milling, and high-efficiency mixing, further improve purity and microstructure.

Looking ahead, the industry is likely to adopt more automation and continuous monitoring systems for real-time quality control, along with advanced eco-friendly additives that reduce firing temperatures and energy consumption. The integration of AI-driven optimization, renewable energy in drying and firing processes, and closed-loop recycling of clay waste offers promising pathways toward higher efficiency, reduced environmental impact, and enhanced product performance. Such developments will support the dual goals of technical excellence and sustainability in the future of ceramic manufacturing.

References

[1]   A.K.A. Khalil, A. Elgamouz, S. Nazir, M.A. Atieh, H. Alawadhi, T. Laoui, Preparation and characterization of clay based ceramic porous membranes and their use for the removal of lead ions from synthetic wastewater with an insight into the removal mechanism, Heliyon 10 (2024).

[2]   Y. Al-Hadeethi, M.I. Sayyed, A.Z. Barasheed, M. Ahmed, M. Elsafi, Preparation and radiation attenuation properties of ceramic ball clay enhanced with micro and nano ZnO particles, J. Mater. Res. Technol. 17 (2022) 223–233.

[3]   Y. Ji, E. Li, G. Zhu, R. Wang, Q. Sha, Preparation and performance of ceramic tiles with steel slag and waste clay bricks, Materials (Basel). 17 (2024) 1755.

[4]   W. Guo, K. Guo, Y. Xing, X. Gui, A comprehensive review on evolution behavior of particle size distribution during fine grinding process for optimized separation purposes, Miner. Process. Extr. Metall. Rev. (2024) 1–20.

[5]   X. Fang, C. Wu, N. Liao, C. Yuan, B. Xie, J. Tong, The first attempt of applying ceramic balls in industrial tumbling mill: A case study, Miner. Eng. 180 (2022) 107504.

[6]   I. Ortiz de Landazuri Suárez, M.-J. Oliveros Colay, Design of Comminution Plants in the Ceramic Industry Using a Simulation-based Optimization Approach, Ing. e Investig. 41 (2021).

[7]   B.C. O'Kelly, Review of recent developments and understanding of Atterberg limits determinations, Geotechnics 1 (2021) 59–75.

[8]   G.E. Barnes, A multi-linear approach to strength and plasticity states between the Atterberg limits, Proc. Inst. Civ. Eng. Eng. 174 (2021) 102–117.

[9]   G. Spagnoli, S. Shimobe, Statistics of Atterberg limit values of some pure kaolinitic clays, Geomech. Geoengin. 18 (2023) 105–120.

[10]   B.P. Kagonbé, D. Tsozué, A.N. Nzeukou, S. Ngos, Mineralogical, physico-chemical and ceramic properties of clay materials from Sekandé and Gashiga (North, Cameroon) and their suitability in earthenware production, Heliyon 7 (2021).

[11]   E.C. Casella, Ceramic Production, in: Oxford Handb. Ind. Archaeol., Oxford University Press, 2022: p. 1951.

[12]   M.S. Rahman, H.S.C. Metselaar, B.B.A. Razak, Mineralogical and Thermal Analysis of Ancient Ceramic Artifacts Based on Modern Techniques of Ceramic Studies, Mediterr. Archaeol. Archaeom. 24 (2024) 38–56.

[13]   J.H. Choi, S.M. Kim, K.S. Han, U.S. Kim, M.S. Kim, Role of earthenware in food processing applications, J. Ceram. Process. Res. 22 (2021) 91–97.

[14]   F.N. Nordin, N. Ayob, Earthenware clay as a base material for ash glaze, IJARBSS 13 (2023) 4732–4743.

[15]   G. Pfaff, Ceramic colors, Phys. Sci. Rev. 7 (2022) 95–102.

[16]   G.N. Aernyi, A.M. Ahuwan, C. V Alkali, The Development Of Red Slip Clay From Dogo Benue State For The Decoration Of Ceramic Wares, Reson. J. Contemp. Scholarsh. 1 (2024).

[17]   L. Bloomfield, Colour in glazes, Bloomsbury Publishing, 2022.

[18] U.D.E.C. Michael, Industrial Materials And Their Processing, (n.d.).

[19] A.C.X. Ren, Elemental Compositions of Kota Cina and Singapore Stoneware Glazes: A Preliminary Study, Arch. Études Interdiscip. Sur Le Monde Insulind. (2025) 203–225.

[20] Q. Yuan, D. Robert, A. Mohajerani, P. Tran, B.K. Pramanik, Sustainable ceramic tiles incorporated with waste fly ash from recycled paper production, J. Clean. Prod. 425 (2023) 138814.

[21] S. Saif, W. Abbass, S. Mubin, F. Aslam, R. Alyousef, Artificial intelligence-based predictive model for utilization of industrial coal ash in the production of sustainable ceramic tiles, Arch. Civ. Mech. Eng. 24 (2024) 222.

[22] H. Ma, J. Henderson, J. Evans, Q. Ma, J. Cui, Searching for the earliest use of limestone as a flux in Chinese high-fired ceramic glazes—evidence from Sr isotopic analysis of Chinese northern porcelain, J. Archaeol. Sci. 131 (2021) 105395.

[23] Q. Yuan, A. Mohajerani, H. Kurmus, J. V Smith, Possible recycling options of waste materials in manufacturing ceramic tiles, Geomate J. 20 (2021) 73–80.

[24] M. Bustillo Revuelta, Ceramic Products, in: Constr. Mater. Geol. Prod. Appl., Springer, 2021: pp. 339–374.

[25] A.G. Adeniyi, K.O. Iwuozor, E.C. Emenike, Material development potential of Nigeria's Kaolin, Chem. Africa 6 (2023) 1709–1725.

[26] A. Kalendova, J. Kupkova, M. Urbaskova, D. Merinska, Applications of clays in nanocomposites and ceramics, Minerals 14 (2024) 93.

[27] C. Elmi, Physical-chemical properties of nano-sized phyllosilicates: recent environmental and industrial advancements, Encyclopedia 3 (2023) 1439–1460.

[28] E.M. Serwicka, Titania-clay mineral composites for environmental catalysis and photocatalysis, Catalysts 11 (2021) 1087.

[29] A. Hein, V. Kilikoglou, Ceramic raw materials: how to recognize them and locate the supply basins: chemistry, Archaeol. Anthropol. Sci. 12 (2020) 180.

[30] G. Montana, Ceramic raw materials: how to recognize them and locate the supply basins—mineralogy, petrography, Archaeol. Anthropol. Sci. 12 (2020) 175.

[31] S. Seifert, S. Dittrich, J. Bach, Recovery of raw materials from ceramic waste materials for the refractory industry, Processes 9 (2021) 228.

[32] T. Zanatta, R.A.A.B. Santa, N. Padoin, C. Soares, H.G. Riella, Eco-friendly ceramic tiles: development based on technical and market demands, J. Mater. Res. Technol. 11 (2021) 121–134.

[33] N.F. Luiz, D. Cecchin, A.R.G. Azevedo, J. Alexandre, M.T. Marvila, F.C. Da Silva, A.L.C. Paes, D.F. Do Carmo, P.F.P. Ferraz, C.M. Huther, Characterization of materials used in the manufacture of ceramic tile with incorporation of ornamental rock waste, (2020).

[34] K. Boulaiche, Recycling and Valorisation of Industrial Wastes In the Formulation of a Sanitary Ceramic Body, (2023).

[35] C. Ferreira, A. Silva, J. de Brito, I. S. Dias, I. Flores-Colen, Maintenance modelling of ceramic claddings in pitched roofs based on the evaluation of their in situ degradation condition, Infrastructures 5 (2020) 77.

Materials Research Forum LLC
https://doi.org/21741/9781644903834

[36] M.S. Medina-Salgado, F.E. García-Muiña, M. Cucchi, D. Settembre-Blundo, Adaptive life cycle costing (LCC) modeling and applying to Italy ceramic tile manufacturing sector: Its implication of open innovation, J. Open Innov. Technol. Mark. Complex. 7 (2021) 101.

[37] V. Lang, S. Weingarten, H. Wiemer, U. Scheithauer, F. Glausch, R. Johne, A. Michaelis, S. Ihlenfeldt, Process data-based knowledge discovery in additive manufacturing of ceramic materials by multi-material jetting (CerAM MMJ), J. Manuf. Mater. Process. 4 (2020) 74.

[38] G. Eramo, Ceramic technology: how to recognize clay processing, Archaeol. Anthropol. Sci. 12 (2020) 164.

[39] M. Dondi, G.P. Bertolotti, Basic guidelines for prospecting and technological assessment of clays for the ceramic industry, part 1, Interceram-International Ceram. Rev. 70 (2021) 36–46.

[40] S.R. Christofoletti, A. Batezelli, M.M.T. Moreno, Facies, geochemistry, and ceramic properties of corumbataí formation, upper permian of paraná basin, and its application in the ceramic industry, Brazil, Clays Clay Miner. 70 (2022) 712–732.

[41] H.A. de Oliveira, C.P. dos Santos, Limestone clays for ceramic industry, in: Clay Sci. Technol., IntechOpen, 2020.

[42] L. Vaiani, A. Boccaccio, A.E. Uva, G. Palumbo, A. Piccininni, P. Guglielmi, S. Cantore, L. Santacroce, I.A. Charitos, A. Ballini, Ceramic materials for biomedical applications: an overview on properties and fabrication processes, J. Funct. Biomater. 14 (2023) 146.

[43] J. Binner, M. Porter, B. Baker, J. Zou, V. Venkatachalam, V.R. Diaz, A. D'Angio, P. Ramanujam, T. Zhang, T. Murthy, Selection, processing, properties and applications of ultra-high temperature ceramic matrix composites, UHTCMCs–a review, Int. Mater. Rev. 65 (2020) 389–444.

[44] M. Hoorizad, S. Valizadeh, H. Heshmat, S.F. Tabatabaei, T. Shakeri, Influence of resin cement on color stability of ceramic veneers: in vitro study, Biomater. Investig. Dent. 8 (2021) 11–17.

[45] J. Götze, M. Göbbels, Ceramic materials, in: Introd. to Appl. Mineral., Springer, 2023: pp. 79–95.

[46] A. Zharmenov, S. Yefremova, B. Satbaev, N. Shalabaev, S. Satbaev, S. Yermishin, A. Kablanbekov, Production of refractory materials using a renewable source of silicon dioxide, Minerals 12 (2022) 1010.

[47] M.Y. Balde, C. Njiomou Djangang, A. Bah, P. Blanchart, D. Njopwouo, Effect of physicochemical characteristics on the use of clays from Kindia (Guinea) in ceramic compositions, Int. J. Appl. Ceram. Technol. 18 (2021) 1033–1042.

[48] H. Slimanou, A. Baziz, N. Bouzidi, D.E. Quesada, A. Tahakourt, Thermal, physical, mechanical and microstructural properties of dredged sediment-based ceramic tiles as substituent of kaolin, Environ. Sci. Pollut. Res. 29 (2022) 26792–26809.

[49] J. Mulinari, J.V. Oliveira, D. Hotza, Lipase immobilization on ceramic supports: An overview on techniques and materials, Biotechnol. Adv. 42 (2020) 107581.

[50] S. Deng, C. Li, X. Huang, H. Guo, W. Zhao, B. Yan, P. Li, Effect of phase evolution and pyroplastic formation behavior on glass-ceramic foam derived from silicomanganese slag

https://doi.org/21741/9781644903834

and feldspar tailings, Chem. Eng. J. 489 (2024) 151486.

[51] M. Rodchom, P. Wimuktiwan, K. Soongprasit, D. Atong, S. Vichaphund, Preparation and characterization of ceramic materials with low thermal conductivity and high strength using high-calcium fly ash, Int. J. Miner. Metall. Mater. 29 (2022) 1635–1645.

[52] V. Xanthopoulou, I. Iliopoulos, I. Liritzis, Characterization techniques of clays for the archaeometric study of ancient ceramics: A review, Sci. Cult. 6 (2020) 73–86.

[53] F. Azaman, M. Nor, W.R.W. Abdullah, M.H. Razali, R.C. Zulkifli, M.A.A. Zaini, A. Ali, Review on natural clay ceramic membrane: Fabrication and application in water and wastewater treatment, Malays. J. Fundam. Appl. Sci 17 (2021) 62–78.

[54] A.W. Vieira, L.S. Rosso, A. Demarch, D. Pasini, S.P. Ruzza, S. Arcaro, M.J. Ribeiro, E. Angioletto, Life cycle assessment in the ceramic tile industry: a review, J. Mater. Res. Technol. 23 (2023) 3904–3915.

[55] Y. Qin, Y. Tian, Y. Peng, L. Luo, X. Zan, Q. Xu, Y. Wu, Research status and development trend of preparation technology of ceramic particle dispersion strengthened copper-matrix composites, J. Alloys Compd. 848 (2020) 156475.

[56] L. Treccani, Processing Methods for Advanced Ceramics, Surface-Functionalized Ceram. Biotechnol. Environ. Appl. (2023) 47–84.

[57] S.R. Kandavalli, S.R. Kandavalli, R.S. Ruban, C.H. Lo, R. Kumar, A.B. Elshalakany, C.I. Pruncu, A conceptual analysis on ceramic materials used for dental practices: manufacturing techniques and microstructure, ECS J. Solid State Sci. Technol. 11 (2022) 53005.

[58] T.C. Chukwueke, I. Okonkwo, CLAY Essential Material For Ceramic Production, Environ. Rev. 9 (2024).

[59] F. Andreola, I. Lancellotti, R. Sergi, V. Cannillo, L. Barbieri, Comparison of three manufacturing techniques for sustainable porous clay ceramics, Materials (Basel). 14 (2020) 167.

[60] O. V Suvorova, E.A. Selivanova, J.A. Mikhailova, V.A. Masloboev, D. V Makarov, Ceramic products from mining and metallurgical waste, Appl. Sci. 10 (2020) 3515.

[61] S.S. Hossain, P.K. Roy, Sustainable ceramics derived from solid wastes: a review, J. Asian Ceram. Soc. 8 (2020) 984–1009.

[62] L. Shchukina, E. Fedorenko, Y. Galushka, S. Lihezin, Technology of obtaining building ceramics with the use of large-tonnage waste of fuel and energy industry, (2021).

[63] M. Karhu, Utilization of Mineral Side Streams in High Temperature Ceramic Materials, Tampere Univ. Fac. Eng. Nat. Sci. Tampere Univ. Diss. 327 (2020).

[64] Y. Xiong, C. Liu, W. Wang, R. Jiang, B. Huang, D. Wang, S. Zhang, Assessment of machined surface for SiCf/SiC ceramic matrix composite during ultrasonic vibration-assisted milling-grinding, Ceram. Int. 49 (2023) 5345–5356.

[65] X. Zhang, Y. Qin, J. Jin, J. Li, P. Gao, High-efficiency and energy-conservation grinding technology using a special ceramic-medium stirred mill: A pilot-scale study, Powder Technol. 396 (2022) 354–365.

[66] Y. Xiong, W. Wang, R. Jiang, B. Huang, C. Liu, Feasibility and tool performance of ultrasonic vibration-assisted milling-grinding SiCf/SiC ceramic matrix composite, J.

Mater. Res. Technol. 19 (2022) 3018–3033.

[67] J. Yin, J. Xu, W. Ding, H. Su, Effects of grinding speed on the material removal mechanism in single grain grinding of SiCf/SiC ceramic matrix composite, Ceram. Int. 47 (2021) 12795–12802.

[68] S. Liu, J. Lu, H. Xu, C. Yuan, J. Xu, E. Dianyu, S. Liu, Z. Zhou, Performance analysis of mixtures of ceramic and steel grinding balls in an industrial scale semi-autogenous mill, Powder Technol. (2025) 121428.

[69] S. Wang, Q. Zhao, P. Xing, Y. Zhuang, L. Wang, Influence of dispersant on the microstructure and performance of the hot-pressed B4C-YB4 ceramics, J. Aust. Ceram. Soc. 59 (2023) 1065–1077.

[70] F.A. Shilar, S. V Ganachari, V.B. Patil, T.M.Y. Khan, N.M. Almakayeel, S. Alghamdi, Review on the relationship between nano modifications of geopolymer concrete and their structural characteristics, Polymers (Basel). 14 (2022) 1421.

[71] J.M. Migliore, P. Hewitt, T.J. Dingemans, D.L. Simone, W.J. Monzel, Effect of Water-Soluble Polymers on the Rheology and Microstructure of Polymer-Modified Geopolymer Glass-Ceramics, Materials (Basel). 17 (2024) 2856.

[72] L.E. Dörner, Highly energetic Al/CuO thermite coatings through nanoparticle composites, (2020).

[73] B. Wei, W. Wu, D. Xie, M. Nastasi, J. Wang, Strength, plasticity, thermal stability and strain rate sensitivity of nanograined nickel with amorphous ceramic grain boundaries, Acta Mater. 212 (2021) 116918.

[74] A. Liens, H. Reveron, T. Douillard, N. Blanchard, V. Lughi, V. Sergo, R. Laquai, B.R. Müller, G. Bruno, S. Schomer, Phase transformation induces plasticity with negligible damage in ceria-stabilized zirconia-based ceramics, Acta Mater. 183 (2020) 261–273.

[75] J.L. Bombazaro, A.M. Bernardin, Improving plasticity of kaolins by high-energy milling for use in porcelain tile compositions, Open Ceram. 10 (2022) 100256.

[76] H. Mgbemere, E. Obidiegwu, A. Oginni, Production and Characterisation of Porcelain Insulator Modified with Talc and Bentonite, (2020).

[77] N.R. Rakhimova, A review of calcined clays and ceramic wastes as sources for alkali-activated materials, Geosystem Eng. 23 (2020) 287–298.

[78] N. Lauro, S. Oummadi, A. Alzina, B. Nait-Ali, D.S. Smith, Computer model of drying behaviour of ceramic green bodies with particular reference to moisture content dependent properties, J. Eur. Ceram. Soc. 41 (2021) 7321–7329.

[79] J. Castellano, V. Sanz, E. Cañas, E. Sánchez, Assessment of humidity self-regulation functionality for ceramic tiles, J. Eur. Ceram. Soc. 42 (2022) 716–723.

[80] K.-D. Kim, J.-T. Suh, D.-L. Yoo, C.-S. Kim, S.-N. Jee, S.-C. Hong, S.K. Kim, Validation on Performance of Soil Moisture Sensor using Weighing Porous Ceramic Cone Filled with Soil, J. Bio-Environment Control 30 (2021) 157–164.

[81] B. Liang, M. Zhang, H. Li, M. Zhao, P. Xu, L. Deng, Preparation of ceramic foams from ceramic tile polishing waste and fly ash without added foaming agent, Ceram. Int. 47 (2021) 23338–23349.

[82] M. Moghadasi, W. Du, M. Li, Z. Pei, C. Ma, Ceramic binder jetting additive

manufacturing: Effects of particle size on feedstock powder and final part properties, Ceram. Int. 46 (2020) 16966–16972.

[83] A. Vetser, Cold Sintering of Na2Mo2O7+ Hexagonal BN Composites for MLCC Applications, (2023).

[84] S. Zago, Fe-NC electrocatalysts from waste biomass for the oxygen reduction reaction, (2022).

[85] U. Önen, T. Boyraz, The Relationship Between Bending Strength And Porosity In Ceramic Brick That Contains Vermiculite, (2024).

[86] T. Boyraz, U. Önen, Ş. Tapik, Investigation Of The Effect Of Vermiculite (Yıldızeli/Sivas) Addition On The Properties Of Sanitaryware Ceramic, Osmaniye Korkut Ata Üniversitesi Fen Bilim. Enstitüsü Derg. 5 (2022) 632–642.

[87] H. Xing, B. Zou, X. Liu, X. Wang, Q. Chen, X. Fu, Y. Li, Effect of particle size distribution on the preparation of ZTA ceramic paste applying for stereolithography 3D printing, Powder Technol. 359 (2020) 314–322.

[88] S. Manotham, P. Tesavibul, Effect of particle size on mechanical properties of alumina ceramic processed by photosensitive binder jetting with powder spattering technique, J. Eur. Ceram. Soc. 42 (2022) 1608–1617.

[89] T. Rijwani, P. Ramkumar, Thermal debinding for binder burnout in metal and ceramic processing, Heat Transf. Eng. 46 (2025) 615–626.

[90] L. Briest, R. Wagner, A. Tretau, E. Tsotsas, N. Vorhauer-Huget, Microwave-assisted drying of clay roof tiles, Dry. Technol. 40 (2022) 1804–1818.

[91] C. Ratti, A.S. Mujumdar, Infrared drying, in: Handb. Ind. Dry., CRC Press, 2020: pp. 567–588.

[92] M.J. Ribeiro, D. Tulyaganov, Traditional ceramics manufacturing, in: Ceram. Glas. Glas. From Early Manuf. Steps Towar. Mod. Front., Springer, 2021: pp. 75–118.

[93] S. Yu, P. Ji, Z. Dai, L. Xu, H. Xu, L. Zeng, W. Han, A novel bionic design method of groove-shaped pressure head for ceramic de-airing extruder: Improving surface quality and forming performance of extruded clay, J. Manuf. Process. 149 (2025) 383–398.

[94] T. Chekai, R. Wijnhorst, P. Sénéchal, D. Grégoire, N. Shahidzadeh, H. Derluyn, Salt weathering of antique Dutch ceramic tiles, J. Cult. Herit. 72 (2025) 121–130.

[95] S. Basak, S. Majumdar, S. Barma, S. Ghosh, Remediation of Polycyclic Aromatic Hydrocarbons (PAHs) in Coastal Subsurface Water by an Optimized Multilayer Ceramic Composite Superhydrophobic Membrane, ACS ES&T Water (2025).

Materials Research Forum LLC

https://doi.org/21741/9781644903834

Chapter 5

Masse Preparation Methods in Ceramic Industry

Hussein Elaibi[1*], Farah Mutlag[1,2], Idris Kaynak[3], Ebru Halvacı[1], Selcuk Erdogan[1], Fatih Sen[1*]

[1]Sen Research Group, Department of Biochemistry, Kutahya Dumlupinar University, Kutahya 43000, Türkiye

[2]Ministry of Education, Karbala Education Directorate, Karbala, 56001, Iraq

[3]Machinery and Metal Technologies, Vocational School of Technical Sciences, Usak University, 1 Eylul Campus, 64200 Usak, Türkiye

hussein.elaibi@ogr.dpu.edu.tr, fatihsen1980@gmail.com

Abstract

Masse preparation is a foundational stage in ceramic manufacturing, transforming raw materials into a homogeneous, processable ceramic body. It includes critical steps such as weighing, grinding, mixing, filtration, rheological adjustment, water removal, and aging. Each step is essential to achieving structural integrity, dimensional accuracy, and performance in the final product. Preparation techniques vary by product type dry for tiles, wet for porcelain, plastic for extrusion, and slip for casting, each with specific rheological and physical demands. Modern advancements like spray drying, high shear mixing, and Industry 4.0 technologies (e.g., digital twins, AI analytics, and SCADA systems) enhance efficiency, consistency, and sustainability. Rheological behavior and granule structure are tightly monitored through real-time sensors and AI, ensuring consistent quality and minimizing waste. Compared to traditional methods, modern approaches offer superior energy efficiency, automation, and reproducibility. Challenges such as particle agglomeration or moisture imbalance still exist but are mitigated by data-driven monitoring and predictive analytics. Tailored preparation strategies support diverse applications from tiles and sanitary ware to electronics and medical ceramics by aligning material properties with functional requirements.

Keywords

Ceramic Mass Preparation, Rheology, Spray Drying, Smart Manufacturing, Industry 4.0

1. Introduction

The preparation of the ceramic body, commonly referred to as masse preparation is one of the most critical and technically complex stages in ceramic manufacturing. It goes far beyond the simple mechanical mixing of raw materials, serving instead as a strategically engineered process where principles of materials science, mineralogy, and rheology intersect to produce a homogeneous, stable, and performance-optimized body [1–3]. While the chemical composition of raw materials defines their theoretical potential, it is the precision of the preparation process that transforms this potential into tangible properties. Any flaws introduced at this stage such as uneven particle distribution, uncontrolled moisture content, or poor plasticity can result in serious structural or aesthetic defects in the final product, including warping, cracking, or insufficient densification during firing [4,5].

As the ceramics industry has evolved from traditional craftsmanship to high-precision manufacturing, body preparation has become increasingly data-driven and reliant on advanced process control [3,6]. Modern technologies such as spray drying, high-shear mixing, and real-time rheological monitoring now enable manufacturers to engineer ceramic bodies with tailored performance characteristics whether for advanced applications like electrical insulators and medical ceramics, or more conventional products like tiles and tableware [7–10]. This chapter does not revisit the raw materials themselves, as they have been discussed previously, but rather focuses on the transformation processes that make these materials functionally viable. By examining the stages, methods, and innovations involved in masse preparation, we aim to understand how this pivotal step forms the foundation for structural integrity, functional performance, and overall product excellence in ceramic manufacturing [11,12].

2. Stages of Masse Preparation

The preparation of a ceramic body is not a linear or routine task, but a multifaceted process that plays a decisive role in determining the structural integrity, processability, and ultimate functionality of the final ceramic product. Each stage in masse preparation is interlinked with the next, forming a closed-loop system where variations introduced at any point whether in particle size, chemical uniformity, moisture content, or rheology can propagate through subsequent operations and manifest as critical defects after firing.

Figure 1 presents the analytical sequence of ceramic body preparation, essential for ensuring final product quality. The process includes raw material selection and weighing, grinding, mixing, filtration, rheology adjustment, water removal, and aging. Table 1 summarizes these stages in detail.

2.1 Selection and precise weighing of raw materials

Every ceramic formulation begins with defining the target properties thermal resistance, mechanical strength, whiteness, electrical insulation and then selecting raw materials accordingly [13–15]. Kaolin, ball clay, feldspar, quartz, and functional additives are weighed with high accuracy using calibrated systems. Deviations as small as 0.5% in component ratios can significantly alter sintering behavior, shrinkage profile, or even aesthetic outcome. As such, this step establishes the chemical identity and functional trajectory of the ceramic body [16,17].

2.2 Particle size reduction: crushing and grinding

The transformation from geological material to ceramic-grade powder necessitates careful control over particle size distribution. Initial crushing breaks down aggregates; grinding, whether wet or dry, refines the material to specific size ranges optimized for packing density and sintering kinetics. Wet grinding is preferred for high-end ceramics due to better particle dispersion and minimization of agglomeration, while dry grinding offers energy efficiency for large-scale tile production. In both cases, surface area control is key to reaction uniformity during thermal processing [18–20].

Stages of Masse Preparation

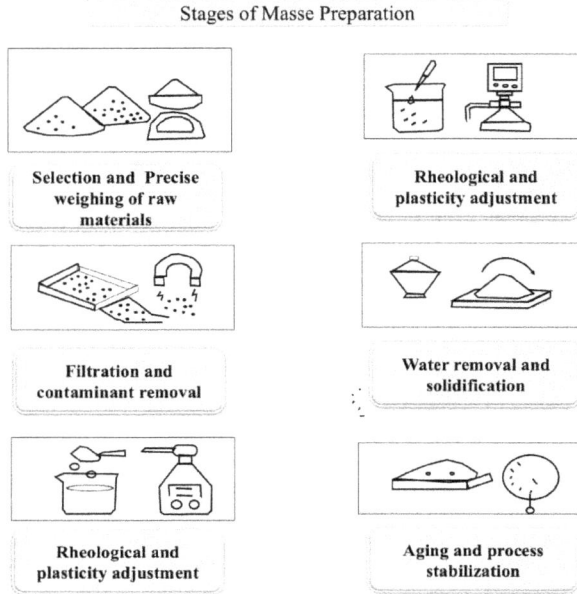

Figure 1. *Stages of masse preparation.*

2.3 Mixing and chemical homogenization

At this stage, the prepared components are blended using high-efficiency mixers, such as ribbon blenders or blungers, to ensure uniform dispersion of particles and consistent chemical composition across the mass. Homogenization directly influences not only the physical behavior during forming but also the thermal behavior during firing. Advanced systems may employ in-line sensors to monitor and adjust parameters in real time, eliminating batch-to-batch variability a critical factor in high-precision applications [21–23].

2.4 Filtration and contaminant removal

Before shaping, it is essential to eliminate coarse particles, magnetic impurities (e.g., iron oxides), and foreign inclusions. Sieving, filtering, and magnetic separation safeguard the visual and structural quality of the final product. Even trace amounts of ferromagnetic particles can cause localized melting or discoloration during firing, making this step vital in high-purity ceramics like porcelain or electronic substrates [24–27].

Table 1. Analytical of ceramic body preparation stages.

Stage	Operations & Tools	Criteria & Process Control	Impact on Final Product	Advanced Measurement & Monitoring Tools
Selection & Precise Weighing of Raw Materials	- Selection of raw materials: kaolin, feldspar, quartz, functional additives - High-precision weighing (±0.5% or less)	- Predefined chemical and physical specifications - Verified source quality	- Defines chemical composition, sintering behavior, shrinkage, and end-use properties (e.g., insulation, whiteness)	- Electronic precision balances - RFID-based formulation tracking - Formulation software
Grinding & Particle Size Reduction	- Dry grinding for high-volume production (e.g., tiles) - Wet grinding for high-end ceramics	- Particle size distribution (PSD) analysis - Energy input and time optimization	- Influences packing density, sintering rate, and surface quality	- Laser diffraction particle analyzers - BET surface area analysis - Power consumption meters
Mixing & Chemical Homogenization	- Ribbon blenders or blungers - Addition of water or chemical modifiers	- Real-time monitoring of homogeneity - Sample testing to verify uniformity	- Reduces structural defects and ensures even thermal reaction during firing	- In-line NIR sensors - Automated blending systems - Sample testing protocols
Filtration & Impurity Removal	- Sieving and mesh screening - Magnetic separation (iron oxides) - Filtration of slurries	- Microscopic inspection or metal detectors - High-purity standards for technical ceramics	- Prevents discoloration, microcracks, and mechanical or dielectric failures	- High-sensitivity magnetic separators - Microscopy (SEM, optical) - Infrared impurity detectors
Rheology & Plasticity Adjustment	- Addition of plasticizers (e.g., bentonite), deflocculants, binders - Consistency tuning for forming method	- Viscosity measurements - Plasticity and flow testing under shear	- Determines formability (casting, extrusion, pressing), reduces cracks and warping	- Rotational viscometers - Shear behavior analyzers - Rheometers
Water Removal & Forming Preparation	- Filter pressing for plastic cakes - Spray drying to produce granules	- Moisture content analysis - Granule flow rate testing	- Affects green strength, dimensional accuracy, and shaping consistency	- Moisture analyzers - TGA (Thermogravimetric Analysis) - Granule flowability testers
Aging & Stabilization	- Controlled storage under specific humidity (60–80%) and temperature - Intermittent mixing for moisture redistribution	- Aging time control (24 hours to 7 days) - Monitoring of rheological stabilization	- Enhances plasticity, minimizes warping, and ensures predictable shrinkage	- Environmental data loggers (humidity/temp) - post-aging forming tests - Plasticity evaluation tools

2.5 Rheological and plasticity adjustment

This phase is the core of processability engineering. Through the controlled addition of plasticizers (e.g., bentonite), deflocculants, binders, and rheology modifiers, the mixture is tuned for optimal forming behavior whether by casting, extrusion, or pressing. The goal is to strike a balance between fluidity and cohesion, minimizing internal stress and cracking during drying, while ensuring stability during shaping. Rheological profiling is often conducted using viscometers or flow curve analysis, allowing fine-tuning of behavior under shear stress [28–31].

2.6 Water removal and solidification

Depending on the shaping method, the ceramic body must be brought to a suitable consistency plastic, semi-dry, or granulated. For slip-based processes, filter pressing concentrates the slurry into a workable plastic cake. For isostatic or uniaxial pressing, spray drying converts the slurry into free-flowing granules with controlled residual moisture. This phase is crucial for dimensional accuracy and green strength in the forming process [32–34].

2.7 Aging and process stabilization

Before forming, the mass may undergo a controlled aging period to stabilize interparticle interactions and allow for moisture redistribution, which improves plasticity and forming consistency. In high-precision ceramics, this period is carefully timed and conducted under specific environmental conditions (temperature, humidity) to avoid premature drying or microbial growth. Proper aging leads to predictable shrinkage, uniform drying, and fewer warping tendencies [13,32,34].

3. Preparation Techniques Based on Product

In ceramic manufacturing, the selection of a suitable preparation method is closely linked to the type of product being manufactured and the forming technology it requires. Each preparation technique yields a ceramic body with distinct physical, chemical, and rheological properties, tailored to meet specific functional and production demands. Figure 2 illustrates the preparation methods, forming technologies, and the resulting material properties, providing a visual overview of how each method aligns with specific ceramic applications.

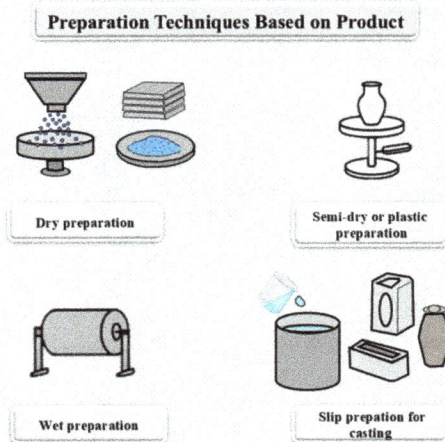

Figure 2. Preparation techniques based on product.

3.1 Dry preparation

Dry preparation is commonly used in mass production of ceramics, particularly in tile manufacturing and standard tableware. The process involves dry milling of raw materials followed by homogenization and granulation through spray drying, which produces free-flowing granules with a controlled moisture level. This method is highly efficient in terms of energy consumption and automation potential. However, it may exhibit lower chemical uniformity compared to wet processing and often requires additives to adjust plasticity and binding characteristics [7,25,35].

3.2 Wet preparation

For high-performance products such as porcelain, sanitary ware, and electrical ceramics wet preparation is preferred due to its superior mixing and particle dispersion. Raw materials are milled in a water-based system to form a slurry, which enhances uniformity at the microscopic level. After filtration and adjustment, the slurry is dried to obtain the desired consistency for shaping. Though energy- and time-intensive, wet preparation offers exceptional quality control and is ideal for applications requiring refined texture, whiteness, and defect-free surfaces [24,36–38].

3.3 Semi-dry or plastic preparation

When extrusion or hand-forming is the chosen shaping method, a semi-dry or plastic body is prepared by incorporating a moderate amount of water. This results in a moldable, cohesive mass that retains its shape during manipulation. Such preparation is especially common in the production of ceramic bricks, pipes, and traditional pottery, where high plasticity is required without the need for slurry processing [22,39,40].

3.4 Slip preparation for casting

In slip casting, which is widely used for forming intricate and hollow shapes, the ceramic mass is prepared as a liquid suspension. A finely milled slurry is formulated with deflocculants to control viscosity and improve flow behavior. This technique allows for precise shaping with minimal mechanical stress, making it ideal for delicate geometries and thin-walled components [24,35,41].

Each of these methods reflects a deliberate balance between material behavior, production scalability, and final product performance. As ceramic applications diversify and demand greater technical precision, the ability to tailor preparation methods to suit specific requirements becomes not only advantageous but essential. Table 2 summarizes the key elements discussed in Section 3, offering a structured overview of the methods and their relevance to modern ceramic manufacturing.

Table 2. Ceramic preparation techniques.

Technique	Process Description (Simplified)	Technical Specifications	Advantages	Limitations	Operational Tips	Common Applications
Dry Preparation	- Dry grinding - Spray drying for free-flowing granules	- Moisture: 4-7% - Angle of repose: 25-35° - Binders: 2-5%	- 30% energy savings - High throughput	- Moderate homogeneity - Requires binders	- Monitor granule moisture daily - Test flowability per batch	- Floor tiles - Mass-produced tableware
Wet Preparation	- Wet milling to slurry - Filtration and drying	- Viscosity: 800-1200 cP - Particle size: ≤1μm - Aging: 24-72h	- Superior surface finish (Ra ≤0.1μm) - Excellent homogeneity	- High water usage (3-5m³/ton) - Slow process	- Maintain pH 8-9 - Inspect slurry microscopically	-Sanitaryware - Electronic substrates
Plastic Preparation	- Kneading with 15-25% water - Extrusion forming	- Green strength: ≥1 MPa - Drying shrinkage: 5-8%	- No molds needed - Hand-formable	- Limited shape complexity	- Control paste viscosity - Store at 60% RH	- Ceramic pipes - Art pottery
Slip Casting	- Deflocculated slurry (1-3% additives) - Plaster mold pouring	- Casting rate: 2-4 mm/min - Demolding: 2-6h	- Ultra-fine details (50μm) - Hollow shapes	- Mold consumption high - Slow drying	- Stir slurry regularly - Clean molds after use	- Fine china - Decorative pieces

4. Equipment Used in Masse Preparation

Modern ceramic manufacturing demands not only precision in formulation but also consistency and efficiency in processing. As such, the equipment used in masse preparation is no longer considered merely supportive; it is a critical component in material engineering, directly influencing particle behavior, rheological control, and ultimately, product performance. Figure 3 highlights the core technologies that form the backbone of contemporary ceramic body preparation systems, illustrating the key machinery and their functional roles within the production line.

4.1 Ball mills

Ball milling remains one of the most essential operations in ceramic processing. Whether in dry or wet systems, these mills enable precise control over particle size distribution, surface area, and powder reactivity. In wet milling, they facilitate not only grinding but also the initial homogenization of the slurry, promoting molecular-scale interaction between components. The degree of fineness achieved in this step significantly influences sintering kinetics and final microstructure [39,42,43].

88

4.2 High efficiency blungers and mixers

These machines ensure uniform dispersion of raw materials, correcting for variations in moisture content, density, and composition. Unlike basic mixing tools, modern blungers integrate automated control systems, allowing dynamic adjustment of mixing speed, shear rate, and cycle timing. This ensures that the mixture achieves optimal plasticity and rheological stability parameters that are critical for shaping precision and drying behavior [28,44–46].

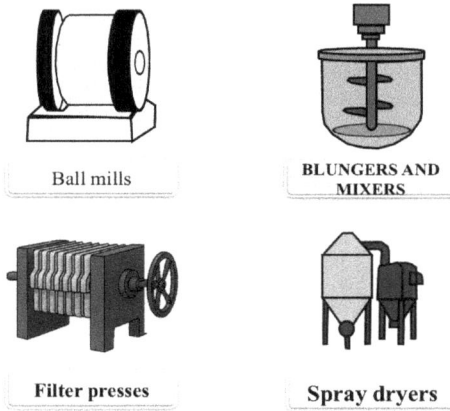

Equipment used in masse preparation

Ball mills

BLUNGERS AND MIXERS

Filter presses

Spray dryers

Figure 3. Equipment used in masse preparation.

4.3 Filter presses

In wet preparation systems, filter presses are used to dewater slurries into plastic cakes with tightly controlled moisture content. The resulting material is highly moldable and ideal for extrusion and manual forming processes. The role of the filter press is not limited to water removal; it also enhances packing density and ensures that the body maintains internal cohesion during handling and shaping [47,48].

4.4 Spray dryers

Spray drying is among the most transformative advancements in ceramic body preparation. It converts liquid slurry into flowable granules with uniform size and residual moisture properties essential for high-pressure forming techniques such as isostatic or uniaxial pressing. The morphology and density of the granules can be tailored through parameters such as atomization pressure and inlet temperature, offering a high level of control over green body behavior [49–51].

The relationship between mass preparation equipment and the final properties of ceramics is a vital aspect of modern material design. Equipment choices directly influence factors such as particle dispersion, moisture control, and rheological stability which contribute to the mechanical strength, surface quality, and functional performance of the final ceramic product. Table 3 summarizes the key points discussed in Section 4, illustrating how specific types of equipment impact critical material characteristics throughout the preparation process.

Table 3. *Relationship between mass preparation equipment and its effect on the final properties of ceramics.*

Equipment	Controlled Parameter	Effect on Ceramic Body	Impact on Final Product	Relevant Application / Forming Method
Ball Mill	Particle size distribution & surface area	Improves powder reactivity and sintering uniformity; reduces agglomeration	Enhances microstructural integrity and mechanical strength post-sintering	Wet milling for porcelain and technical ceramics
High-Efficiency Mixer	Homogeneity, plasticity, rheological stability	Ensures plastic consistency and uniform material behavior during shaping	Reduces warping and cracks, improves shape retention in complex forms	Body preparation for tiles, insulators, and sanitaryware
Filter Press	Moisture content and cake density	Produces a cohesive, moldable body with controlled drying behavior	Minimizes drying shrinkage, boosts forming efficiency	Extrusion-based forming, traditional plastic shaping
Spray Dryer	Granule morphology & residual moisture	Enables high green density and uniform compaction in pressing	Improves dimensional stability and surface quality after firing	Pressing of tiles, isostatic pressing of technical bodies
Magnetic Separator	Ferromagnetic impurities (e.g., Fe_2O_3)	Prevents defect initiation from metallic contamination during firing	Critical for porcelain aesthetics and dielectric purity in technical ceramics	Glazed tableware, electrical ceramics, whitewares
Precision Sieve	Oversized particle removal, granulometry control	Promotes even packing, improves slurry behavior and surface finish	Optimizes texture and gloss; avoids surface blemishes in glazing	Casting slips, fine-grained decorative ceramics

5. Rheological and Physical Control in Mass Preparation

In modern ceramics, ceramic mass is not regarded as a mere rigid mixture of raw materials, but as a complex dynamic system with rheological (behavior under shear and pressure) and physical properties (such as density, shrinkage, and moisture) that directly control its behavior during forming, drying, and sintering. Precise management of these properties during preparation is critical to avoiding structural defects and achieving stable performance in the final product. As illustrated in Figure 4, these interrelated properties influence each stage of the production process, highlighting the need for integrated monitoring and control systems to ensure consistency and reliability in ceramic manufacturing.

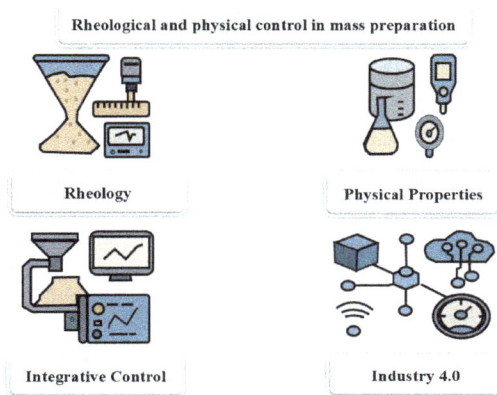

Figure 4. *Rheological and physical control in mass preparation.*

5.1 Understanding rheology in ceramic systems

Rheology in ceramic body preparation refers to how the mass flows or deforms under applied stress. It affects processes such as casting, extrusion, and pressing. A body with optimized plasticity and shear resistance will respond predictably during shaping, maintaining dimensional stability and reducing internal stresses. Parameters like yield stress, viscosity, and thixotropy must be tailored through precise control of particle interactions, moisture content, and additive selection (e.g., bentonite as a plasticizer or dispersants in slip casting) [13,39,52,53].

5.2 Physical Properties; Density, Moisture, and Granule Structure

The physical characteristics of the ceramic mass, especially green density, moisture content, and granule morphology determine how it compacts and behaves thermally. For instance, low green density often leads to weak mechanical strength post-firing, while high residual moisture can trigger warping or bloating during drying. Spray drying techniques, filter pressing, and controlled humidity environments allow engineers to precisely modulate these variables. In particular, particle packing efficiency plays a key role in shrinkage control and microstructural homogeneity.[35,54–56].

5.3 Integrative control and real-time monitoring

With the advent of smart manufacturing, real-time rheological and physical property monitoring has become integral to quality control. Techniques such as in-line viscometry, moisture probes, and laser diffraction particle analysis allow feedback-driven adjustments in milling or mixing. This ensures each batch adheres to tight specifications, essential in high-end applications like technical ceramics, where even micro-level variations can compromise function [57–59].

5.4 Industry 4.0 applications in ceramic mass monitoring and control

With the global shift towards Industry 4.0, ceramic mass preparation processes have become part of smart manufacturing systems that rely on data, cloud computing, and the Industrial Internet of Things (IIoT). By integrating technologies such as Digital Twin, SCADA systems, and MES systems, it is now possible to achieve real-time control over ceramic body properties, improve productivity, and reduce waste [60–62].

5.4.1 Digital twin

An advanced virtual model that instantly reflects the behavior of ceramic mass within production lines by integrating sensory data from various sensors such as moisture, viscosity, and temperature sensors, along with machine vision cameras and particle distribution analysis tools. This data is collected and analyzed to provide an accurate and dynamic picture of the material's behavior during different stages of manufacturing. Sacmi Imola, an Italian company, has developed the "Smart Factory 4.0" platform, which includes a digital twin of the entire ceramic powder preparation line. This innovative simulation system is used to analyze the impact of changes in the composition of materials on properties like density and plasticity. This technology enables real-time monitoring and optimization of production, enhancing efficiency, reducing costs, and increasing productivity [63,64].

5.4.2 Supervisory Control and Data Acquisition (SCADA)

A software system is used to collect data from grinding, mixing, and drying units and analyzing it in real-time. In the tile factories of RAK Ceramics Group, SCADA is used to automatically monitor the moisture characteristics in spray dryers and adjust them based on changes in water salinity or kaolin content. This ensures optimal production conditions and enhances process control [65–69].

5.4.3 Manufacturing Execution Systems (MES)

The MES (Manufacturing Execution System) links preparation processes (such as grinding and drying) with ERP (Enterprise Resource Planning) systems, ensuring real-time tracking of each batch in terms of: Raw material source, Grinding transactions, Moisture and density values, Storage time and the allowed aging window of the mass. System Ceramics offers an MES solution called Prime Line, which connects powder preparation lines with ERP and SAP systems. This integration enables decision-making based on actual performance and product quality at every moment, ensuring optimal production management [70–73].

5.4.4 AI-based analytics

Machine learning is used to analyze patterns in viscosity, mixing time, and resulting firing defects, predicting potential issues before they occur. In the Villeroy & Boch sanitary ceramics factories, machine learning algorithms are employed to identify mixtures that cause the highest crack rates after drying. The material ratios are automatically adjusted to prevent defects, ensuring better product quality and reducing waste [74–76].

Achieving comprehensive control in ceramic body preparation is essential for ensuring repeatability, quality assurance, and process optimization across industrial production. Section 5 outlines the critical parameters that must be monitored and regulated throughout each stage ranging from raw material quality to slurry rheology and moisture content. Table 4 summarizes the content of Section 5, presenting the key control points and their impact on process stability and product performance.

Table 4. Comprehensive control in ceramic body preparation.

Parameter	Scientific Standards	Measurement Tools	Control Techniques	Industry 4.0 Applications	Consequences of Poor Control
Rheology	- Yield stress: 50-200 Pa - Plastic viscosity: 500-1500 cP - Thixotropy index: 1.2-1.8	- Rotational rheometer - Brookfield viscometer - 3-interval thixotropy test	- Deflocculants (0.1-0.5%) - Bentonite (2-4%) - Moisture control ($\pm 0.3\%$)	- Digital twin (viscosity tracking) - SCADA for additive adjustment	- Forming deformations - post-firing surface defects
Physical Properties	- Green density: 1.8-2.4 g/cm³ - Moisture content: 5-22% - Granule flow: 25-30°	- NIR moisture probes - Gamma density gauges - Granule morphology analysis	- Spray drying (400-450°C inlet) - Bimodal particle distribution - Aging time control	- MES for particle monitoring - AI density prediction	- Differential shrinkage - Weak mechanical strength
Smart Monitoring	- Particle size: 0.1-875 µm - Moisture stability: $\pm 0.1\%$ - Mix homogeneity: $\pm 2\%$	- Laser diffraction (HELOS/KR) - IR thermography - NIR spectroscopy	- Self-adjusting systems - Closed-loop control	- AI molecular analysis - Early deviation alerts	- Batch inconsistency - Material waste
Digital Integration	- Aging time: 24±2 hrs - Raw material traceability - Energy efficiency: +15%	- IoT platforms (Siemens MindSphere) - ERP integration - Vibration sensors	- Digital twin process modeling - Dynamic simulation	- SCADA for milling - MES (Prime Line) - AI defect prediction	- Material loss - Unplanned downtime

6. Comparison Between Traditional and Modern Methods

As the ceramic industry evolves from craft-based production to data-driven precision engineering, the transition from traditional to modern methods of masse preparation represents a fundamental shift. This section compares the two approaches across three key dimensions: energy efficiency, product quality, and process control.

6.1 Efficiency and energy consumption

Traditional preparation techniques such as manual mixing, open-air drying, and batch-based milling tend to be labor-intensive and energy inefficient. The lack of process optimization often results in excessive material waste, prolonged processing times, and inconsistent moisture removal, all of which translate to higher energy demands [77–79].

In contrast, modern processes emphasize closed-loop control and energy optimization. Technologies like high-efficiency spray dryers, continuous ball mills, and inline moisture sensors dramatically reduce energy input by tailoring conditions in real-time. Furthermore, advanced scheduling algorithms and lean manufacturing principles minimize idle time and thermal losses, leading to measurable gains in sustainability and operational cost [80–82].

6.2 Product quality and consistency

Product consistency is one of the most defining challenges in traditional systems, where heterogeneity in raw materials, manual weighing errors, and inconsistent mixing often result in variable green body properties. Such variability may cause differential shrinkage, deformation, or cracking during sintering, and these defects are often only visible at the end of production [83–85].

Modern methods integrate precision dosing, automated homogenization, and feedback-driven adjustments, ensuring uniform particle size distribution, controlled rheology, and optimal plasticity. These controls directly translate into improved surface finish, dimensional accuracy, and structural integrity across batches, which is especially critical for technical ceramics and applications requiring tight tolerances [86–88].

6.3 Automation and process control

Traditional preparation relies heavily on operator expertise, with minimal instrumentation. Decision-making is often empirical and reactive responding to defects after they occur. In contrast, advanced preparation lines feature automated control systems, real-time data acquisition, and predictive modeling. These systems monitor parameters like viscosity, moisture content, and granule structure, allowing proactive intervention before deviations lead to defects. Moreover, machine learning and AI-based models are increasingly being used to optimize mixing cycles and forecast behavior under different firing conditions [89–91].

This shift towards smart manufacturing not only reduces human error and waste but also enables reproducibility and traceability key requirements in regulated industries such as electronics and medical ceramics. Section 6 compares traditional and modern approaches in ceramic masse preparation, emphasizing the evolution of technology and its role in improving precision and control. Table 5 summarizes this comparison, highlighting the key differences between conventional methods and advanced, digitally integrated processes.

The versatility of ceramic materials allows their integration into a wide range of industrial sectors, each with specific requirements for body formulation, processing precision, and end-use performance. Section 6 provides an overview of practical applications and the corresponding preparation strategies adopted to meet functional demands. Table 6 presents selected examples of industrial applications and their relevance to specific ceramic body preparation techniques, illustrating how tailored processes contribute to performance optimization across different fields.

Table 5. Radiational and modern methods in ceramic masse preparation.

	Traditional Methods	Modern Methods	Real-World Example
Energy Efficiency	High energy use due to manual drying and intermittent kilns.	Optimized with spray dryers and continuous flow lines with thermal recovery.	Use of spray drying in porcelain tile production reduces drying energy by up to 30%.
Mixing & Homogeneity	Manual or low-shear mixing often leads to poor dispersion of fine additives.	Automated high-shear mixers ensure complete dispersion of kaolin, fluxes, and binders.	Uniform mixing of zirconia in dental ceramics enhances translucency and strength.
Moisture Management	Judged by operator feel; prone to variation during seasons or shifts.	Moisture sensors and closed-loop drying systems enable consistent drying profiles.	Controlled dewatering via filter press ensures stable extrusion of ceramic pipes.
Rheological Behavior	Plasticity inconsistencies affect casting and extrusion, especially in humid climates.	Real-time rheology monitoring allows adjustment of dispersants and binders to maintain flow curves.	Slip casting for sanitary ware improved by on-the-fly viscosity control.
Process Control	Empirical control based on end-of-line inspection.	Inline sensors with digital feedback loops maintain setpoints for moisture, particle size, and pH.	Technical ceramics (e.g. substrates) require real-time control of granule uniformity.
Labor Dependency	Skilled workers are essential to detect anomalies manually.	Machine-integrated diagnostics reduce need for operator judgment and error.	Tile plants using SCADA systems reduced operator count by 40%.
Product Consistency	Variable shrinkage, cracks, or surface finish due to batch variation.	Granular control allows tolerance bands within ±1% for dimensions and strength.	Electronic insulators require near-zero batch variability to meet IEC standards.
Automation Level	Minimal to no automation, difficult to integrate with MES/ERP systems.	Fully automated lines with sensor arrays, PLCs, and AI-based optimization algorithms.	Smart kilns and AI-adjusted batching systems used in aerospace ceramic part production.

Table 6. Examples of industrial applications and their relevance.

Example in the Table	Target Industry	Example
Porcelain tile production	Porcelain tile manufacturing	Spray drying is widely used to enhance efficiency and reduce energy consumption.
Zirconia in dental ceramics	Medical ceramics / Dental applications	Highlights the importance of homogeneity in blending high-cost materials like zirconia.
Sanitary ware slip casting	Sanitary ware production (e.g., sinks, toilets)	A classic example of how rheological control impacts slip casting performance.
Ceramic pipes extrusion	Ceramic piping for abrasion and heat resistance	Demonstrates the critical role of moisture control in extrusion stability.
Electronic insulators	Electrical ceramics (e.g., insulators)	High product consistency is essential to meet international standards such as IEC.

7. Challenges and Common Defects

Despite the advancements in modern processing technologies, the preparation of ceramic bodies continues to face several inherent challenges. These issues can compromise the structural integrity, dimensional accuracy, and aesthetic quality of the final product if not properly controlled. The most recurrent problems are linked to material variability, process sensitivity, and uncontrolled rheological behavior.

7.1 Inhomogeneity

Inhomogeneous distribution of raw materials or additives due to insufficient mixing, poor weighing, or inadequate particle dispersion often leads to defects such as warping, delamination, or localized shrinkage during firing [92–94].

7.2 Over- or under-plasticization

Improper control of plasticity (either excess or deficiency) affects the formability and shape retention of the ceramic body. Over-plasticized batches may deform or collapse, while under-plasticized ones are prone to cracking and reduced green strength [95–97].

7.3 Particle agglomeration

Insufficient deagglomeration leads to large clusters of particles that disrupt sintering behavior, reduce density, and introduce internal porosity or weak points within the structure [98–100].

8. Conclusion

Masse preparation stands as the fundamental pillar of ceramic manufacturing, where the marriage of precision engineering and technological innovation drives the production of exceptional quality ceramics. The transformative integration of Industry 4.0 technologies, particularly artificial intelligence and industrial automation is reshaping traditional processes by enabling predictive quality control, real-time parameter optimization, and intelligent waste reduction. This technological evolution is not only elevating product performance and consistency but also establishing new benchmarks for sustainable manufacturing practices. As the ceramic industry continues its digital transformation, the convergence of material science expertise with smart manufacturing capabilities promises to unlock unprecedented levels of efficiency, reliability, and innovation for both conventional and advanced technical applications. The future of ceramic production lies in this harmonious blend of time-honored craftsmanship and cutting-edge digital solutions, ensuring the industry remains at the forefront of material innovation while meeting increasingly stringent quality and environmental standards.

References

[1] V. Occari, C. Fenwick, H. Möller, P.S. Quinn, I.C. Freestone, V. Sacco, P. von Rummel, M. Chaouali, Kairouan and the medieval ceramic industry in Tunisia: New data from compositional analysis of glazed and unglazed wares from the sites of Bulla Regia and Chimtou, J. Archaeol. Sci. Reports 66 (2025) 105245. https://doi.org/10.1016/j.jasrep.2025.105245

[2] R.B. Heimann, On the nature of ceramics technology: from Empedocles to Dawkins, Archaeometry 67 (2025) 55–71. https://doi.org/10.1111/arcm.12997

[3] M. Kurian, S. Thankachan, Introduction: ceramics classification and applications, in: Ceram. Catal. Mater. Synth. Appl., Elsevier, 2023: pp. 1–17. https://doi.org/10.1016/B978-0-323-85746-8.00009-6

[4] Z. Ye, P. Ruangchewin, P. Panthupakorn, The Historical Development and Site Value of the Imperial Porcelain Factory in Jingdezhen, China, Int. J. Multidiscip. Manag. Tour. 8 (2024) 13–28.

[5] M. Tanguler-Bayramtan, C.B. Aktas, I.O. Yaman, Environmental Assessment of Calcium Sulfoaluminate Cement: A Monte Carlo Simulation in an Industrial Symbiosis Framework, Buildings 14 (2024) 3673. https://doi.org/10.3390/buildings14113673

[6] S.S. Anandakrishnan, S. Yadav, M. Tabeshfar, V. Balanov, T. Kaushalya, M. Nelo, J. Peräntie, J. Juuti, Y. Bai, Toward Ecofriendly Piezoelectric Ceramics—Reduction of Energy and Environmental Footprint from Conceptualization to Deployment, Glob. Challenges 7 (2023) 2300061. https://doi.org/10.1002/gch2.202300061

[7] M.F. Zawrah, M.A. Taha, R.A. Youness, Advanced ceramics: stages of development, in: Adv. Ceram., Springer, 2023: pp. 1–46.

[8] Q. He, X. Zheng, X. Xiao, L. Luo, H. Lin, S. He, The Spatiotemporal Evolution and Influencing Factors of the Ceramics Industry in Jingdezhen in the Last 40 Years, Land 12 (2023) 1554. https://doi.org/10.3390/land12081554

[9] A. Maged, S.A. Abu El-Magd, A.E. Radwan, S. Kharbish, S. Zamzam, Evaluation insight into Abu Zenima clay deposits as a prospective raw material source for ceramics industry: Remote Sensing and Characterization, Sci. Rep. 13 (2023) 58. https://doi.org/10.1038/s41598-022-26484-5

[10] H. Blaifi, M. Guendouz, A.E. Belhadj, D. Boukhelkhal, M. Hadjadj, Sustainable Use of Recycled Plastic and Ceramic Industrial Wastes in Eco-Friendly Construction Materials, Environ. Eng. Manag. J. 22 (2023) 1471–1486. https://doi.org/10.30638/eemj.2023.124

[11] R.S. Mahale, V. Shamanth, P.C. Sharath, K. Goggi, V. Mahesh, Comparative Studies on Dry and Wet Milling: Uncovering the Secrets of Material Processing, in: Adv. Struct. Mater., Springer, 2024: pp. 11–40. https://doi.org/10.1007/978-981-97-6504-1_2

[12] S.E. Avram, L.B. Tudoran, G. Borodi, I. Petean, Microstructural Characterization of the Mn Lepidolite Distribution in Dark Red Clay Soils, Appl. Sci. 15 (2025) 6445.

https://doi.org/10.3390/app15126445

[13] Giacomo Boschi, A. Tugnoli, M. Chiara Bignozzi, Sustainability and environmental impact of the Italian ceramic tile industry, (2023).

[14] M.A. Ali, H.A.M. Ahmed, H.M. Ahmed, M. Hefni, Pyrophyllite: An economic mineral for different industrial applications, Appl. Sci. 11 (2021) 11357. https://doi.org/10.3390/app112311357

[15] F. Gol, A. Yilmaz, E. Kacar, S. Simsek, Z.G. Sarıtas, C. Ture, M. Arslan, M. Bekmezci, H. Burhan, F. Sen, Reuse of glass waste in the manufacture of ceramic tableware glazes, Ceram. Int. 47 (2021) 21061–21068. https://doi.org/10.1016/j.ceramint.2021.04.108

[16] H. Alves de Oliveira, C. Pereira dos Santos, Limestone Clays for Ceramic Industry, Clay Sci. Technol. (2021) 83. https://doi.org/10.5772/intechopen.92506

[17] M. Dondi, B. Gian Paolo, Basic Guidelines for Prospecting and Technological Assessment of Clays for the Ceramic Industry. Part 2, InterCeram Int. Ceram. Rev. 71 (2022) 28–37. https://doi.org/10.1007/s42411-022-0484-1

[18] W. Guo, K. Guo, Effect of Solid Concentration on Particle Size Distribution and Grinding Kinetics in Stirred Mills, Minerals 14 (2024) 720. https://doi.org/10.3390/min14070720

[19] S. Shojaee Barjoee, V. Rodionov, A.M. Vaziri Sereshk, Noise climate assessment in ceramic industries (Iran) using acoustic indices and its control solutions, Adv. Environ. Technol. 11 (2025) 91–115. https://doi.org/10.22104/aet.2024.6922.1899

[20] I.O. de Landazuri Suárez, Design of comminution in ceramic plants using a simulation-based optimization approach, Ing. e Investig. 41 (2021). https://doi.org/10.15446/ing.investig.v41n3.87761

[21] E. Durgut, M. Çınar, M. Terzi, İ. Kurşun Ünver, Y. Yıldırım, F. Boylu, O. Özdemir, Improvement of Sintering Properties of Halloysit-Kaolinite Mixture Low Quality Ceramic Raw Material by Mechanical Dispersion and Wet Magnetic Separation, (n.d.).

[22] A.A. Mansor, M.A. Al-Nuaimy, Study and Evaluating the Physical and Chemical Properties of Clay of Injana Formation in Al-Anbar Governorate - Western Iraq for the Purposes Some Ceramic Industries, J. Univ. Anbar Pure Sci. 17 (2023) 219–228. https://doi.org/10.37652/juaps.2023.178928

[23] I. Djobov, E. Karamanova, G. Avdeev, A. Karamanov, Characterization of Clays From "Mines Maritsa Iztok" As Raw Materials for Ceramic Industry, J. Chem. Technol. Metall. 59 (2024) 1331–1340. https://doi.org/10.59957/jctm.v59.i6.2024.7

[24] S. Bose, C. Das, Introduction to Ceramics: Fabrication, Characterizations, and Applications, CRC Press, 2024.

[25] M.J. Ribeiro, D. Tulyaganov, Traditional Ceramics Manufacturing, in: PoliTO Springer Ser., Springer, 2021: pp. 75–118. https://doi.org/10.1007/978-3-030-85776-9_3

[26] A.G. Adeniyi, K.O. Iwuozor, E.C. Emenike, Material Development Potential of Nigeria's Kaolin, Chem. Africa 6 (2023) 1709–1725. https://doi.org/10.1007/s42250-023-00642-2

[27] A. Venkateswara Rao, K. Srinivasa Rao, Effect of fly ash on strength of concrete, in: Circ. Econ. Fly Ash Manag., Materials Research Forum LLC, 2019: pp. 125–134. https://doi.org/10.1007/978-981-15-0014-5_9

[28] T. Palomäki, Measurement of electrostatic forces for in-line quality control of ceramic

slip, (2021).

[29] J. Wang, J. Zhao, J. Mao, W. Liu, H. Ji, J. Zhang, S. Wang, Enhanced plasticity of spontaneous coagulation cast oxide ceramic green bodies, J. Adv. Ceram. 13 (2024) 568–578. https://doi.org/10.26599/JAC.2024.9220879

[30] L.S.O. Pires, J. Luís, M.H.V. Fernandes, M. Oliveira, Controlling properties of ceramic formulations for porcelain robocasting, Ceram. Int. 49 (2023) 4764–4774. https://doi.org/10.1016/j.ceramint.2022.09.366

[31] T. Zat, S.R.C. Matos, P.R. de Matos, J. Georgin, L.P. Specht, J.P. Gonçalves, G. Mohamad, E.D. Rodríguez, Sustainable red ceramic block: recycling of a sewage sludge as raw material, J. Mater. Cycles Waste Manag. 25 (2023) 1019–1034. https://doi.org/10.1007/s10163-022-01587-3

[32] Y. Zhang, Y. Tan, R. Sun, W. Zhang, Preparation of Ceramic Membranes and Their Application in Wastewater and Water Treatment, Water (Switzerland) 15 (2023) 3344. https://doi.org/10.3390/w15193344

[33] H. Heidari Forozabadi, M. Mehdi Khabiri, K. Barkhordari Bafghi, D.A. Miti Ku, Use of Ceramic-Industry Wastewater in Sandy Subgrade Stabilization by Deep Mixing Method, J. Rehabil. Civ. Eng. 11 (2023) 1–17. https://doi.org/10.22075/JRCE.2022.26315.1614

[34] M. Fukushima, T. Ohji, Macroporous ceramics for the sustainable development goals (SDGs): Review, Int. J. Appl. Ceram. Technol. 20 (2023) 660–680. https://doi.org/10.1111/ijac.14261

[35] A. Lyons, Ceramic products, in: Mater. Archit. Build., Springer, 2020: pp. 258–264. https://doi.org/10.4324/9780080465791-19

[36] T.L. Akinbogun, Abundant Resources, Waning Growth; The Paradox of Ceramic Industry in Nigeria, Inaug. Lect. Ser. 122 (2021).

[37] Y. Sawadogo, M. Sawadogo, N. Sory, M. Ouedraogo, K. Dao, M. Seynou, P. Blanchart, L. Zerbo, Porcelain: raw materials, technological properties and applications--a review., J. La Société Ouest-Africaine Chim. 53 (2024).

[38] G. Sariisik, Performance Analysis in Ceramic Porcelain Production with the Use of Alternative Raw Materials and Composition Improvement, Int. J. Curr. Nat. Adv. Phytochem. 4 (2024).

[39] C. Lourenco Alves, J. de Oliveira Martins Müller, A. de Noni, S. Heinrich, Challenges and opportunities for increase sustainability and energy efficiency in ceramic tile industry, Int. J. Appl. Ceram. Technol. 22 (2025) e15097. https://doi.org/10.1111/ijac.15097

[40] J. dos Santos Silva, M.S.A. Leite, Analysis of a supply chain in the ceramic sector: a look at business processes, Gest. e Prod. 31 (2024) e12021. https://doi.org/10.1590/1806-9649-2024v31e12021

[41] E. Cumbajin, N. Rodrigues, P. Costa, R. Miragaia, L. Frazão, N. Costa, A. Fernández-Caballero, J. Carneiro, L.H. Buruberri, A. Pereira, A Real-Time Automated Defect Detection System for Ceramic Pieces Manufacturing Process Based on Computer Vision with Deep Learning, Sensors 24 (2024) 232. https://doi.org/10.3390/s24010232

[42] J. Yu, S.H. Jin, K. Raju, Y. Lee, H.K. Lee, Analysis of individual and interaction effects of processing parameters on wet grinding performance in ball milling of alumina ceramics

Materials Research Forum LLC
https://doi.org/21741/9781644903834

using statistical methods, Ceram. Int. 47 (2021) 31202–31213.
https://doi.org/10.1016/j.ceramint.2021.07.296

[43] M. Hawryluk, J. Marzec, Problems related to the operation of machines and devices for the production of ceramic roof tiles with a special consideration of the durability of tools for band extrusion, Arch. Civ. Mech. Eng. 25 (2025) 60. https://doi.org/10.1007/s43452-024-01106-1

[44] J. Qian, B. Sun, Q. Zhao, J. Sun, H. Zhang, Study on Rheology and Stability of Light Curing Silicon Oxide Ceramic Slurry, in: Adv. Transdiscipl. Eng., IOS Press, 2022: pp. 80–88. https://doi.org/10.3233/ATDE220421

[45] C.F. Revelo, C.M.F. Vieira, H.A. Colorado, Innovative Use of Glass Waste and Additives for Additive Manufacturing of Ceramic Bricks, in: Miner. Met. Mater. Ser., Springer, 2025: pp. 136–146. https://doi.org/10.1007/978-3-031-80748-0_12

[46] N. Acharjee, S. Ganguly, P. Biswas, B. Sarangi, Uses of black pigmented CAS glass in conventional ceramic industry: production, impediments to metallization, and startling approaches for color evaluation in ceramic tech-artistry, Pigment Resin Technol. 54 (2024) 689–701. https://doi.org/10.1108/PRT-02-2024-0018

[47] M.U. Taşkıran, K. Kayacı, E.S. Vilches, V.S. Solana, Ş.K. Türk, Ş.C. Genç, javier C. Brull, Development of alternative wall tile compositions by using different industrial wastes, Physicochem. Probl. Miner. Process. 61 (2025). https://doi.org/10.37190/ppmp/203420

[48] N. Matsumoto, T. Watanabe, M. Maruyama, Y. Horimoto, T. Maeda, K. Kato, Development of mass measurement equipment using an electronic mass-comparator for gravimetric preparation of reference gas mixtures, Metrologia 41 (2004) 178–188. https://doi.org/10.1088/0026-1394/41/3/011

[49] T.L. Akinbogun, I.B. Kashim, T.P. Ologunwa, Adoption of Sublimation Digital Printing Technology for Innovative Surface Decoration of Ceramic Wares in Nigeria: A Review, Editors (2025) 1.

[50] Y. Tabak, The Utilization of Wet Silica Sand Sludge as an Additive in Different Temperature Sustainable Brick Production, Buildings 15 (2025) 849. https://doi.org/10.3390/buildings15060849

[51] Z.J. Ting, X. Meng, Z. Yang, S.A. Jiskani, L. Hu, W. Dong, M. Zhao, Solid Recovered Fuel (SRF): A Comprehensive Review of Its Origins, Production, and Industrial Utilization, Energy and Fuels 39 (2025) 9726–9761. https://doi.org/10.1021/acs.energyfuels.5c01674

[52] S. Sremac, Z. Backalic, V. Petrovic, D. Jovanovic, M. Vasic, Shortening of fast firing in the ceramic tile production by inorganic additives - an upscale to the industry, Sci. Sinter. (2024) 4–4. https://doi.org/10.2298/sos231213004s

[53] M. Hawryluk, J. Marzec, T. Leśniewski, J. Krawczyk, Ł. Madej, K. Perzyński, Analysis of the Wear of Forming Tools in the Process of Extruding Ceramic Bands Using Selected Research Methods for Evaluating Operational Durability, Materials (Basel). 18 (2025) 1994. https://doi.org/10.3390/ma18091994

[54] Z.T. Yaseen, A.M. Aqrawi, I.N. Ahmed, An Industrial Evaluation and Chemical and Physical Properties of the Clay from the Taq Taq Area in Northern Iraq for some Ceramic

Applications, Iraqi Natl. J. Earth Sci. 22 (2022) 47–66.
https://doi.org/10.33899/earth.2022.134616.1018

[55] M.G.P. Cherene, G. de C. Xavier, L. da S. Barroso, J. de S.M. Oliveira, A.R.G. de
      Azevedo, C.M. Vieira, J. Alexandre, S.N. Monteiro, Technological and microstructural
      perspective of the use of ceramic waste in cement-based mortars, Constr. Build. Mater.
      367 (2023) 130256. https://doi.org/10.1016/j.conbuildmat.2022.130256

[56] L. Gautam, J.K. Jain, A. Jain, P. Kalla, Recycling of bone china ceramic waste as cement
      replacement to produce sustainable self-compacting concrete, in: Structures, Elsevier,
      2022: pp. 364–378. https://doi.org/10.1016/j.istruc.2022.01.019

[57] A. Chen, K. Liu, C. Yan, Editorial: Additive manufacturing of advanced ceramic materials
      and its applications, Front. Mater. 11 (2024) 1519909.
      https://doi.org/10.3389/fmats.2024.1519909

[58] S. Chen, C. Wu, B. Han, Z. Liu, Z. Mi, W. Hao, J. Zhao, X. Wang, Q. Zhang, K. Liu, J.
      Qi, J. Cao, J. Feng, D. Yu, J. Li, P. Gao, Atomic-scale imaging of CH3NH3PbI3 structure
      and its decomposition pathway, Nat. Commun. 12 (2021) 5516.
      https://doi.org/10.1038/s41467-021-25832-9

[59] X. Wang, N. Wei, Y. Zhang, Y. Fang, Y. Li, S. Li, Z. Wang, C. Sun, Nanozyme-mediated
      glutathione depletion for enhanced ROS-based cancer therapies: a comprehensive review,
      Nanomedicine 20 (2025) 279–290. https://doi.org/10.1080/17435889.2024.2446138

[60] J. Bhadu, J. Bhamu, D. Singh, P. Saraswat, R. Agrawal, Implementation of an Integrated
      Framework of Lean Manufacturing and Industry 4.0 toward Sustainability: A Case Study
      of the Indian Ceramic Industry, EMJ - Eng. Manag. J. (2025) 1–20.
      https://doi.org/10.1080/10429247.2024.2440248

[61] S. Ozcan, U. Sengul, The use of supervised artificial intelligence methods in quality
      determination in continuous production lines: a case study of ceramic industry, Int. J.
      Interact. Des. Manuf. (2025) 1–20. https://doi.org/10.1007/s12008-025-02310-w

[62] L. Ruivo, M. Russo, R. Lourenço, D. Pio, Energy management in the Portuguese ceramic
      industry: Analysis of real-world factories, Energy 237 (2021) 121628.
      https://doi.org/10.1016/j.energy.2021.121628

[63] R. Raffaeli, L. Pazzi, M. Pellicciari, Industry 4.0 Solutions as Enablers for the
      Sustainability of the Italian Ceramic Tiles Sector, Sustain. 16 (2024) 4301.
      https://doi.org/10.3390/su16104301

[64] C.L. Alves, V. Skorych, A. De Noni, D. Hotza, S.Y.G. González, S. Heinrich, Application
      of Flowsheet Simulation Methodology to Improve Productivity and Sustainability of
      Porcelain Tile Manufacturing, Machines 11 (2023) 137.
      https://doi.org/10.3390/machines11020137

[65] V. Varta, R.A. Kantale, V. Kumar Sharma, P. Bhagat, M. Lahamge, SCADA-Automation
      Key Concept of Dairy Industrial Control System, Vigyan Varta 3 (2022) 45–50.

[66] N.M. Nail Mammadov, Problems of Managing Technological Regimes in Gas
      Distribution Networks and Modern Approaches To Their Solution, PAHTEI-Proceedings
      Azerbaijan High Tech. Educ. Institutions 44 (2024) 122–130.
      https://doi.org/10.36962/pahtei44092024-14

[67] A. Rubio-Rico, F. Mengod-Bautista, A. Lluna-Arriaga, B. Arroyo-Torres, V. Fuster-Roig,

The Industrial Digital Energy Twin as a Tool for the Comprehensive Optimization of Industrial Processes, Processes 11 (2023) 2353. https://doi.org/10.3390/pr11082353

[68] A. Vlasopoulos, Assessment of life cycle impacts of sustainable technologies in industrial environments, (2024).

[69] T.D.K. Saradha, Implementace Průmyslu 4.0, (2022).

[70] L. Zhang, X. Yang, Y. Xu, Q. Yu, Research on the models and implementation techniques of ceramic MES system Based on ISA95, in: E3S Web Conf., EDP Sciences, 2021: p. 2030. https://doi.org/10.1051/e3sconf/202123602030

[71] S.H. Devjibhai, Performance Improvement Through Implementation Of MFCA Based System Framework In Ceramic Tiles Manufacturing, (2023).

[72] R.D.Q. de Jesus Silva, Optimization of Color Adjustment in the Ceramic Industry using Genetic Algorithms, (2023).

[73] S. Ramesan, Digitization and optimization of production flow in a manufacturing company, (2021).

[74] O.B. Al-Ameri, M. Alzuhairi, E. Bailón-García, F. Carrasco-Marín, J. Amaro-Gahete, Transforming Petrochemical Processes: Cutting-Edge Advances in Kaolin Catalyst Fabrication, Appl. Sci. 14 (2024) 9080. https://doi.org/10.3390/app14199080

[75] L. Meng, Y. Hu, J. Guo, S. Shen, Intelligent Design of Biotechnology Ceramic Structures Using 3D Structured Light Simulation Technology for Commercial Applications, J. Commer. Biotechnol. 30 (2025) 234–247.

[76] S. Ma, Y. Huang, Y. Chen, Q. Xiao, J. Xu, J. Leng, Edge-Cloud Cooperation-Driven Intelligent Sustainability Evaluation Strategy Based on IoT and CPS for Energy-Intensive Manufacturing Industries, IEEE Internet Things J. 12 (2025) 12287–12297. https://doi.org/10.1109/JIOT.2024.3520612

[77] K.H. Mohd, S. Yahya, The significance of ceramic products in Malaysian culture, Int. J. Heritage, Art Multimed. 4 (2015) 75–82.

[78] M. Sharma, V.D. Vani, J.V.H. Raj, A. Dutt, J.S. Kumar, M. Almusawi, N. Gupta, R. Goyal, Sustainable Ceramics: Creating Effective Key Performance Indicators for Industry Monitoring, in: E3S Web Conf., EDP Sciences, 2024: p. 1045. https://doi.org/10.1051/e3sconf/202459601045

[79] M.C. Oliveira, R.C. Oliveira, H.A. Matos, Dynamic simulation and optimisation of water and energy consumption in a ceramic plant: Application of the customised ThermWatt computational tool, in: Comput. Aided Chem. Eng., Elsevier, 2024: pp. 2377–2382. https://doi.org/10.1016/B978-0-443-28824-1.50397-5

[80] J. Singh, Nanomaterials and Nanotechnology, Springer, 2006. https://doi.org/10.1016/b978-0-12-805457-4.50007-0

[81] L. Branchini, M.C. Bignozzi, B. Ferrari, B. Mazzanti, S. Ottaviano, M. Salvio, C. Toro, F. Martini, A. Canetti, Cogeneration supporting the energy transition in the italian ceramic tile industry, Sustain. 13 (2021) 4006. https://doi.org/10.3390/su13074006

[82] D.R. Macedo Ramos De Afonso Matias, Optimization of energy consumption in the ceramics industry: a techno-economic analysis, (2023).

[83] L. Silvestri, A. Forcina, G. Di Bona, C. Silvestri, Circular economy strategy of reusing

olive mill wastewater in the ceramic industry: How the plant location can benefit environmental and economic performance, J. Clean. Prod. 326 (2021) 129388. https://doi.org/10.1016/j.jclepro.2021.129388

[84] T. Lansdorf, Hydrogen - a Game Changer for the Ceramic Industry, InterCeram Int. Ceram. Rev. 71 (2022) 48–54. https://doi.org/10.1007/s42411-022-0497-9

[85] Z. Gu, Research About the Influence of Artificial Intelligence Technology on the Development of Ceramic Industry, Art Soc. 2 (2023) 74–76. https://doi.org/10.56397/as.2023.02.12

[86] A. Ay, The self-standing, malleable doughs of advanced ceramics facilitate low-number production and prototyping on a benchtop, (2024).

[87] C.F. Revelo, C.M.F. Vieira, H.A. Colorado, Manufacturing of Ceramic Bricks, in: TMS 2025 154th Annu. Meet. Exhib. Suppl. Proc., Springer Nature, 2025: p. 136.

[88] A.M. Laptev, M. Bram, D. Garbiec, J. Räthel, A. van der Laan, Y. Beynet, J. Huber, M. Küster, M. Cologna, O. Guillon, Tooling in Spark Plasma Sintering Technology: Design, Optimization, and Application, Adv. Eng. Mater. 26 (2024) 2301391. https://doi.org/10.1002/adem.202301391

[89] M. Alvarez, L. Brancalião, D. Gomes, V. Pinto, J. Carneiro, J. Santos, J.P. Coelho, J. Gonçalves, Prototyping and Control of an Automatic Ceramic Tableware Finishing Device, in: Lect. Notes Electr. Eng., Springer, 2022: pp. 167–178. https://doi.org/10.1007/978-3-031-10047-5_15

[90] A. Pan, Application of Multi-Objective Genetic Algorithm in Ceramic Image Segmentation Technology, in: Proc. - 2023 Int. Conf. Comput. Simul. Model. Inf. Secur. CSMIS 2023, IEEE, 2023: pp. 337–342. https://doi.org/10.1109/CSMIS60634.2023.00067

[91] G. Contini, F. Grandi, M. Peruzzini, Human-Centric Green Design for automatic production lines: Using virtual and augmented reality to integrate industrial data and promote sustainability, J. Ind. Inf. Integr. 44 (2025) 100801. https://doi.org/10.1016/j.jii.2025.100801

[92] M. Huang, C.Z. Chen, Traditional Villages, Corp. Manag. Digit. Age (2025) 1.

[93] A. Darkhan, A. Anarbayev, B. Yessimov, T. VAKALOVA, V. Stanevich, A. Molodykh, Optimization of the Process of Producing Ceramic Granite From Mineral Raw Materials and Secondary Raw Material, (2025).

[94] A. Naderi, Characterization of Application-Driven Sol-Gel Coated Meshes Characterization of Application-Driven Sol-Gel Coated Meshes and Composites and Composites, Louisiana State University and Agricultural & Mechanical College, 2021.

[95] O.E. Abamwa, Clay Properties and The Challenges to Ceramic Cottage Industries in Nigeria., (2024)

[96] A. Tazzini, F. Gambino, M. Casale, G.A. Dino, Managing Marble Quarry Waste: Opportunities and Challenges for Circular Economy Implementation, Sustain. 16 (2024) 3056. https://doi.org/10.3390/su16073056.

[97] C. Matter, H.B. Khanh, V.O. Mittova, N.A. Tien, P. Thi, H. Duyen, Condensed Matter and Interphases, Mater. Sci. 24 (2022) 109–115.

[98] V.N. Mehta, J.B. Raval, S.R. Patel, V.S. Prajapati, R. M. Patel, Bio-functionalized Silver Nanoparticles: A Versatile Candidate for the Ceramic Industry, in: Handb. Polym. Eramic Nanotechnol. Vol. 1,2, Springer, 2021: pp. 83–98. https://doi.org/10.1007/978-3-030-40513-7_52.

[99] N. Quaranta, M. Caligaris, M. Unsen, H. López, G. Pelozo, A. Cristóbal, Ceramic bricks using pistachio shells as controlled porosity former, J. Build. Mater. Struct. 10 (2023) 16–26. https://doi.org/10.34118/jbms.v10i1.2614

[100] A.S. Shathi, M.G. Mostafa, M.A. Rahman, P.K. Biswas, M.S. Alam, M.S. Rana, M.R. Uddin, M. Nuruzzaman, M.S. Shahriar, M.N. Zaman, Iron removal from red clay using oxalic acid leaching for enhanced ceramic industry applications, Heliyon 10 (2024).

A Journey from Raw Materials to Ceramics
Materials Research Foundations 184 (2025)

Materials Research Forum LLC
https://doi.org/21741/9781644903834

Chapter 6

Model Design and Pattern Preparation Methods

Farah Mutlag [1,2*], Hussein Elaibi[1], Idris Kaynak[3], Ebru Halvaci[1], Cumhur Eren Isık[4], Fatih Sen[1*]

[1]Sen Research Group, Department of Biochemistry, Kutahya Dumlupinar University, Kutahya 43000, Türkiye

[2]Ministry of Education, Karbala Education Directorate, Karbala, 56001, Iraq

[3]Machinery and Metal Technologies, Vocational School of Technical Sciences, Usak University, 1 Eylul Campus, 64200 Usak, Türkiye

[4]Department of Handicrafts, Kutahya Fine Arts Vocational School, Kutahya Dumlupınar University, Evliya Çelebi Campus, 43000 Kutahya, Türkiye

farahfakhir28@gmail.com, fatihsen1980@gmail.com

Abstract

The process of model design and pattern preparation is fundamental to ceramic production, whether in artisanal or industrial settings. It begins with model design, where conceptual ideas are transformed into physical prototypes. This phase blends artistic vision with technical analysis, using materials such as clay, plaster, resin, and foam each selected for specific advantages in flexibility, cost, and precision. The integration of digital technologies like 3-dimensional printing, Computer-aided design (CAD) modeling, and numerical control (CNC) milling has enhanced accuracy and accelerated development, allowing for detailed and efficient prototyping. In the pattern preparation stage, models are converted into molds for repeatable production. Traditional plaster molding remains valuable for small-scale or customized items, while techniques such as press molding and slip casting offer consistency and efficiency in mass production. Contemporary methods combine manual craftsmanship with digital tools to achieve complex shapes and reliable reproduction. Choosing the right mold material plaster, silicone, or high-temperature resin depends on product requirements, clay body characteristics, and production volume. The forming stage defines the final geometry and structure of the ceramic cost. Techniques like hand shaping, slip casting, pressing, extrusion, and isostatic pressing vary in precision, scalability, and use case. Hand shaping is prized for creative freedom, while isostatic pressing is ideal for technical ceramics requiring high strength. The selected forming method must align with the material's state solid, liquid, or powder and the intended design and production scale.

Keywords

Pattern Preparation, Model Design, Ceramic

1. Introduction

The design of models and preparation of patterns represent a foundational and critical stage in the ceramic manufacturing process whether in traditional craftsmanship or industrial-scale production. This phase establishes the physical and aesthetic blueprint of the ceramic product, ensuring dimensional accuracy, structural integrity, and reproducibility across batches [1–3].

Clay and plaster were the materials of choice for the manual shaping of models in traditional industries. Furthermore, as the shapes and requirements became more complicated, there was a growing demand for more modern technologies, like 3D printing, parametric modeling, and CNC milling. These computer technologies have made it possible to design accurate geometric patterns, generate models that can be customized in a short amount of time, and even reach internal compositions that were previously unattainable [4–7].

For visualizing and refining the final form before beginning mass production, producers might construct a precise model. After that, this model is used as the foundation for the development of mold, which in turn makes it possible to shape ceramic materials in an effective manner. Model and pattern preparation has developed into a sophisticated process that uses artisanal expertise in conjunction with technical precision as a result of the growing integration of digital technologies such as 3D printing and CNC milling [8–11].

The integration of craftsmanship and engineering precision at this stage highlights how mastering pattern design and pattern preparation is not just a technical step, but a crucial factor in achieving ceramic products that meet both functional and aesthetic quality standards.

2. Model Design

Model design is the first critical step in ceramic production, where the creative vision integrates with material reality. It involves shaping a prototype that anticipates both the visual form and technical behavior of the final ceramic object. Figure 1 illustrates the various model designs, from manual shaping to digital fabrication tools.

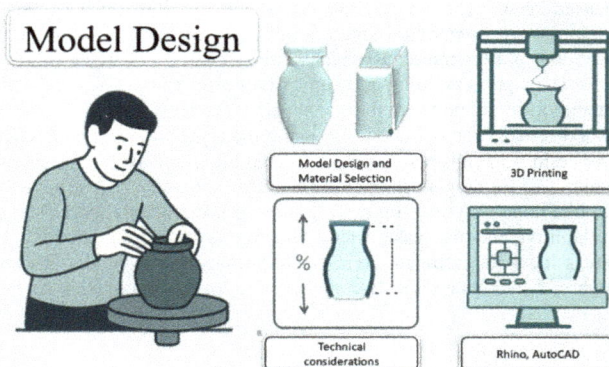

Figure 1. Model design in ceramics industry.

2.1 Conceptual foundations

Model design in the ceramics industry is an integrated scientific creative process, where abstract ideas are transformed into precise physical models through a methodology based on a deep understanding of material behavior. A ceramic model acts as a simulation-based prototype, not just a mold. It analyzes physical transformations (e.g., clay shrinkage during firing) to preemptively address defects. For instance, a vase design must account for 12% vertical shrinkage to avoid warping [12–15].

This process is based on three fundamental principles: first, proactive modeling that predicts the behavior of clay under the influence of thermal and mechanical factors. Second, precise physical analysis of the material's responses during the drying and firing stages. Third, functional integration between the aesthetic aspect and the technical requirements of the final piece [16–18]. A summary of the key technical and methodological insights derived from these references is presented in Table 1.

Table 1. Technical framework of ceramic modeling principles.

Concept	Tools/Methodology	Studied Properties	Mechanisms/Results
Predictive Modeling	- Finite Element Analysis (FEA) using ANSYS. - Bingham plastic models for clay.	- Elasticity. - Yield behavior. - Material deformation properties.	- Simulation of plastic deformation under thermal and mechanical loads. - Identification of critical stress zones.
Physical Laboratory Analysis	- Thermogravimetric and calorimetric analysis (TGA/DSC). - CT scanning for internal defect detection.	- Mass loss. - Thermal response. - Porosity levels.	- Determination of glass transition temperature. - Correlation between shrinkage and clay composition.
Functional-Aesthetic Integration	- Topology optimization software (e.g., nTopology). - Colorimetric analysis (CIE L*a*b* system).	- Surface finish quality. - Color uniformity.	- Achieving a balance between aesthetic and structural requirements. - Conformance to ISO aesthetic and durability standards.

This methodology requires a comprehensive understanding of material properties, whereby the designer becomes an engineer capable of anticipating and guiding physical transformations to achieve the desired results. A successful model depends on a delicate balance between artistic vision and scientific knowledge, making it a vital tool that bridges the gap between initial design and final product [19–21].

2.2 Material selection and influence

The choice of material plays a pivotal role in shaping both the design process and the final outcome of ceramic models. Each material offers distinct advantages and limitations, influencing the feasibility, precision, and cost of production.

Traditional clay remains a favored medium for its tactile flexibility, allowing for intuitive shaping and rapid iteration. However, its tendency to warp and shrink during drying poses challenges for projects requiring dimensional accuracy[4,20,22].

In contrast, plaster provides superior dimensional stability, with minimal shrinkage (0.5%), making it ideal for master models that demand sharp edges and precise replication. Yet, its brittle nature limits its use in dynamic or flexible applications.[23–25].

The advent of digital fabrication has introduced advanced materials such as UV-cured acrylic resin (used in 3D printing), which enables the creation of intricate geometries with exceptional detail. Despite its high cost approximately three times that of plaster its ability to produce complex, high-resolution models make it indispensable for modern precision-driven designs [26–28].

Similarly, high-density foam, often utilized in CNC milling, offers a lightweight and machinable alternative for large-scale models. While it excels in producing rough forms quickly, its porous surface and lower durability may restrict its use in fine-detail applications [29–31].

2.3 Technical considerations in design

Designing for ceramic production requires a meticulous balance between artistic vision and material science. Unlike other media, ceramics impose unique constraints that demand anticipatory adjustments throughout the design process.

Material Shrinkage and Dimensional Compensation; Ceramic bodies undergo significant volumetric reduction, typically 8–15% during drying and firing. This nonlinear transformation must be preemptively calculated to ensure final dimensions match design intent. For example, a 10 mm wall designed in clay with 12% shrinkage will fire to 8.8 mm. Failure to account for these yields misshapen or undersized results [32–35].

Structural Integrity & Wall Thickness; Optimal wall thickness (5–8 mm) mitigates warping and cracking during firing. Excessively thick; sections risk uneven drying (leading to fractures), while overly thin walls compromise durability. Uniformity is sudden transitions in thickness create stress concentrations [36,37].

2.4 Digital tools and their impact

The integration of computational tools has fundamentally altered the landscape of ceramic design, merging traditional craftsmanship with algorithmic precision. Advanced software now enables designers to navigate complex material behaviors and fabrication constraints before physical production begins ushering in an era of predictive creativity [38–40].

In the design phase, software like Rhinoceros with Grasshopper has become indispensable for parametric modeling, enabling the creation of intricate forms such as lattice structures and organic tessellations with micrometer-level precision and effortless adjustability. Meanwhile, simulation tools like ANSYS elevate the process further by analyzing the design's behavior under firing conditions, predicting potential weak points and cracks before physical production begins [41–43].

When transitioning to fabrication, options range from 3D printing technologies capable of producing models with resolutions up to 25 microns to CNC milling, ideal for large-scale, precision components. 3D printing encompasses SLA/DLP for fine details and direct clay extrusion, which liberates designers from traditional mold constraints [8,44,45].

Table 2 delineates the fundamental steps integral to ceramic model creation, from conceptual development through to digital manufacture. It emphasizes the technical and creative goals, the instruments and materials employed, and the influence each phase exerts on the quality of the result. This organized analysis elucidates the contribution of each phase to both utility and aesthetic value in the production cycle. The table functions as a consolidated review of the principal literature and references examined in Section 2, providing a comparative perspective on the translation of theoretical principles into practical design workflows.

Table 2. Core processes in ceramic model design analytical.

Process	Technical / Artistic Objective	Tools / Materials Used	Impact on Final Product Quality	Design Stage in Production Cycle
Conceptual Foundation	Generating the initial idea and shaping the preliminary form	Hand sketches, clay modeling, freehand drawing	Guides the design direction and defines initial features	Pre-production Phase
Material Selection	Choosing materials that ensure accuracy and efficiency	Clay, plaster, wax, resin, foam	Affects detail resolution and ease of processing	Experimental Phase
Technical Considerations	Managing thickness, shrinkage, draft angles, and stability	Simulation software, measuring tools, expert input	Reduces structural flaws during shaping and firing	Final Design Phase
Digital Tools & Methods	Creating precise, replicable models for industrial production	Rhino, AutoCAD, 3D printing, CNC milling	Enhances geometric control and reduces development time	Pre-production & Direct Manufacturing Phase

3. Pattern Preparation Methods

Pattern preparation constitutes a critical phase in ceramic manufacturing, facilitating the transition from a unique prototype to a scalable and repeatable production workflow. By means of meticulously engineered molds, the intended design is accurately and consistently reproduced. This phase serves as an essential link between artistic vision and technical implementation, ensuring that each ceramic piece maintains its structural integrity and aesthetic quality throughout repeated production cycles. An illustrative representation of this stage is provided in Figure 2, which highlights the Pattern preparation methods.

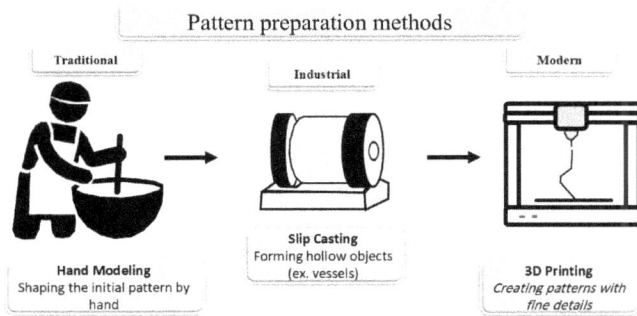

Pattern preparation methods

Traditional

Industrial

Modern

Hand Modeling
Shaping the initial pattern by hand

Slip Casting
Forming hollow objects
(ex. vessels)

3D Printing
Creating patterns with
fine details

Figure 2. Pattern preparation methods.

3.1 Traditional methods of pattern making

Traditional pattern preparation in ceramics is rooted in manual craftsmanship and relies on centuries-old techniques of forming and mold-making. These methods typically involve sculpting an initial prototype using clay or wax, which is then used to create a negative impression in plaster. Plaster remains the most widely used material for mold making due to its affordability, porosity, and ease of shaping [46–48].

Hand-modeling allows the artisan to directly shape expressive forms, but it also introduces a degree of variation that may be undesirable in mass production. Nevertheless, this process continues to play a vital role in ceramic art, prototyping, and small-batch studio production [49–51].

The process is intuitive but technically demanding; it requires careful timing, moisture control, and a high sensitivity to material behavior. Poorly executed traditional molds can result in surface defects, dimensional distortion, or even failure during casting [4,52,53].

3.2 Industrial and semi-automated methods

As ceramic production expanded to meet industrial demand, pattern preparation evolved into more standardized and scalable formats. In this context, molds are no longer single-use but are designed for durability and repeatability [54–56].

One of the most common industrial methods is plaster press molding, used extensively for flat or semi-flat products like tiles. Here, the model is translated into a split mold capable of withstanding high mechanical pressure [57,58].

For hollow or complex forms, slip casting is the preferred method. In this technique, a liquid clay body (slip) is poured into a plaster mold that absorbs moisture and forms a solid shell along the interior surface. The mold must be carefully engineered to allow for drying, demolding, and shrinkage without deformation [59–61].

More complex products may require multi-part molds, especially when undercuts or internal cavities are involved. These molds demand precise alignment systems and dimensional accuracy.

3.3 Contemporary digital and hybrid techniques

In recent years, digital fabrication has revolutionized pattern preparation. 3D printing of molds or master patterns using high-temperature resins or printable ceramic pastes allows for intricate detailing that would be impossible with traditional carving or casting [62–64].

Computer numerical control milling is another hybrid method, used to sculpt master models out of rigid materials like polyurethane foam or high-density plastics. These digitally produced models ensure precision and symmetry, and are ideal for applications requiring consistency across thousands of units [65–67].

Furthermore, laser scanning and reverse engineering allow existing ceramic pieces to be digitized and replicated with minimal loss in fidelity. These tools enhance archival work, restoration efforts, and experimental design [68–70].

Digital methods combine with traditional techniques in hybrid workflows - for instance, 3D-printed master models used to create plaster molds.

3.4 Material and technical considerations

The selection of mold material, such as plaster, silicone, or composite polymer, is primarily dictated by three interrelated factors; the intended production scale, the characteristics of the clay body, and the chosen forming technique (e.g., slip casting or press molding). Each material exhibits distinct properties in terms of moisture absorption, thermal expansion, and surface finish, all of which critically influence both the dimensional stability of the ceramic product and the durability of the mold itself [71–73].

In parallel, technical design, elements must be integrated at the early stages of pattern preparation to ensure process efficiency and product fidelity. These include the incorporation of appropriate draft angles (to facilitate mold release), venting channels (to eliminate trapped air and enhance flow), and gating systems (to guide the slip or clay into the mold cavity uniformly). Neglecting these aspects can lead to defects such as warping, incomplete filling, or surface blemishes, thereby compromising the accuracy and reproducibility of the final ceramic form [60,74–77].

Table 3 presents a comparative overview of traditional, industrial, and digital/hybrid methods used in ceramic model and mold fabrication. It highlights each approach's general features, advantages, limitations, and typical applications. This classification clarifies how production goals, precision requirements, and available resources influence the selection of an appropriate fabrication method.

Table 3. Comparison of modeling and molding methods in ceramic design.

Category	General Description	Key Advantages	Challenges/Limitations	Typical Applications
Traditional Methods	Hand-shaped models and plaster mold casting	-Low cost -Easy modification -Ideal for artworks	- Inconsistent accuracy -Time-consuming -Limited mass production	-Art pieces -Prototypes -Custom individual works
Industrial Methods	Multi-part or compression molds for repetitive production	-Good precision -High durability -Efficient mass production	- High initial mold cost - Limited design flexibility	-Tiles - Sanitary ware (toilets, sinks) - Standardized tableware
Digital/Hybrid Methods	3D printing or CNC for direct model/mold production	-Extreme precision -High repeatability -Rapid development	- High initial investment -Material limitations (clay types)	-Complex architectural designs -Precision industrial components

4. Forming Methods in the Ceramic Industry

Forming is the stage in ceramic production where the raw material, whether in plastic, semi-fluid, or dry form, is shaped into a defined geometry using molds, mechanical force, or casting techniques. This process is critical because it determines not only the shape of the ceramic object but also its structural density, mechanical strength, and surface integrity. Each forming method is selected based on the type of ceramic body, the intended application, and the production scale. A

Materials Research Forum LLC

https://doi.org/21741/9781644903834

schematic representation of common forming techniques and their material-flow behavior is illustrated in Figure 3, providing a visual reference for understanding the selection criteria and process implications.

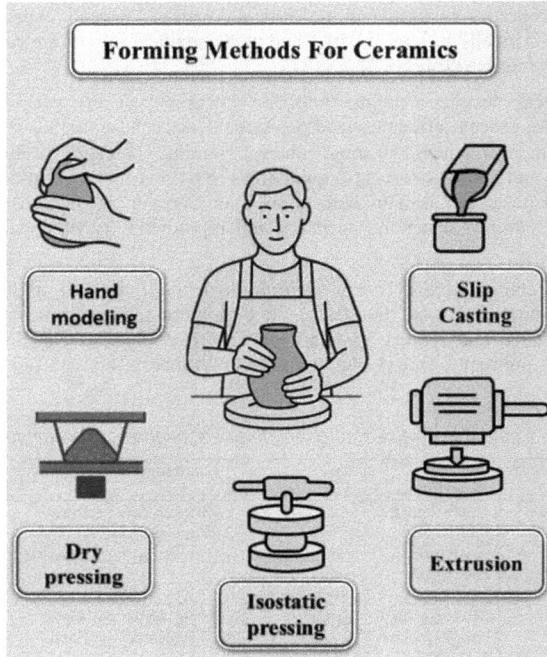

Figure 3. *Forming methods in the ceramic industry.*

4.1 Hand forming techniques

Hand forming represents the oldest and most intuitive method of shaping ceramics. Techniques such as pinching, coiling, and wheel throwing allow artisans to shape soft clay manually, offering complete control over form and texture. Although these methods are not suited for high-volume industrial production, they remain essential in ceramic art, studio pottery, and prototype development due to their expressive flexibility [78,79].

Despite their creative richness, hand-forming techniques require high levels of skill and consistency, and the results may vary slightly from piece to piece.

4.2 Slip casting

Slip casting is a widely used method for producing hollow or intricate forms with thin and uniform walls. It involves pouring liquid clay (slip) into a porous plaster mold. The plaster absorbs moisture from the slip, causing a solid layer to form along the inner walls of the mold [59,60].

This technique is ideal for complex shapes like vases, figurines, and sanitary ware. The final thickness of the wall is controlled by the casting time, and the excess slip is drained out once the desired shell has formed. Slip casting allows for high repeatability and relatively low-cost tooling for medium to large batch production [80,81].

4.3 Pressing (Dry and Semi-Dry Pressing)

Pressing is one of the most common industrial forming techniques, particularly for flat or symmetrical products like tiles, electrical insulators, and tableware blanks. In dry pressing, fine ceramic powder with minimal moisture is compacted into a steel mold under high pressure [82–84].

Semi-dry pressing involves slightly more moisture and may yield better mechanical strength, but still requires precision in controlling compaction and ejection. Pressing offers excellent dimensional accuracy and high production speed, but it's typically limited to simpler shapes [85,86].

4.4 Extrusion

Extrusion is a continuous forming method used to produce long, uniform cross-sectional profiles such as tubes, bricks, rods, or tiles. A plastic ceramic body is forced through a shaped die using mechanical or hydraulic pressure [87–89].

The main advantage of extrusion lies in its efficiency for mass production and its ability to produce products with consistent shapes and minimal material waste. However, it often requires post-forming operations such as cutting, drying, and sometimes surface finishing [88–90].

4.5 Isostatic pressing

Isostatic pressing is an advanced forming technique where pressure is applied uniformly in all directions using a flexible mold within a fluid-filled chamber. This method ensures even compaction of the ceramic powder, resulting in high-density and low-porosity parts [91–93].

It is used especially for technical ceramics and products that require superior strength and dimensional stability, such as ceramic bearings, cutting tools, and bioceramics. Although it is more expensive and slower than uniaxial pressing, it offers unmatched material uniformity [94–96].

Table 4 summarizes the main ceramic forming methods, comparing them based on material type, geometric accuracy, production scalability, and typical applications. The table illustrates how each technique is suited to specific functional and industrial needs, ranging from artisanal pottery to advanced technical ceramics. This comparative analysis supports informed decision-making in selecting the appropriate forming method depending on the design and manufacturing requirements.

Table 4. Comparative analysis of ceramic forming methods.

Forming Method	Material Used	Geometric Accuracy / Shape Control	Mass Production Capability	Common Applications
Hand Forming	Plastic clay	Low to medium depends on craftsmanship	Low unique, non-uniform	Artistic pottery, prototypes, sculptural works
Slip Casting	Liquid slip (clay + water)	High excellent for detailed surfaces	Medium to high	Tableware, hollowware, sanitary ceramics (sinks, toilets)
Dry / Semi-Dry Pressing	Low-moisture ceramic powder	High excellent dimensional control	Very high	Tiles, electrical insulators, semi-finished dinnerware blanks
Extrusion	High-plasticity clay	Medium excellent for continuous forms	High	Tubes, rods, bricks, elongated tiles
Isostatic Pressing	Ultra-fine ceramic powder	**Very high** uniform density and shape	Medium	Technical ceramics, medical tools, high-strength engineering parts

5. Conclusion

The synergy of model design, pattern preparation, and forming methods determines the functional integrity, aesthetic appeal, and reproducibility of ceramic products. With the integration of digital technologies, ceramic design has evolved from a purely manual craft to a multi-disciplinary process that blends engineering precision with creative expression. This evolution not only increases efficiency and quality but also opens new possibilities for complex designs and advanced applications. Ultimately, mastering these foundational stages ensures that ceramic products fulfill both artistic and technical expectations, whether in a studio setting or a high-volume production line.

References

[1] Y. Ye, Y. Du, T. Hu, J. You, B. Bao, Y. Wang, T. Wang, 3D Printing of Integrated Ceramic Membranes by the DLP Method, Ind. Eng. Chem. Res. 60 (2021) 9368–9377. https://doi.org/10.1021/acs.iecr.1c02224

[2] A. Guo, D. Kong, X. Zhou, H. Kong, P. Qu, S. Wang, H. Wang, Y. Hu, Method for preparing damage-resistant 3D-printed ceramics via interior-to-exterior strengthening and toughening, Addit. Manuf. 60 (2022) 103272. https://doi.org/10.1016/j.addma.2022.103272

[3] G. Chen, Y. Zeng, F. Zhao, C. Wu, X. Pan, F. Lin, L. Xu, Y. He, G. He, Q. Chen, D. Sun, Z. Hai, Conformal fabrication of functional polymer-derived ceramics thin films, Surf. Coatings Technol. 464 (2023) 129536. https://doi.org/10.1016/j.surfcoat.2023.129536

[4] M.J. Ribeiro, D. Tulyaganov, Traditional Ceramics Manufacturing, in: PoliTO Springer Ser., Springer, 2021: pp. 75–118. https://doi.org/10.1007/978-3-030-85776-9_3

[5] M. Toka, S. Bourgault, C. Friedman-Gerlicz, J. Jacobs, An Adaptable Workflow for Manual-Computational Ceramic Surface Ornamentation, in: UIST 2023 - Proc. 36th Annu. ACM Symp. User Interface Softw. Technol., 2023: pp. 1–15. https://doi.org/10.1145/3586183.3606726

[6] S. Dashti, F. Hussain, F. Carroll, E. Prakash, A. Navarro-Newball, Role of Intricate

Pottery Visualization in Ceramic Manufacturing, IEEE Comput. Graph. Appl. 42 (2022) 90–97. https://doi.org/10.1109/MCG.2022.3190951

[7]   L. Zhang, G. Liu, Mathematical Modeling for Ceramic Shape 3D Image Based on Deep Learning Algorithm, Adv. Math. Phys. 2021 (2021) 4343255. https://doi.org/10.1155/2021/4343255

[8]   M. Ly, S. Spinelli, S. Hays, D. Zhu, 3D Printing of Ceramic Biomaterials, Eng. Regen. 3 (2022) 41–52. https://doi.org/10.1016/j.engreg.2022.01.006

[9]   W.T. Chou, C.C. Chuang, Y.B. Wang, H.C. Chiu, Comparison of the internal fit of metal crowns fabricated by traditional casting, computer numerical control milling, and threedimensional printing, PLoS One 16 (2021) e0257158. https://doi.org/10.1371/journal.pone.0257158

[10] M.S. Kumar, R. Kumar, R.S. Saini, R. Vyas, S.K. Vaddamanu, A Comparative Analysis of the Accuracy of Crown Fabricated from Conventional Wax Patterns and CAD/CAM Technology – Milling and 3D Printing, J. Pharm. Bioallied Sci. 16 (2024) S3821–S3823. https://doi.org/10.4103/jpbs.jpbs_985_24

[11] F.H. Alhamoudi, Comparing the accuracy of crown fitting between milling and 3D printing techniques using CAD/CAM technologies, Technol. Heal. Care 32 (2024) 2755–2768. https://doi.org/10.3233/THC-231961

[12] R.K. Porwal, U. Chandra, R. Misra, Comprehending and optimising slurry behaviour characteristics of ceramics in additive manufacturing, Adv. Mater. Process. Technol. 8 (2022) 1664–1698. https://doi.org/10.1080/2374068X.2021.1948703

[13] K. Stover, Intangible Property: Architecture, Artificial Stone Ornament, and Insurance from 1769–1821, University of California, Los Angeles, 2022.

[14] A. Maksoud, S.I.A.R. Alawneh, A. Hussien, Integration of computational design and virtual prototyping in optimizing and verifying clay 3D-printed Islamic pattern brick walls for indoor applications, Constr. Innov. (2025). https://doi.org/10.1108/CI-09-2024-0270

[15] B. Bawono, T. Yuniarto, C.E.P. Sanusi, S. Felasari, O.K.W. Widyanarka, P.W. Anggoro, Optimization of virtual design and machining time of the mold master ceramic jewelry products with Indonesian batik motifs, Front. Mech. Eng. 9 (2023) 1276063. https://doi.org/10.3389/fmech.2023.1276063

[16] T. Nenov, S.P. Yordanov, Ceramic sensors: technology and applications, CRC press, 2020.

[17] F. Wu, S. Qiang, X. Zhang, F. Wang, X. Yin, L. Liu, J. Yu, Y.T. Liu, B. Ding, The Rising of Flexible and Elastic Ceramic Fiber Materials: Fundamental Concept, Design Principle, and Toughening Mechanism, Adv. Funct. Mater. 32 (2022) 2207130. https://doi.org/10.1002/adfm.202207130

[18] A. Goldstein, A. Krell, Z. Burshtein, Transparent ceramics: materials, engineering, and applications, John Wiley & Sons, 2020.

[19] E. Gliozzo, Ceramic technology. How to reconstruct the firing process, Archaeol. Anthropol. Sci. 12 (2020) 260. https://doi.org/10.1007/s12520-020-01133-y

[20] G. Eramo, Ceramic technology: how to recognize clay processing, Archaeol. Anthropol. Sci. 12 (2020) 164. https://doi.org/10.1007/s12520-020-01132-z

[21]   Y. Chen, N. Wang, O. Ola, Y. Xia, Y. Zhu, Porous ceramics: Light in weight but heavy in energy and environment technologies, Mater. Sci. Eng. R Reports 143 (2021) 100589. https://doi.org/10.1016/j.mser.2020.100589

[22]   H. Alves de Oliveira, C. Pereira dos Santos, Limestone Clays for Ceramic Industry, Clay Sci. Technol. (2021) 83. https://doi.org/10.5772/intechopen.92506

[23]   D. Tălpeanu, L.E. Geambazu, A.I. Dumitru, F. Clicinschi, V. Tsakiris, A. Cucoş, V. Marinescu, Development of Ceramic Structures With Complex Shapes Based on Hydroxyapatite Using the Technique of Casting in Plaster Molds With Possible Medical Applications, Rev. Rom. Mater. Rom. J. Mater. 53 (2023) 3–12.

[24]   A. Lyons, Ceramic products, in: Mater. Archit. Build., Springer, 2020: pp. 258–264. https://doi.org/10.4324/9780080465791-19

[25]   Y. Lu, H. Shi, K. Lü, J. Li, Y. Li, W. Chen, Properties and Fracture Mechanism of Composite Plaster Mold Covered with Multiple Adhesion Layer for Investment Casting, Int. J. Met. 18 (2024) 1770–1782. https://doi.org/10.1007/s40962-023-01152-0

[26]   Y. Zhang, Y. Xu, A. Simon-Masseron, J. Lalevée, Radical photoinitiation with LEDs and applications in the 3D printing of composites, Chem. Soc. Rev. 50 (2021) 3824–3841. https://doi.org/10.1039/d0cs01411g

[27]   L. Pezzana, Towards a More Sustainable World: UV-Curing & 3D Printing of Bio-Based Monomers Synthesis and Characterization of Bio-Derived Monomers for Cationic and Radical UV-Curing, (2024).

[28]   M. Rank, A. Sigel, Y. Bauckhage, S. Suresh-Nair, M. Dohmen, C. Eder, C. Berge, A. Heinrich, 3D Printing of Optics Based on Conventional Printing Technologies, in: Springer Ser. Opt. Sci., Springer, 2021: pp. 45–167. https://doi.org/10.1007/978-3-030-58960-8_3

[29]   A. Sharma, A. Babbar, Y. Tian, B.P. Pathri, M. Gupta, R. Singh, Machining of ceramic materials: a state-of-the-art review, Int. J. Interact. Des. Manuf. 17 (2023) 2891–2911.

[30]   P. Fernández-Lucio, O.P. Neto, G. Gómez-Escudero, F.J.A. Fuertes, A.F. Valdivielso, L.N.L. de L. Marcaide, Roughing milling with ceramic tools in comparison with sintered carbide on nickel-based alloys, Coatings 11 (2021) 734. https://doi.org/10.3390/coatings11060734

[31]   J. Meher, B.B. Nayak, A. Panda, R. Kumar, A.K. Sahoo, Recent research development of CNC based milling machining conditions: A comprehensive review, Mater. Today Proc. 62 (2022) 3983–3989. https://doi.org/10.1016/j.matpr.2022.04.582

[32]   P. Arora, K.G. Mostafa, E. Russell, S. Dehgahi, S.U. Butt, D. Talamona, A.J. Qureshi, Shrinkage Compensation and Effect of Building Orientation on Mechanical Properties of Ceramic Stereolithography Parts, Polymers (Basel). 15 (2023) 3877. https://doi.org/10.3390/polym15193877

[33]   M. Strano, K. Rane, G. Herve, A. Tosi, Determination of process induced dimensional variations of ceramic parts, 3d printed by extrusion of a powder-binder feedstock, Procedia Manuf. 34 (2019) 560–565. https://doi.org/10.1016/j.promfg.2019.06.220

[34]   N. Lauro, A. Alzina, B. Nait-Ali, D.S. Smith, Non-Uniform Drying Shrinkage in Robocasted Green Body Ceramic Products, Ceramics 7 (2024) 1122–1136. https://doi.org/10.3390/ceramics7030073

[35] X. Gu, Y. Ling, Characterization and properties of Chinese red clay for use as ceramic and construction materials, Sci. Prog. 107 (2024) 00368504241232534. https://doi.org/10.1177/00368504241232534

[36] D. Kong, A. Guo, H. Wu, X. Li, J. Wu, P. Qu, S. Wang, Method for preparing biomimetic ceramic structures with high strength and high toughness, Ceram. Int. 49 (2023) 40284–40296. https://doi.org/10.1016/j.ceramint.2023.10.001

[37] T. Doozandeh, P. Jindal, J. Botchu, Sustainable Design and Wall Thickness Optimization for Enhanced Lifetime of Ultra-High Temperature Ceramic Matrix Composite Thruster for Use in Green Propulsion Systems, Materials (Basel). 18 (2025) 3196. https://doi.org/10.3390/ma18133196

[38] R. Thér, Ceramic technology. How to reconstruct and describe pottery-forming practices, Archaeol. Anthropol. Sci. 12 (2020) 172. https://doi.org/10.1007/s12520-020-01131-0.

[39] Y. Zhao, Y. Ke, Improved style transfer algorithm in decorative design of ceramic painting, Cogent Arts Humanit. 10 (2023) 2247650. https://doi.org/10.1080/23311983.2023.2247650

[40] E. Peters, D. Charlotte Tam-george, P. Edem Peters, Contemporary Design and Ergonomics: Implication in Ceramics Design and Technology, (n.d.). https://www.researchgate.net/publication/380694483

[41] K. Jiang, J. Li, X. Zhang, S. Liu, The Development and Application of One Thermal–Hydraulic Program Based on ANSYS for Design of Ceramic Breeder Blanket of CFETR, J. Fusion Energy 34 (2015) 1088–1093. https://doi.org/10.1007/s10894-015-9923-6

[42] T. Wei, G. Wu, Q. Tan, T. Luo, S. Tang, D. Shen, C. Li, J. Xiong, Modeling, simulation and coupling experiment for integrated passive wireless multi-parameters ceramic sensor, Sens. Rev. 36 (2016) 98–106. https://doi.org/10.1108/SR-03-2015-0045

[43] K.C. Lim, J. Lawrence, L. Li, R.E. Edwards, A.W. Gale, Finite element analysis of stress distribution and the effects of geometry in a laser-generated single-stage ceramic tile grout seal using ANSYS, Proc. Inst. Mech. Eng. Part B J. Eng. Manuf. 218 (2004) 1227–1237. https://doi.org/10.1243/0954405042323603

[44] S. Bose, E.K. Akdogan, V.K. Balla, S. Ciliveri, P. Colombo, G. Franchin, N. Ku, P. Kushram, F. Niu, J. Pelz, A. Rosenberger, A. Safari, Z. Seeley, R.W. Trice, L. Vargas-Gonzalez, J.P. Youngblood, A. Bandyopadhyay, 3D printing of ceramics: Advantages, challenges, applications, and perspectives, J. Am. Ceram. Soc. 107 (2024) 7879–7920. https://doi.org/10.1111/jace.20043

[45] L.C. Hwa, S. Rajoo, A.M. Noor, N. Ahmad, M.B. Uday, Recent advances in 3D printing of porous ceramics: A review, Curr. Opin. Solid State Mater. Sci. 21 (2017) 323–347. https://doi.org/10.1016/j.cossms.2017.08.002

[46] W.H. Holmes, Origin and Development of Form and Ornament in Ceramic Art, DigiCat, 1886.

[47] T. Lin, V.V. Vermol, J. Yu, H. Jiang, Cultural inheritance and technological innovation in modern ceramics: A historical study based on the evolution of individual practice and aesthetic consciousness of ceramic artists, Herança 8 (2025)

[48] S. Chakraborty, ceramic Analysis on selected Potsherds recovered from the excavation at Kunal during 2018-2019, (n.d.).

[49] N.N. Kilchitskaya, Technologies of using colored clay in the manufacture of ceramic products, in: E3S Web Conf., EDP Sciences, 2023: p. 11009. https://doi.org/10.1051/e3sconf/202340211009

[50] Y. Ming, R.C. Me, J.K. Chen, R.W.O.K. Rahmat, A Systematic Review on Virtual Reality Technology for Ancient Ceramic Restoration, Appl. Sci. 13 (2023) 8991. https://doi.org/10.3390/app13158991

[51] Z. Luo, X. Hu, Y. Zhou, Y. Ding, W. Zhang, T. Li, M. Liu, Harnessing High-Throughput Computational Methods to Accelerate the Discovery of Optimal Proton Conductors for High-Performance and Durable Protonic Ceramic Electrochemical Cells, Adv. Mater. 36 (2024) 2311159. https://doi.org/10.1002/adma.202311159

[52] A. Elgazzar, S.J. Zhou, J.H. Ouyang, Z.G. Liu, Y.J. Wang, Y.M. Wang, A Critical Review of High-Temperature Tribology and Cutting Performance of Cermet and Ceramic Tool Materials, Lubricants 11 (2023) 122. https://doi.org/10.3390/lubricants11030122

[53] Y. Zhang, Y. Wang, J. Hu, L. Zhong, Z. Su, B. Qian, X. Liu, Y. Dai, J. Song, S. Sun, J. Qiu, Rapid high-precision machining of LTCC utilizing water-jet guided laser technology: Expanding the horizons of functional ceramic processing, Mater. Today Commun. 41 (2024) 111013. https://doi.org/10.1016/j.mtcomm.2024.111013

[54] D. Carloni, B. Šegvić, G. Zanoni, M. Sartori, M. Besse, a Critical Account on the Automated Sem-Eds Usage in Ceramic Analyses At the Example of Prehistoric Pottery From the Site of Petit-Chasseur (3100-1600 Bc), Southwestern Switzerland, Mediterr. Archaeol. Archaeom. 22 (2022) 237–237. https://www.maajournal.com/index.php/maa/article/view/741

[55] A.N. Abolela, The impact of technology on the ceramic industry and its relationship to the circular economy, (2024).

[56] N.N. Nekmat, S.N. Ab Rahim, A. Abdul Aziz, Innovative Semi-Automated Kuih Kapit Production: Leveraging Infrared Heating for Consistent Quality and Sustainability, J. Adv. Res. Technol. Innov. Manag. 14 (2025) 10–19. https://doi.org/10.37934/jartim.14.1.1019

[57] Y. Goren, L. Friedlander, O. Marder, N. Shalev, M. Teng, D. Tu, G. Shelach-Lavi, Ceramic production and the transition to agriculture in Northeast China: Neolithic pottery technology in the Fuxin Region, Archaeol. Anthropol. Sci. 16 (2024) 9. https://doi.org/10.1007/s12520-023-01912-3

[58] A.M. Carreño, M.E. Torres, M.S. Ramirez Rojas, J. Núñez Rodriguez, Comparative Study To Evaluate the Subtractive and Additive Process in Dentistry: a Systematic Review, Ing. USBMed 15 (2024) 70–91. https://doi.org/10.21500/20275846.6104

[59] E. Baseta, M. Palma, F. Heher, T. Konegger, M. Kaftan, Differentiated Slip Casting: Producing Variable Thickness Ceramic Tiles with Functionally Graded Plaster Moulds, Ceramics 8 (2025) 6. https://doi.org/10.3390/ceramics8010006

[60] K. Lyons, Crafting the Curve: Automating Plaster Mold Design for Ceramic Slip Casting with Shape Cast, in: Conf. Hum. Factors Comput. Syst. - Proc. , 2025: pp. 1–11. https://doi.org/10.1145/3706598.3713866

[61] K. Lyons, Shape Cast: Automating 3D Design for Plaster Molds in Ceramic Slip Casting, in: Conf. Hum. Factors Comput. Syst. - Proc., 2024: pp. 1–7. https://doi.org/10.1145/3613905.3651020

[62] J. Carvalho, B. Figueiredo, P.J.S. Cruz, Ceramic AM and beyond: The potential of hybrid construction systems, in: 3D Print. Constr. Transform. Build. Ind., CRC Press, 2024: pp. 96–120. https://doi.org/10.1201/9781003403890-5

[63] R. Zhang, F. Peng, I. Gwilt, Exploring the role of immersive technology in digitally representing contemporary crafts within hybrid museum exhibitions: a scoping review, Digit. Creat. 35 (2024) 355–377. https://doi.org/10.1080/14626268.2024.2398457

[64] M. Albrecht, F. Schmidt, F. Menzel, J. Yassine, F. Beuer, A. Unkovskiy, Comparative Analysis of Modern 3D-Printed Hybrid Resin-Ceramic Materials for Indirect Restorations: An In Vitro Study, Polymers (Basel). 16 (2024) 3161. https://doi.org/10.3390/polym16223161

[65] N.K. Eid, A Review on the Power of CAD/CAM Technology and the Material Science in Modern Manufacturing, ERU Res. J. 4 (2025) 2223–2250. https://doi.org/10.21608/erurj.2025.299610.1167

[66] M. Soori, F. Karimi, G. Jough, R. Dastres, B. Arezoo, A Review in Capabilities and Challenges of 5-Axis CNC Milling Machine Tool Operations, Preprint (2024).

[67] B. Han, J. Lim, K. Lim, A. Choo, C.C. Yen, G. Ang, C. Zheng, Slip Casting as a Machine for Making Textured Ceramic Interfaces, in: Conf. Hum. Factors Comput. Syst. - Proc. , 2025: pp. 1–18. https://doi.org/10.1145/3706598.3714396

[68] M.A. Mihalache, A. Sover, O. Dodun-Des-Perrieres, G. Nagit, V. Merticaru, A. Hrituc, L. Slatineanu, 2 Modern Trends of Reverse Engineering in Digital Manufacturing, Digit. Prod. Des. Manuf. (2025) 20.

[69] Y. Wakjira, N.S. Kurukkal, H.G. Lemu, Reverse engineering in medical application: literature review, proof of concept and future perspectives, Sci. Rep. 14 (2024) 23621. https://doi.org/10.1038/s41598-024-74176-z

[70] X. Zhang, Y. Yuan, Integrating Ceramic Design CAD and Virtual Reality Based on 3D Printing, Comput. Aided. Des. Appl. 21 (2024) 139–152. https://doi.org/10.14733/cadaps.2024.S28.139-152

[71] G. Pelin, M. Sonmez, C.E. Pelin, The Use of Additive Manufacturing Techniques in the Development of Polymeric Molds: A Review, Polymers (Basel). 16 (2024) 1055. https://doi.org/10.3390/polym16081055

[72] A. Kumar, T. Debica, S. Priyadarshini, R. Nongthombam, M. Das, B. Ningthoujam, Material and Technological Advancements in the Recontouring of Maxillofacial, Somatic, and Dental Structures and Their Implications for Auricular Prosthesis: A Case Report, Cureus 17 (2025).

[73] V. Eklund, Ceramics 3D printing: State-of-the-Art, (2025).

[74] M.B. Islam, H.J. Sumona, M.J. Haque, A. Al Mahmood, A progressive overview of the mainstream additive manufacturing of ceramic components for industrial advancement, Mater. Open Res. 3 (2024) 8. https://doi.org/10.12688/materialsopenres.17669.1

[75] I. Aguiar, S. Cunha, J. Aguiar, Application of Foundry Wastes in Eco-Efficient Construction Materials: A Review, Appl. Sci. 15 (2025) 10. https://doi.org/10.3390/app15010010

[76] Y. Xu, Cultural Expression and Inheritance in Ceramic Art Design Based on Cross-border

Integration, Art Perform. Lett. 5 (2024) 39–46. https://doi.org/10.23977/artpl.2024.050307

[77] Q. Wang, J. Zheng, Research on the thermal energy environment and process parameters of milling process in ceramic art design, Therm. Sci. Eng. Prog. 55 (2024) 102925. https://doi.org/10.1016/j.tsep.2024.102925

[78] M.G. Perna, F. Falcone, C. Casolino, E. Metalla, G. Rosatelli, S. Antonelli, F. Stoppa, Analysing the glaze of a medieval ceramic fragment from the Durres Amphitheater in Albania, Herit. Sci. 12 (2024) 82. https://doi.org/10.1186/s40494-024-01175-8

[79] F. Bell, E. McClure, C. Friedman-Gerlicz, R. Ta, L. Buechley, Shape-Changing Clay-Dough: Taking a Material-Oriented Approach to 3D Printing Ceramic Forms, in: Conf. Hum. Factors Comput. Syst. - Proc., 2024: pp. 1–19. https://doi.org/10.1145/3613904.3642246

[80] J.L. Allan, Clay: A Human History, Simon and Schuster, 2024.

[81] Z. Jwaida, A. Dulaimi, L.F.A. Bernardo, The Use of Waste Ceramic in Concrete: A Review, CivilEng 5 (2024) 482–500. https://doi.org/10.3390/civileng5020024

[82] S. Shakirov, B. Bektemirov, S. Sadaddinova, U. Umirov, M. Abdurakhmonova, K. Urokov, Z. Mirzarakhimova, Mathematical Modelling Concerning Compressibility of Air In Porosity During Semi-Dry Pressing Process Of Ceramic Powder, Int. J. Integr. Eng. 17 (2025) 1–16.

[83] K. Akishev, K. Aryngazin, B. Biybosynov, the Use of Information Technologies in Calculating the Productivity of Technological Equipment for the Production of Ceramic Products Based on Man-Made Raw Materials, Казутб 3 (2024). https://doi.org/10.58805/kazutb.v.3.24-523

[84] R. V. Kryvobok, K. V. Bilogubkina, O.Y. Fedorenko, G. V. Lisachuk, V. V. Voloshuk, O.M. Tarasenko, Y.L. Tataryntseva, M.O. Tarasenko, Technology of production of radio-transparent ceramic materials, in: 2024 IEEE 5th KhPI Week Adv. Technol. KhPIWeek 2024 - Conf. Proc., IEEE, 2024: pp. 1–6. https://doi.org/10.1109/KHPIWEEK61434.2024.10878028

[85] S. Qu, Y. Yang, P. Yao, L. Li, Y. Sun, D. Chu, Fiber reinforced ceramic matrix composites: from the controlled fabrication to precision machining, Int. J. Extrem. Manuf. (2025).

[86] L. Sawunyama, O.C. Olatunde, O.A. Oyewo, M.F. Bopape, D.C. Onwudiwe, Application of coal fly ash based ceramic membranes in wastewater treatment: A sustainable alternative to commercial materials, Heliyon 10 (2024). https://doi.org/10.1016/j.heliyon.2024.e24344

[87] X. Shuisheng, Continuous Extrusion, in: ECPH Encycl. Min. Metall., Springer, 2024: pp. 355–356. https://doi.org/10.1007/978-981-99-2086-0_1073

[88] R. Spina, L. Morfini, Material Extrusion Additive Manufacturing of Ceramics: A Review on Filament-Based Process, Materials (Basel). 17 (2024) 2779. https://doi.org/10.3390/ma17112779

[89] M. Hawryluk, J. Marzec, Problems related to the operation of machines and devices for the production of ceramic roof tiles with a special consideration of the durability of tools for band extrusion, Arch. Civ. Mech. Eng. 25 (2025) 60. https://doi.org/10.1007/s43452-024-01106-1

[90] M. Hölzgen, P. Quirmbach, Additives for Extrusion, in: Art Ceram. Extrus., Springer, 2009: pp. 211–220. https://doi.org/10.1007/978-3-540-27102-4_11

[91] P. Dehghani, S. Salman Seyed Afghahi, F. Soleimani, Hot Isostatic Pressing (HIP) in Advanced Ceramics Production, in: Adv. Ceram. Mater. Technol., IntechOpen, 2025. https://doi.org/10.5772/intechopen.1007176

[92] D. Richter, G. Haour, D. Richon, Hot isostatic pressing (HIP), Mater. Des. 6 (1985) 303–305. https://doi.org/10.1016/0261-3069(85)90012-3

[93] A. Magnusson, J. Shipley, M. Karlsson, C. Beamer, Latest in Hot Isostatic Pressing for Advanced Ceramics, Funtai Oyobi Fummatsu Yakin/Journal Japan Soc. Powder Powder Metall. 72 (2025) S1649–S1653. https://doi.org/10.2497/jjspm.17C-T4-10

[94] B.J. McEntire, Processing of Silicon Nitride Bioceramics, in: Silicon Nitride Bioceram., Springer, 2024: pp. 53–99. https://doi.org/10.1007/978-3-031-67047-3_3

[95] G. Velu Kaliyannan, R. Rathanasamy, R. Gunasekaran, S. Sivaraj, S. Kandasamy, S. Krishna Rao, Preparation of Ceramics: Different Approaches, in: Eng. Mater., Springer, 2024: pp. 53–86. https://doi.org/10.1007/978-981-97-9018-0_3

[96] W.J. Miao, S.Q. Wang, Z.H. Wang, F. Bin Wu, Y.Z. Zhang, J.H. Ouyang, Y.M. Wang, Y.C. Zou, Additive Manufacturing of Advanced Structural Ceramics for Tribological Applications: Principles, Techniques, Microstructure and Properties, Lubricants 13 (2025) 112. https://doi.org/10.3390/lubricants13030112

Materials Research Forum LLC

https://doi.org/21741/9781644903834

Chapter 7

Current Shaping Methods Used in the Production of Technical Ceramic Materials Forming Methods for the Ceramic Industry

Tuna Aydın[1*], Farah Mutlag[2,3], Hussein Elaibi[2], Irem Turk[2], Iskender Isık[4], Fatih Sen[2*]

[1]Kırıkkale University, Faculty of Engineering and Architecture, Department of Metallurgy and Materials Engineering, Kırıkkale, Türkiye

[2]Sen Research Group, Department of Biochemistry, Kutahya Dumlupinar University, Kutahya 43000, Türkiye

[3]Ministry of Education, Karbala Education Directorate, Karbala, 56001, Iraq

[4]Department of Materials Science & Engineering, Faculty of Engineering, Kutahya Dumlupınar University, Evliya Çelebi Campus, 43100 Kutahya, Türkiye

Abstract

Ceramics are the oldest materials developed by humans. With the discovery of fire, humans discovered how to bake the forms they made from clay, and ceramics have played a very important role in human life ever since. From the simplest ceramic decorative items in our homes to computers developed in the modern age and even the hulls of space shuttles, ceramic materials are used in many industries. Ceramic materials are divided into two main groups: "Traditional Ceramics" and "Technical (Engineering) Ceramics." Regardless of the group, all ceramic materials must undergo a shaping process. The dimensions, microstructure properties, reproducibility, cost, and shape of the product to be produced influence the choice of shaping method. Ceramics are shaped using three main methods: dry, semi-wet, and wet shaping.

Keywords

Ceramic Materials, Dry, Semi-Wet, Wet Shaping

1. Introduction

Ceramic materials cannot be shaped using known manufacturing processes compared to metallic or polymer materials. The energy required to melt and cast ceramic raw materials is very costly. In addition, the fragile nature of ceramic end products requires as little shaping as possible after the sintering process. Therefore, ceramic parts must be shaped before the sintering process. Shaping transforms an unstable powder mixture into a consistent and compact structure with the selected geometry [1–4].

After the raw material preparation stage, the shaping stage is necessary for the ceramic product to take its final form. The shaping process results in a semi-finished product. Ceramic materials can be shaped using many different methods depending on the composition, structure, application area, intended use, and shape of the product to be produced [5–7].

When selecting a shaping technique, important factors to consider include the product's dimensions and dimensional tolerances, required microstructural properties, repeatability, cost, and the desired final product shape. Considering these factors, we can generally categorize ceramic material shaping methods into three main categories [8]; Methods of molding ceramic materials fall into three main categories. Dry molding includes Compression, injection molding, 3D printing, and isotropic cold pressing. Semi-wet molding (plastic molding) includes Forming at skin hardness, and vacuum pressing (extrusion). Wet forming includes: Slip casting, belt casting, and gel casting [9–11].

1.1 Dry forming method

The dry forming method is based on the principle that the material to be formed must be dry. Ceramic powders (mass) are formed using the dry method with high-tonnage single-axis hydraulic presses, injection molding, 3D printers, isostatic presses, and hot pressing methods. Traditional ceramics, such as ceramic coating materials (floor tiles, wall tiles, porcelain tiles, etc.), and some technical ceramics (B4C-based ballistic armor) can be shaped using the pressing method, while most oxide or non-oxide engineering ceramics are shaped using injection molding, 3D printers, isostatic presses, and hot pressing methods [12–14].

1.1.1 Pressing

It is a shaping method widely used in the production of ceramic coating materials (ceramic floor/wall and porcelain floor/wall tiles), refractory brick production, and the production of certain technical ceramics [13,15].

In the production of ceramic coating materials, the mass prepared with a specific moisture content (~5-6%) in the slurry preparation section is molded under a certain pressure. This method is called pressing. Ceramic semi-finished products are produced by pressing. This process ensures the shaping of semi-finished products, granule pressing, and increased density of granules. The shaping of the semi-finished product provides the necessary geometry for the final product. Granule pressing also provides the semi-finished product with mechanical characteristics (such as raw strength). Increasing the density of the granules also limits the voids (raw density) in the semi-finished product [16,17].

The pressing process takes place in three main stages: Filling the mold, compression, and removal of the compressed part from the mold. Two main factors influence the compression process: The properties of the compressed granules, and the properties of the compression process itself. The microstructure and pressed cake properties of ceramic coating materials are determined by several factors such as: Particle size distribution and shapes, agglomeration properties, moisture content, binders and plasticizers used in the compaction process. These properties affect the volumetric density inside the mold, the flowability of the granules, the compression ratio, and the friction angle between the granules, and flowability is one of the most important properties to consider to ensure that the mold cavity is properly filled. To obtain a high volumetric density, it must be ensured that the air trapped between the granules is removed during compression. Volumetric density is controlled by factors such as voids between granules and the density of granules or agglomerated particles. Agglomeration refers to the aggregation of particles into small clumps as a result of surface forces or bonds between particles [14,18,19].

If the particles have different diameters, the smaller particles fill the voids, resulting in a higher mold filling degree and higher apparent density. The mold filling degree (ε) is calculated as the ratio of particle volumes to prism volume. If the particles are perfectly spherical, the mold filling degree (ε) is 0.74. However, in reality, particles are never perfectly spherical. Therefore, this value is much lower ($\varepsilon \sim 0.5$). Depending on the mold filling degree, the apparent density of a pressed raw carbon is approximately 2 kg/dm³. The specific volume is around 0.5 dm³/kg, meaning that approximately 80% of the mold is filled with solid material and 20% with air [20,21].

During pressing, the visible volume decreases and the visible density increases. The grains are compressed, allowing each grain to settle into more voids. Plastic components (clay, kaolin, etc.) deform until they come into contact with hard components (quartz, feldspar). The corners of hard particles are broken down, resulting in a higher degree of compression [22,23].

There are a range of important parameters to consider during the pressing stage for both semi-finished and finished products, including: Fracture strength or fracture modulus of the raw material, fracture strength or fracture modulus of the dry material, fracture strength or fracture modulus of the material after firing, as well as firing shrinkage, water absorption, edge flatness, dimensional stability and deviation from square shape, as well as the overall flatness of the surface [24,25].

During the pressing stage, as the compression (specific pressure) increases, the material density (compactness) also increases. The degree of compression is a very important factor affecting the final product characteristics. Soft granules are in contact with each other, which causes larger voids where air can enter. During compression, hard particles deform elastically to a minimum extent, while plastic particles change their shape and begin to fill the spaces occupied by air. As pressure increases further, more shape changes occur, and plastic particles can settle into smaller voids. This reduces the number of pores and the volume of air in the raw material. However, it should be noted that while increasing the press pressure for high compression reduces pore volume, it ultimately also reduces shrinkage during firing [20].

1.1.1.1 Mold systems for ceramic coating materials

There are three types of mold systems used to form ceramic floor and wall tiles and porcelain tiles, which are part of the ceramic coating materials group, using presses: Standard Mold System (SMS), Mirror Mold System (MMS), and Top Forming System (SFS).

The standard mold system is the most commonly used mold system in tile shaping. It consists of an upper mold attached to a movable beam during the pressing process and a lower mold system designed to fit this mold and attached to the lower side. After pressing, the lower mold is raised to lift the pressed tile. These pressing molds are an extremely suitable mold system for traditional single-fired, double-fired, and porcelain tile products. There are two versions of this mold system: face down for traditional shapes and face up for large tiles that undergo subsequent processes (e.g., chamfering) [20,26,27].

The mirror mold system is a more expensive and difficult-to-install mold system compared to the standard mold system. In this system, the mold is pressed in the correct direction, so there is no need to turn it again. The mold cavity moves with the push of the upper mold. It is suitable for tile shaping in all sizes, especially very thick ones, where chamfering is required. It is a mold system that allows expansion control by gradually releasing the traditional or mold removal pressure force [20].

The Upper Shaping Mold System (SFS) has been developed to meet the growing demand for pressed decoration. It is a suitable system for producing high-quality porcelain tiles. The tiles are produced flat in this system. In the SFS mold system, the tiles always face upward, allowing manufacturers to shape and process the tiles without causing any damage or wear. Thanks to hydraulic circuits equipped with dual action, the movements of the upper hydraulic ejector, mold box, lower ejector (SPE), and press beam are coordinated, enabling tile ejection from the upper cavity with extreme precision. When compared to other mold systems, the SFS mold system offers the following advantages: These molds have the advantage that the tiles are always facing upwards and are protected from corrosion, tiles with beveled edges can be produced directly on the press, and there is no maximum applied pressure.

1.1.2 Injection molding

Production speed is high due to the use of continuous production methods, resulting in lower costs. The particle size must be below the micron level. It is widely used in the production of engineering ceramics. Material temperature, flow rate, cavity pressure, and cooling rate are the primary variables and the main factors influencing injection molding and processing [28]. Injection pressure, piston speed, mold geometry, and mold temperature are factors that depend on the machine. The injection molding method is a compression method based on the rapid injection of powder fed by a rotating screw system inside a cylindrical tube made of steel into the mold cavity. The method is similar to the injection system used in the production of plastic materials. The process consists of four main stages: mixing, injection, cooling, removal from the mold, and removal of the binder (Figure 1) [24,29,30].

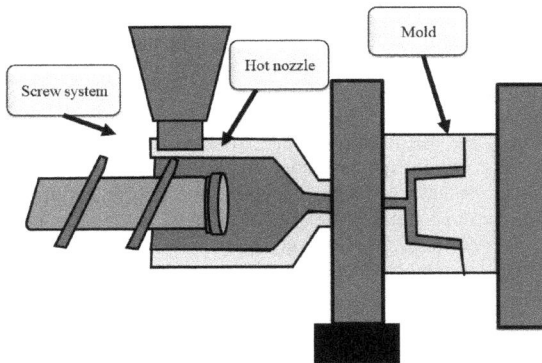

Figure 1. Schematic injection molding.

Materials Research Forum LLC

https://doi.org/21741/9781644903834

Ceramic powders are mixed with a polymer with a low melting point of approximately 40% to enable them to move within the screw and to obtain a compact structure. During the injection stage, the hot powder and molten binder are pressed into the mold via the screw system. After injection into the mold, the part is allowed to cool for a period of time, and once cooling is complete, the part is removed from the mold. In the final stage, the binder is removed from the structure either using a solvent or water, or by applying heat above the binder's vaporization temperature [24,28,31].

1.1.3 3D printing

3D printing is a series of advanced production technologies used to produce physical parts by separating 3D CAD models, which have been digitally sliced into 2D sections, point by point, line by line, or layer by layer, with additive layers. The 3D printing method is a unique production technology that enables the preparation of extremely complex and precise products in a much more flexible manner than traditional manufacturing methods such as casting and pressing. For this reason, this technology has been developing very rapidly in recent years.

There are three different methods preferred in 3D printing production: Forming methods fall into three main categories: Liquid suspension-based molding, powder-based molding, and bulk solids-based molding [32,33].

In liquid slurry-based 3D printing methods, technologies such as Stereolithography (SL), Digital Light Processing (DLP), Two-Photon Polymerization (TPP), Inkjet Printing (IJP), and Direct Ink Writing (DIW) are used. In powder-based 3D printing methods, Three-dimensional printing (3DP), Selective laser sintering (SLS), and Selective laser melting (SLM) technologies are used. Solid-based 3D printing methods use laminated object manufacturing (LOM) and fused deposition modeling (FDM) technologies [34], [35].

The 3DP (3D printing) method, one of the powder-based 3D printing methods, was first designed by Sachs and his colleagues at the Massachusetts Institute of Technology (MIT) in 1989 and a patent application was also filed. In the 3DP process, the organic binder solution in droplet form is sprayed onto selected areas of the powder bed surface through print heads. Solid layers are formed by the solidification (i.e., bonding) of the liquid binder that penetrates and surrounds the powder. A new layer of powder is then applied and spread over the previous layer, repeating the construction process until the part is fully formed. Afterward, the loose powder is removed to reveal the part. The most significant advantage of the 3DP printing method over other 3D powder printing methods is that it does not require a substrate material and enables printing without sintering. It is the most suitable 3D printing method for the production of bioceramics (Figure 2). However, its significant disadvantages include low surface quality, the possibility of unbound powder particles, low mechanical strength, and a challenging powder removal process [34], [35].

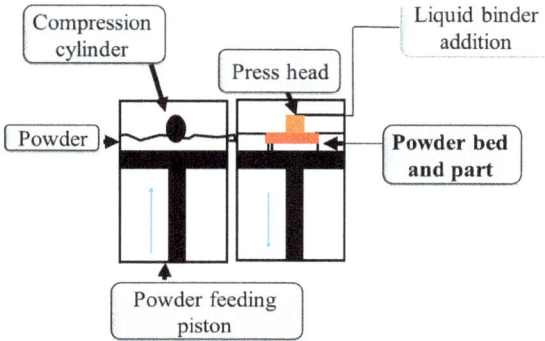

Figure 2. Schematic 3DP among powder-based 3D printing methods.

1.1.4 Isostatic pressing

There are two types of isostatic pressing: hot isostatic pressing (HIP) and cold isostatic pressing (CIP). In this method, powders are prepared in the same way as in the dry pressing method. In isostatic pressing, granules are transferred to a rubber mold that does not contain liquid. One of the most important points here is the removal of air from the system. The liquid surrounding the mold is compressed to approximately 400MN/m2. The bulk density of the compressed granules can reach up to 50%. As a result of the firing process, this value can reach the theoretical density [36,37].

If it is desired to increase the density of the ceramic product, hot isostatic pressing is preferred. Densification through sintering can only reach a maximum of 90%. Since the sintering of ceramics is based on the densification process, full densification can only be achieved at temperatures close to the melting point, where the diffusion rate is fastest. Since the melting point of ceramics is very high, this is both technically and economically challenging [38].

1.1.4.1 Cold isostatic pressing (CIP)

Cold isostatic pressing is a method in which a rubber mold is filled with a powder to be shaped and molded using a homogeneous hydraulic press. The reason for using a rubber mold is that it allows the hydraulic press to be applied in a much more homogeneous manner. For this reason, it is also referred to as a "hydraulic press" or "rubber press." This method is based on the principle of molding the powder in the rubber mold under high hydraulic pressure within a liquid-filled container [39,40]. Unlike the single-axis dry pressing method, this method provides unlimited multi-axis pressing and homogeneous molding independent of size and shape. A typical cold isostatic press pressure ranges from 100 to 500 MPa [20], [28].

1.1.4.2 Hot isostatic pressing (HIP)

The hot isostatic pressing (HIP) method involves the simultaneous application of high pressure inert gas (nitrogen or argon) and high temperature inside a container. As mentioned earlier, the hot isostatic pressing method is preferred when increasing the densification of the ceramic product is desired. In this shaping method, densification occurs not primarily through the diffusion of atoms but through the movement of the powder to be shaped. As a result, densification can be achieved at lower temperatures or in shorter time frames. In the hot isostatic pressing method, there are three important powder behaviors to consider—rearrangement, plastic deformation, and creep—along with important parameters that must be continuously monitored, such as temperature, pressure, powder size, and the mechanical behavior of the material. With this method, sudden densification is achieved through the combined effects of increased pressure and hot deformation. Approximately 90% of the densification occurs due to the effect of hot deformation, while the remaining portion occurs through creep-type densification. With this method, 100% densification is achieved within approximately 15 minutes [41]. This method has many advantages over other compression methods. It eliminates all internal voids in ceramic and metal parts, reducing rejection rates, and improves mechanical properties such as fatigue resistance, ductility, and impact resistance, allowing for more unique and aesthetically pleasing designs. It also increases vacuum tightness and treated surface quality, and produces full-density materials from powdered metals, composites, polymers or ceramics without the need for melting. Using powders, solids can be manufactured with superior properties due to fine, homogeneous particle size and isotropic structure, as well as enabling the production of complex-shaped products that cannot be manufactured by other manufacturing methods [37,38,42]. Shows as figure 3

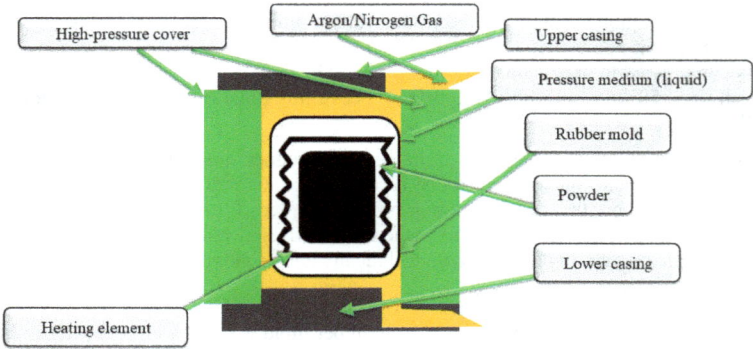

Figure 3. *Schematic hot isostatic pressing.*

1.2 Semi-wet shaping

Another name for this is the plastic shaping method. In the grinding process, sludge containing 50% liquid is first taken from the sludge settling ponds by means of pumps and passed through filter presses to obtain sludge cakes with a water content of around 20-25%. These cakes are then processed through vacuum-free presses (Shineke presses) to form large sludge blocks with a moisture content of 18-19%. These sludge blocks are then transferred to settling chambers and left to settle for approximately 5 days. The rested sludge blocks are then passed through a vacuum press to reduce the moisture content to 15-16%, resulting in a preliminary shape. Semi-wet shaping includes five distinct methods: freehand shaping, shaping on a lathe, shaping on plaster on a lathe, shaping between molds, and shaping using a vacuum press [8,43,44].

1.2.1 Shaping at leather hardness

In order to apply the shaping method at leather hardness, which is one of the semi-wet shaping methods, a preliminary shape must be given to the clay using a vacuum press. It is a shaping method that is also used in the production of porcelain insulators and kitchenware. The plastic clay, which has been given a preliminary shape using a vacuum press, is left to rest in curing rooms at the appropriate humidity and temperature until it reaches the desired hardness. Horizontal and/or vertical lathes are used to process the plastic blocks that have reached the desired hardness. Vertical lathes are used to shape chain-type insulators used in high-voltage lines. A blade moving over a template attached to the lathe is used to give the clay block the required shape [45,46]. In addition, internal or external coating methods are also used, particularly in the shaping of pin-type porcelain insulators and plate and bowl groups of kitchenware [8], [47].

1.2.2 Vacuum press (Extrusion)

The vacuum press method, one of the plastic shaping methods, is used in both pre-shaping and final shaping of ceramic-based materials. It is used for final shaping in the production of brick and tile construction materials, and for pre-shaping in the production of high and low voltage insulators and kitchenware. There are two types of extrusion devices: vacuum and non-vacuum. The non-vacuum ones are also called "shinike presses." Shinike presses are not used to compress mud cakes that have come out of the filter press. They do not have the task of removing air from the mud. Ceramic muds obtained from the shinike press in the form of mud logs called "sucuk" with a certain moisture content are taken to the resting rooms [48,49].

The vacuum press contains an endless screw (auger) system. The slurry, which is compressed and pushed toward the press opening in the screw system, is also deaerated in the vacuum chamber. Depending on the mold opening cross-section, it can be removed from the device in the desired shape [8], [47]. In vacuum pressing, there are a range of important parameters that affect the drying and firing processes, including: The plasticity of the clay and its pH value, screw design and angle, rotation speed, the need to knead the clay homogeneously inside the vacuum press, the strength of the clay in the wet state, the moisture content of the clay entering the press, the vacuum pressure, as well as the particle size distribution after the grinding process [50].

At least 20% plasticizer must be used in vacuum forming. Clay or organic binders are used as plasticizers. If the material to be formed does not have sufficient wet strength and has a very high moisture content, surface defects such as lamination, cracks, and breaks may occur.

1.3 Wet forming

In wet molding, the preparation of the slip slurry used for molding is the most important factor, as it depends on several key parameters including: The physical structure of the raw materials such as particle size distribution and shapes, grinding time, firing color, and hardness; the chemical properties of the raw materials such as their chemical composition; their mineral structure including the minerals and crystal structure they contain; the rheological properties such as literal weight, viscosity, density, thixotropy and others; and finally the water content, selection and ratio of the electrolyte used [24,51].

These are the important parameters required for the successful preparation of casting slurry. The particle size distribution (PSD) of the raw materials used encompasses not only the average grain size but also the entire distribution. PSD has a significant impact on the rheology of the system. PSD also affects casting behavior. As particle size increases, plasticity and shrinkage decrease. The amount of particles smaller than 1 µm also affects casting speed [52,53].

The particle size and structure of the clay used to control drying cracks are very important. As the clay particle size decreases, the plasticity and water absorption capacity of the clay will increase. Conversely, drying shrinkage will also increase proportionally. During the raw material preparation process, when clay and kaolin are blended with inert raw materials such as quartz and feldspar, the average particle size increases. As a result, the amount of water absorbed decreases, thereby minimizing the risk of drying cracks [8], [54].

The rheological properties of the prepared liquid slurry are monitored by measuring viscosity and thixotropy. The resistance of liquids to flow is called viscosity. The desired fluidity of the slip must be achieved using the minimum amount of water. If a slip containing more water is used, the molds become overly saturated, which prolongs the time it takes for the casting to come out of the mold and may cause cracks to form within the mold. For this reason, fluidizers (electrolytes) are used instead of water. The fluidizers used reduce the amount of water and ensure that the desired fluidity is achieved. Fluidizers (electrolytes) change the charges of the particles, enabling them to disperse effectively in water. The rheological properties of the slip also change depending on the type and proportion of the fluidizer used. This affects the casting speed [55–57].

The phenomenon where fluid slurries lose their fluidity when stationary and behave like a static or solid substance, only to regain their fluidity upon re-mixing, is called thixotropy. A certain degree of thixotropic property is necessary for the prepared slip slurry to thicken quickly. Prepared casting slurries may lose their fluidity when left stationary (if they are not continuously stirred in slurry tanks). This phenomenon is called thixotropy, and these structural changes are characterized by the formation of weak bonds between particles. Molds with high thixotropy have a high viscosity gain rate, resulting in a softer semi-finished product, but their drainage behavior is poor. The term "drainage" refers to the mold leaving the mold cavity without defects after the casting process. When the viscosity of the casting slurry is very low, it causes long casting times, uneven casting surfaces, and cracks, while when the viscosity of the slurry is very high, it causes pinhole surface defects and difficulties in the slurry discharge stage. Thixotropic values that are significantly lower than normal can result in brittle structures, prolonged casting times, uneven casting surfaces, and weak structures, while very high values can cause difficulties in draining the slurry, prolonged drying times, and soft castings [16,49,51].

1.3.1 Slip casting (casting method)

Slip casting slurry is a mixture of plastic or absorbent raw materials (such as bentonite and some kaolins) and non-plastic or non-absorbent raw materials (such as quartz and feldspars), water, and a fluidizing agent (electrolyte Na-silicate) that has a high solid content, low viscosity, and is stabilized [58,59].

The slip casting method is a shaping technique used to produce complex-shaped and large-volume products that cannot be manufactured using other ceramic shaping methods, whether they are solid or hollow. This method is widely used in the shaping of ceramic tableware, sanitary ware products such as toilets, sinks, and bathtubs (vitrified), and some engineering (high-tech) ceramics [8], [50], [60].

In casting molding, plaster molds with a porous structure and excellent water absorption properties are commonly used, and synthetic materials with a porous structure may also be preferred. The water content of the casting slurry poured into the plaster mold is quickly absorbed by the mold, starting from the mold surface. As the particles that make up the casting slurry and the water between them decrease over time, the attractive forces between the particles become more dominant than the repulsive forces. This causes the particles to accumulate on top of each other and stick together. As time passes, the number of particles piling up increases, and a solid-liquid transition layer forms starting from the inner surface of the mold. As time passes, the thickness of this layer also increases. This time-dependent process is called thickening rate. Thickening rate is the most important property of casting slurries. The solid-to-water ratio in the casting slurry, the specific properties of the deflocculant type, and the particle size and distribution of the solids are also important parameters that influence the thickening rate. A high thickening rate is the key to obtaining a good casting slurry [8], [50], [54,61].

For a good casting slurry, the slurry should spread easily within the plaster core. As mentioned earlier, a low-viscosity slurry should be prepared, the solid particles in the slurry should not settle, the green casting should be easily removable from the mold after casting, and the wall thickness should be neither too fast nor too slow after the casting process. The post-casting strength should be high. Dry shrinkage should be minimal [13,43].

After a certain period of time, once a certain wall thickness is achieved, the excess slip remaining in the mold is drained, and finally, a semi-finished product that has taken the shape of the mold is obtained. This semi-finished product still contains a very thin layer of water between the grains. Most of the moisture in the obtained semi-finished product is this. The removal of this moisture is called dry shrinkage. During dry shrinkage, as the water layer in the semi-finished product dries, the particles move closer together and begin to touch each other at many points. As the particles move even closer together, the product shrinks. As the product loses water, it shrinks, gains strength, and loses plasticity [62,63].

1.3.1.1 Battery casting

This is a shaping method developed to increase production speed in the shaping of ceramic vitrified (sanitary ware) products. It is a system in which the molds to be used in shaping are connected and compressed in a row on a rail, and the clay is pressed into these molds under a certain pressure and left for a certain period of time to thicken. Excess clay is removed by applying compressed air to the mold. This system is more suitable for the production of simple-shaped products such as sinks. Due to the short lifespan of the plaster molds used, the industry has transitioned to high-pressure casting methods [64,65].

A maximum of two castings can be made per day in a mold. The drying process is carried out on these benches using hot air application with rapid drying systems. The main reason why a maximum of two castings can be made per day is precisely this. The biggest disadvantage of the battery molding method is that it is not suitable for every mold. Additionally, the initial investment cost is high, and the filling marks that form during casting are also significant disadvantages. Mold lifespans are a maximum of 200 castings. An average of around 120 parts can be efficiently obtained per worker [66,67].

1.3.1.2 Mechanized casting

Compared to battery casting, this method allows for the mass casting of more complex parts and is a shaping method similar to both hand casting and battery casting. It is a method that combines the production of complex shapes in hand casting with the mass production of battery casting. As in battery casting, a rail system is used. The production quantity per person is higher than in manual casting but lower than in battery casting.

1.3.1.3 High-pressure casting

The latest technology used in the shaping of ceramic sanitaryware products in recent years is the high-pressure casting method. The molds used are made of synthetic materials and are high-cost systems. While water absorption in the plaster mold used in battery casting and mechanized casting systems is achieved through capillary forces, in the high-pressure casting method, pressure is applied to the clay and vacuum is applied to the mold.

This has reduced the thickness acquisition time, one of the most important parameters in production. The application of vacuum to the mold after casting has shortened the normal casting time, and the application of pressure to the clay has further reduced this time.

One of the most critical stages of the shaping process is the drying stage. Pressurized air is supplied to the drying area to remove water from the mold, and the water is physically removed, thereby saving energy. In methods using plaster molds, some of the water absorbed by the mold can be removed, while in high-pressure casting, all of the water can be removed [60,68], [54].

1.3.2 Belt casting

This is a widely used shaping method for the production of ceramic materials required for the electronics industry. In particular, the shaping of very thin ceramic layers made of oxides such as Al_2O_3, $BaTiO_3$, and ZrO_2, which are often stacked on top of each other and used as circuit carriers, is carried out using the ribbon casting method, especially in the field of microelectronics where there is a very high demand. The method begins with the preparation of a ceramic slip, which is a liquid containing the inorganic ceramic material dispersed in a solvent. The resulting slip is a liquid containing a binder and additives that regulate plasticity. The binder used is in a diluted form. The prepared ceramic slip is spread over a flat surface to allow the solvent to evaporate. After drying, the material is peeled off the surface. The resulting form is as thin as paper. The strips produced using a doctor blade are used in the chip industry, capacitors, ferrite, and piezoelectric materials [69].

The organic additives used enhance the specific properties of the strip. These additives form a film on the surface of the slip that binds inorganic materials. For this purpose, PVB (polyvinyl butyral) or acrylic polymers are used. Theoretically, approximately 5 g of binder is required for 100 g of inorganic material, and this amount should not exceed 8 g. In the strip casting method, the same amount of plasticizer as the binder must be used. Butyl benzyl phthalate and polyethylene glycol are the most commonly used plasticizers [70–72].

The process begins with the grinding stage, where agglomerated parts in the slip are broken down and the particles are uniformly distributed to obtain a viscous slip. The mixing and homogenization stages are carried out in two steps. In this stage, additives (plasticizers and binders) and solvents dissolve in the ceramic slip. The air is removed from the slip in a vacuum chamber. Strip production is carried out in a continuous production machine. This machine contains a moving conveyor system. The doctor blade on the conveyor determines the thickness of the strip. A reel system is used to wind the strip obtained during the drying process.

Strip casting is a process that primarily allows for the production of thin, two-dimensional parts in a raw or uncooked state. These parts can be as thin as 2-3 microns and up to 6 mm thick. However, most strips are between 0.25 mm and 1.27 mm thick. Strips can be cast in widths exceeding 1 meter and lengths defined by the casting time (and depending on the available slip amount). Today, strip casting sheets measuring 1.2 meters x 2.4 meters x 0.5 mm have been produced for use in fuel cells. This enables the production of large-scale fuel cells for energy generation. Fuel cell technology is expected to become one of the most important power sources in the future, and it is believed that strip casting will make it even more efficient. Therefore, the ability to produce large, thin sheets from a wide variety of materials is one of the significant advantages of the strip casting process [13,73,74].

1.3.3 Gel casting

Although the slurry casting method, one of the wet forming methods, is often used for shaping complex-shaped materials, the density variations that arise in the production of thick parts, the heterogeneous distribution of soluble substances such as binders and inorganic ions, and particle separation have led to the rapid development of gel casting. This method is based on the dispersion of ceramic powders in a water-based monomer solution, followed by the solidification of the mixture using an initiator and catalyst to achieve the desired shape. The gel casting method enables the production of high-strength green bodies. Green bodies can be processed using this method [75], [30,76].

Due to the relatively high solvent level and to limit stresses caused by shrinkage during the drying stage and prevent warping of the raw material, the gel casting process must be carefully controlled. Controlling the drying process is crucial for the successful production of ceramics using direct solidification techniques [75].

Pectin is commonly used as a gelling agent. The most important feature of pectin is that it is a hydrocolloid that exhibits gelling properties even at low concentrations in water. In addition to its gelling properties, pectin is also used as a stabilizer and thickener.

The gel casting process begins with the preparation of ceramic slip. Pectin solutions are added for gelation. Following homogeneous mixing, the casting process is performed, and then air bubbles are removed using a vacuum. After the gelation and coagulation process, the demolding process is carried out. The semi-finished product removed from the mold takes its final product form through the sintering process after the drying process [30,75].

2. Conclusions

The most important feature of ceramic materials is that they form the basis of other material groups. They constitute the most important component of many industries such as metallurgy, energy, machine tools, automobiles, glass, construction, electrical-electronics, computers, defense, and space. At this point, their production also requires a great deal of precision. The shaping methods we have discussed in this section are methods developed for both traditional and advanced ceramic materials. In particular, tape casting, gel casting, hot-cold isostatic pressing, and 3D printing methods are more commonly preferred for shaping technical ceramic materials. However, another important point here is the costs and obtaining clean products. While 3D printing is currently the most discussed and debated shaping method, it is more accurate to say that this method is still in the development phase for ceramic materials. At this point, the key factors in determining the shaping method for a ceramic material include the product's dimensions and dimensional tolerances, required microstructural properties, repeatability, cost, and the desired final product shape.

References

[1]   N.F. Astbury, Science in the ceramic industry, Proc. R. Soc. London. Ser. A. Math. Phys. Sci. 258 (1960) 27–46.

[2]   H. Celik, Technological characterization and industrial application of two Turkish clays for the ceramic industry, Appl. Clay Sci. 50 (2010) 245–254.

[3]   K. Ding, A. Li, J. Lv, F. Gu, Decarbonizing ceramic industry: Technological routes and cost assessment, J. Clean. Prod. 419 (2023) 138278

[4]   S.O. Abiodun, M.A. Akintonde, T.E. Akinde, Small scale ceramic industry in Oyo State: Challenges and prospects, Small 4 (2013).

[5]   Y. Lakhdar, C. Tuck, J. Binner, A. Terry, R. Goodridge, Additive manufacturing of advanced ceramic materials, Prog. Mater. Sci. 116 (2021) 100736.

[6]   J. Barata, F. Silva, M. Almeida, Ceramic Industry 4.0, Res. Anthol. Cross-Ind. Chall. Ind 4 (2019).

[7]   C. Ganesh, The environmental impact caused by the ceramic industries and assessment methodologies, Int. J. Qual. Res. 13 (2019) 315–334.

[8]   A. Arcasoy, Seramik teknolojisi , (1983) 277.

[9]   S.R. Kandavalli, S.R. Kandavalli, R.S. Ruban, C.H. Lo, R. Kumar, A.B. Elshalakany, C.I. Pruncu, Review—A Conceptual Analysis on Ceramic Materials Used for Dental Practices: Manufacturing Techniques and Microstructure, ECS J. Solid State Sci. Technol. 11 (2022) 053005. https://doi.org/10.1149/2162-8777/ac6113

[10]   M.B. Islam, H.J. Sumona, M.J. Haque, A. Al Mahmood, A progressive overview of the mainstream additive manufacturing of ceramic components for industrial advancement, Mater. Open Res. 3 (2024) 8.

[11]   R. Ulus, Y. Yıldız, S. Eriş, B. Aday, F. Şen, M. Kaya, Functionalized Multi-Walled Carbon Nanotubes (f -MWCNT) as Highly Efficient and Reusable Heterogeneous Catalysts for the Synthesis of Acridinedione Derivatives, ChemistrySelect 1 (2016) 3861–3865. https://doi.org/10.1002/slct.201600719

Materials Research Forum LLC
https://doi.org/21741/9781644903834

[12] A. Zaccaron, V. de Souza Nandi, M. Dal Bó, S. Arcaro, A.M. Bernardin, The behavior of different clays subjected to a fast-drying cycle for traditional ceramic manufacturing, J. King Saud Univ. Sci. 36 (2024) 339–348.

[13] M.J. Ribeiro, D. Tulyaganov, Traditional Ceramics Manufacturing, in: PoliTO Springer Ser., Springer, 2021: pp. 75–118. https://doi.org/10.1007/978-3-030-85776-9_3

[14] T. OHJI, Additive manufacturing of ceramic components—Towards innovation of ceramic industry—, Synth. English Ed. 11 (2019) 81–92.

[15] M. Karakus, R.E. Moore, Refractories Handbook, The Technical Association of Refractories, Japan, 1998. - References - Scientific Research Publishing, (1998).

[16] P. Saha, D. Sarkar, Whiteware and glazes, in: Ceram. Process. Ind. Pract., CRC Press, 2019: pp. 215–245. https://doi.org/10.1201/9781315145808-7

[17] M. Dondi, J. Ten, J.V. Vieira Lisboa, M. Vicent Cabedo, L. Perez-Villarejo, E. Rambaldi, C. Zanelli, Critical Raw Materials in High-Throughput Manufacturing: Case-Study of the Ceramic Industry, Available SSRN 4641609 (n.d.).

[18] M. Mu, Z. Zuo, X. Liu, C. Jiang, Manufacturing ceramic foams from granite sludge, J. Ceram. Soc. Japan 132 (2024) 641–648.

[19] C. Lourenco Alves, J. de Oliveira Martins Müller, A. de Noni, S. Heinrich, Challenges and opportunities for increase sustainability and energy efficiency in ceramic tile industry, Int. J. Appl. Ceram. Technol. 22 (2025) e15097. https://doi.org/10.1111/ijac.15097

[20] A. Vari, Raw material preparation and forming of ceramic tiles , İtalya, 2004.

[21] S. Mahbuba, Synthesis of zirconia nanopowders from local zircon sand and study the influence of crystal structure on photocatalytic activity, (2022).

[22] A. Ay, The self-standıng, malleable doughs of advanced ceramıcs facılıtate low-number productıon and prototypıng on a benchtop, (2024).

[23] G. Eramo, A. Mangone, Archaeometry of ceramic materials, Chem. Anal. Cult. Herit. 4 (2020) 331–356. https://doi.org/10.1515/9783110457537-014

[24] M. Mandal, D. Sarkar, Manufacturing Excellence in Ceramic Industry, in: Ceram. Process., CRC Press, 2019: pp. 1–36.

[25] S.H. Devjibhai, G.D. Acharya, S.G. Acharya, Performance Improvement Through Implementatıon Of Mfca Based System Framework In Ceramıc Tıles Manufacturıng, (2023).

[26] A. Bernasconi, L. Pellegrino, F. Vergani, F. Campanale, N.M. Marian, L. Galimberti, M. Perotti, C. Viti, G. Capitani, Recycling detoxified cement asbestos slates in the production of ceramic sanitary wares, Ceram. Int. 49 (2023) 1836–1845.

[27] M.Y. Tamar-Agha, M.A.A. Mahdi, A.A.A. Ibrahim, The Kaolin clay deposits in the western desert of Iraq: an overeview, IRAQI Bull. Geol. Min. (2019) 147–173.

[28] A.E. Geçkinli, İleri Teknoloji Malzemeleri, İstanbul Üniversitesi Matbaası (1992).

[29] S. Changsheng, W. Ruiqiang, L.I. Bin, W.E.I. Zhonghua, C. Song, C. Bo, W. Weiguo, Z. Weiru, Injection Molding Process of Silicon Nitride Ceramics., Bull. Chinese Ceram. Soc. 42 (2023).

[30] J. Yang, J. Yu, Y. Huang, Recent developments in gelcasting of ceramics, J. Eur. Ceram.

Soc. 31 (2011) 2569–2591. https://doi.org/10.1016/J.JEURCERAMSOC.2010.12.035

[31] C. Berges, A. Gallego, J.A. Naranjo, G. Herranz, Manufacturing porcelain components by CIM: Viability of processing different ceramic powders, Boletín La Soc. Española Cerámica y Vidr. 60 (2021) 307–317.

[32] Y. Wang, T. Wu, G. Huang, State-of-the-art research progress and challenge of the printing techniques, potential applications for advanced ceramic materials 3D printing, Mater. Today Commun. 40 (2024) 110001.

[33] E. Ordoñez, J.M. Gallego, H.A. Colorado, 3D printing via the direct ink writing technique of ceramic pastes from typical formulations used in traditional ceramics industry, Appl. Clay Sci. 182 (2019) 105285.

[34] J. Cesarano, J. Stuecker, P. Calvert, 3D Printing Of Ceramics: Processes And Constraints, Adv. Mater. Process. 177 (2019) 28–31. https://doi.org/10.31399/ASM.AMP.2019-07.P028

[35] Z. Chen, Z. Li, J. Li, C. Liu, C. Lao, Y. Fu, C. Liu, Y. Li, P. Wang, Y. He, 3D printing of ceramics: A review, J. Eur. Ceram. Soc. 39 (2019) 661–687. https://doi.org/10.1016/j.jeurceramsoc.2018.11.013

[36] S. Riehm, V. Friederici, S. Wieland, Y. Deng, S. Herzog, A. Kaletsch, C. Broeckmann, Tailor-made functional composite components using additive manufacturing and hot isostatic pressing, Powder Metall. 64 (2021) 295–307.

[37] S. Güntop, Effect of Hot Isostatic Pressing on Densification, Microstructure, and Electrical Properties of 0.95 MgTiO3-0.05 CaTiO3Ceramic, (2023).

[38] G. Vadolia, K.P. Singh, B.R. Doshi, M.K. Gupta, Survey on Hot Isostatic Pressing Technique for Development of Tokamak Components, (2018) 642–642.

[39] M.U. Iqbal, A. Hussain, M. Shahid, M.R.A. Karim, M. Akhtar, Processing of Bioceramics by Pressing and Tape Casting, in: Adv. Bioceram., CRC Press, 2023: pp. 74–93.

[40] N. Irani, H.R. Rezaie, R. Naghizadeh, Functional Composites and Structures, (2020).

[41] Q. Zhang, The microstructure and properties of powder HIPped nickel-based superalloy CM247LC, (2011).

[42] I.L. Svetlov, K.K. Khvatskiy, M.A. Gorbovets, M.S. Belyaev, An effect of Hot Isostatic Pressing (HIP) on mechanical properties of casting Ni-based superalloys, Aviation 2023 (2022) 2025.

[43] M. Bustillo Revuelta, Ceramic Products, in: Constr. Mater. Geol. Prod. Appl., Springer, 2021: pp. 339–374.

[44] N.D. Coşkun, E. Çaşın, C.E. Işık, Investigation of Defects Occurring in the Casting Shop Process of Ceramic Sanitaryware on the Final Product, Black Sea J. Eng. Sci. 6 (n.d.) 527–534.

[45] G. Kiradoo, The transition of traditional pottery-making into advanced ceramics in context to the indian ceramic industry, Turkish J. Physiother. Rehabil. 32 (2021) 546–556.

[46] K. Fatima, M. Mohiuddin, Impact of capItal Structure on profItabIlIty and corporate Value of ceramIc InduStry: a Study on Some Selected lISted companIeS In dhaka Stock exchange, J. Bus. Adm. 41 (2020) 47–77.

[47] T. Aydın, Alumina esaslı elektroporselen bünyelerin geliştirilmesi , (2006) 67–88.

[48]   T. Jorgensen, New Tools for Ceramic Extrusion: Developing Craft Experiments into Industrial Applications, Mak. Futur. J. (2019).

[49]   F. Händle, The art of ceramic extrusion, Springer, 2019.

[50]   F. Singer, S.S. Singer, Industrial Ceramics, Ind. Ceram. (1963). https://doi.org/10.1007/978-94-017-5257-2

[51]   M.F. Islam, Characterization of Pin-Holing Defect and Its Preventive Measures in Tableware Ceramic Industry, (2019).

[52]   J. Yu, S.-H. Jin, K. Raju, Y. Lee, H.-K. Lee, Analysis of individual and interaction effects of processing parameters on wet grinding performance in ball milling of alumina ceramics using statistical methods, Ceram. Int. 47 (2021) 31202–31213.

[53]   A.C. Iyasara, J.A. Nduka, G.C. Irogbele, Design and fabrication of mechanical sieve shaker for particle size analysis of ceramics, J. Eng. Res. Reports 24 (2023) 12–19.

[54]   Rheology applied to ceramics | Kairos Media Group, (2006).

[55]   S. Vitali, L. Giorgini, Overview of the Rheological Behaviour of Ceramic Slurries., FME Trans. 47 (2019).

[56]   N. Ercioglu Akdogan, E. Arioz, O.M. Kockar, Enhancing rheology and physico-mechanical properties of ceramic slurries: Effect of the addition of various types of deflocculants, J. Dispers. Sci. Technol. 46 (2024) 119–129.

[57]   S. Dıkmen, T. Mucur, Z. Arsoy, B. Ersoy, The relationship between the flow properties of clay slurry samples and the properties of ceramic green/sintered products, Avrupa Bilim ve Teknol. Derg. (2020) 233–247.

[58]   P.O. Odewole, I.B. Kashim, T.L. Akinbogun, Production of refractory porcelain crucibles from local ceramic raw materials using slip casting, Int. J. Eng. Manuf. 9 (2019) 56–69.

[59]   L.B. Palhares, D.F. Galvão, P.R.P. de Paiva, Effects of plaster to water ratio on physical and mechanical properties of ceramic pieces produced by the slip casting process, Cerâmica 68 (2022) 143–151.

[60]   R. Papitha, M. Suresh, Y. Rao, B. Saha, D. Das, R. Johnson, Pressure slip casting and cold isostatic pressing of aluminum titanate green ceramics: A comparative evaluation, Process. Appl. Ceram. 7 (2013) 159–166. https://doi.org/10.2298/PAC1304159P.

[61]   A.P. Parra, P.A.M. Aguilar, M. Vlasova, R.P. Hernandez, R.G. Tapia, M.F. Perez, Preparation of secondary gypsum binders from waste plaster molds of the ceramic industry and their properties, Int. J. Tech. Sci. Res. Eng. 5 (2022) 1–19.

[62]   S. Carvalheira, M. Oliveira, M. Robaina, J.C.O. Matias, Energy Efficiency Improvements in a Portuguese Ceramic Industry: Case Study, Appl. Sci. 13 (2023) 5028.

[63]   R.S. Gomez, T.R.N. Porto, H.L.F. Magalhães, G. Moreira, A.M. André, R.B.F. Melo, A.G.B. Lima, Natural gas intermittent kiln for the ceramic industry: A transient thermal analysis, Energies 12 (2019) 1568.

[64]   D. Kujanen, Technical Ceramics And Refrac-Tories Applications And Volumes, (2019).

[65]   J. Schnell, F. Tietz, C. Singer, A. Hofer, N. Billot, G. Reinhart, Prospects of production technologies and manufacturing costs of oxide-based all-solid-state lithium batteries, Energy Environ. Sci. 12 (2019) 1818–1833.

[66] D. Brough, H. Jouhara, The aluminium industry: A review on state-of-the-art technologies, environmental impacts and possibilities for waste heat recovery, Int. J. Thermofluids 1–2 (2020) 100007. https://doi.org/10.1016/J.IJFT.2019.100007

[67] N.H. Khand, A.R. Solangi, S. Ameen, A. Fatima, J.A. Buledi, A. Mallah, S.Q. Memon, F. Sen, F. Karimi, Y. Orooji, A new electrochemical method for the detection of quercetin in onion, honey and green tea using Co3O4 modified GCE, J. Food Meas. Charact. 15 (2021) 3720–3730. https://doi.org/10.1007/s11694-021-00956-0

[68] A. Şavk, K. Cellat, K. Arıkan, F. Tezcan, S.K. Gülbay, S. Kızıldağ, E.Ş. Işgın, F. Şen, Highly monodisperse Pd-Ni nanoparticles supported on rGO as arapid, sensitive, reusable and selective enzyme-free glucose sensor, Sci. Rep. 9 (2019) 19228. https://doi.org/10.1038/s41598-019-55746-y

[69] L. Ren, X. Luo, H. Zhou, The tape casting process for manufacturing low-temperature co-fired ceramic green sheets: A review, J. Am. Ceram. Soc. 101 (2018) 3874–3889. https://doi.org/10.1111/JACE.15694

[70] T.Y. Zeybek, Reuse of Vapour Condensate in Ceramic Industry, (2019).

[71] S. Shojaee Barjoee, V. Rodionov, A.M. Vaziri Sereshk, Noise climate assessment in ceramic industries (Iran) using acoustic indices and its control solutions, Adv. Environ. Technol. 11 (2025) 91–115. https://doi.org/10.22104/aet.2024.6922.1899

[72] A. Hojjati-Najafabadi, S. Salmanpour, F. Sen, P.N. Asrami, M. Mahdavian, M.A. Khalilzadeh, A Tramadol Drug Electrochemical Sensor Amplified by Biosynthesized Au Nanoparticle Using Mentha aquatic Extract and Ionic Liquid, Top. Catal. 65 (2022) 587–594. https://doi.org/10.1007/s11244-021-01498-x

[73] R. Akoumeh, M. Al-Ejji, B. Aljaoni, M. Abbas, Advances in ceramic membrane technology: Versatility of fabrication technique, industrial applications, and challenges, Inorg. Chem. Commun. 179 (2025) 114685. https://doi.org/10.1016/j.inoche.2025.114685

[74] L. Treccani, Processing Methods for Advanced Ceramics, Surface-Functionalized Ceram. Biotechnol. Environ. Appl. (2022) 47–84. https://doi.org/10.1002/9783527698042.ch2

[75] G. Tari, Gelcasting ceramics: A review, (2003).

[76] H. Göksu, H. Burhan, S.D. Mustafov, F. Şen, Oxidation of Benzyl Alcohol Compounds in the Presence of CarbonHybrid Supported Platinum Nanoparticles (Pt@CHs) in Oxygen Atmosphere, Sci. Rep. 10 (2020) 5439. https://doi.org/10.1038/s41598-020-62400-5

A Journey from Raw Materials to Ceramics
Materials Research Foundations 184 (2025)

Materials Research Forum LLC
https://doi.org/21741/9781644903834

Chapter 8

Applications of Nanotechnology in Ceramic Industry

Farah Mutlag[1,2*], Hussein Elaibi[1], Abdullah Seyrankaya[3], Ebru Halvaci[1], Fatih Sen[1*]

[1]Sen Research Group, Department of Biochemistry, Kutahya Dumlupinar University, Kutahya 43000, Türkiye

[2]Ministry of Education, Karbala Education Directorate, Karbala, 56001, Iraq

[3]Faculty of Engineering, Department of Mining Engineering, Dokuz Eylul University, Izmir, Türkiye

farahfakhir28@gmail.com, fatihsen1980@gmail.com

Abstract

The integration of nanotechnology into the ceramic industry marks a transformative shift in enhancing material performance across mechanical, thermal, and functional dimensions. By employing nanomaterials such as titanium dioxide, zirconium oxide, and silver nanoparticles, ceramics now exhibit improved resistance to fracture, wear, and thermal shock, making them highly suitable for advanced engineering, medical, and environmental applications. Modern implementations of nanoceramics are diverse and expanding. These include self-cleaning smart tiles, antibacterial surfaces for hospitals and sanitaryware, as well as nanoceramic membranes for water and air filtration. In the biomedical field, nanostructured bioceramics play a critical role in bone regeneration and infection control. Furthermore, the integration of artificial intelligence (AI) into the design and manufacturing of nanoceramics unlocks new possibilities for predictive quality control, process automation, and real-time performance optimization. Despite these benefits, several challenges remain. High production costs, difficulty in achieving uniform nanoparticle dispersion, and environmental health concerns associated with nanoparticle exposure are significant obstacles. However, these issues also present opportunities for innovation in safe, scalable, and sustainable nanoceramic production technologies. In conclusion, nanoceramics represent a promising frontier that elevates the capabilities of traditional ceramics and positions the industry for a future that is smarter, more precise, and aligned with the demands of next-generation technologies.

Keywords

Ceramic, Antibacterial, Thermal, Nanotechnology

1. Introduction

Nanotechnology is the branch of science and technology focused on the manipulation of matter at the scale of the nanometer. In relation to the ceramic industry, nanotechnology breakthroughs have permitted the alteration of material characteristics at even the atomic and molecular levels. As a result, nanoceramics have been developed which are ceramic materials consisting of nano-sized particles or structures that improve the performance of the ceramic materials [1–4].

The main distinction between traditional ceramics and nanoceramics is their microstructure. With traditional ceramics, there is a composition of larger grains and particles, which could lead to greater brittleness, lower toughness, and an inability to perform under extreme physical and environmental conditions. Differently, nanoceramics possess ultra-fine grains which increase the density of the grain boundaries, thereby improving mechanical strength, wear resistance, and thermal and electrical properties [5–7].

With the aid of nanotechnology, the exact engineering of surface characteristics is now possible, which enables advanced functions such as self-cleaning surfaces, antimicrobial action, as well as enhanced chemical resistance. In addition, the high surface areas of nanoparticles make them more reactive. This can be beneficial for filtration, catalysis, and energy storage [6,8–10].

Overall, the integration of nanotechnology into ceramic manufacturing is not merely a refinement of existing materials; it represents a paradigm shift that opens new opportunities in industries ranging from electronics and biomedicine to aerospace and environmental technologies.

2. Mechanical Enhancements

The addition of reinforcing particles at the nanoscale level improves the mechanical and physical aspects of ceramics, like fracture toughness, wear, and hardness. This approach sharpens the grain structure and restricts the spread of cracks when stress is applied, as illustrated in Figure 1, Mechanical enhancement.

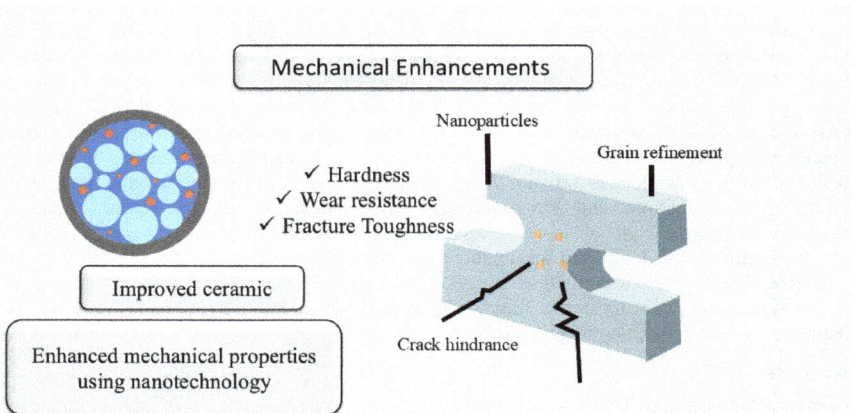

Figure 1. Mechanical enhancement.

Zirconium oxide (ZrO_2) and aluminum oxide (Al_2O_3) nanoparticles are often employed for their outstanding nanoscale mechanical properties. These nanoparticles provide substantial reinforcement when dispersed uniformly within ceramic matrix as they considerably block dislocation movement and enhance grain boundary suffusion, both of which are vital for increasing strength and durability [6,11–14].

Nanoceramics show characterized resilience and fatigue resistance in structural and engineering applications where ceramics experience heavy loads and harsh conditions. For instance, nano-zirconia-toughened alumina serves as a benchmark material in femoral heads and thighs for the orthopedic and dental implants as well, as in industrial applications such as, turbine blades and cutting tools due to its exceptional strength and wear resistance [15–17].

Moreover, nanoscale refinement of microstructure improves compaction and sintering behavior during processing, resulting in denser materials with fewer defects. This not only boosts the reliability of ceramic parts but also extends their service life in critical applications [18,19].

In summary, nanotechnology significantly elevates the mechanical performance of ceramics, making them more viable and competitive in sectors that demand both light weight and high durability. Table 1 shows Nanomaterials for mechanical improvements in ceramics.

Table 1. Nanomaterials for mechanical improvements in ceramics.

Nanomaterial	Targeted Property	Industrial Application	Technical Mechanism	Performance Indicators
Zirconium Oxide (ZrO_2)	Fracture Toughness	Cutting tools, dental prosthetics	Phase transformation toughening increases resistance to crack growth	+30% fracture toughness, reduced micro-cracking
Aluminum Oxide (Al_2O_3)	Wear Resistance	Bearings, valve seats, wear coatings	High hardness reduces surface abrasion	Up to 40% longer service life in high-friction use
Silicon Carbide (SiC)	Thermal Shock Resistance	Furnace linings, heat exchangers	Low thermal expansion and high conductivity	Withstands >1500°C without cracking
Carbon Nanotubes (CNTs)	Flexural Strength	Lightweight structural ceramics, armor systems	Nano-reinforcement bridges crack, distributes stress	+25% flexural strength, +15% impact resistance

3. Electrical and Electronic Applications

Nanotechnology increases the functional prospects of ceramic materials in the broader electrical and electronic sectors. The development of nanoceramics, which possess enhanced electrical, dielectric, and piezoelectric qualities, is made possible through the incorporation of traditional ceramic matrices integrated with nanoparticles, as illustrated in Figure 2, Electrical and electronic applications.

141

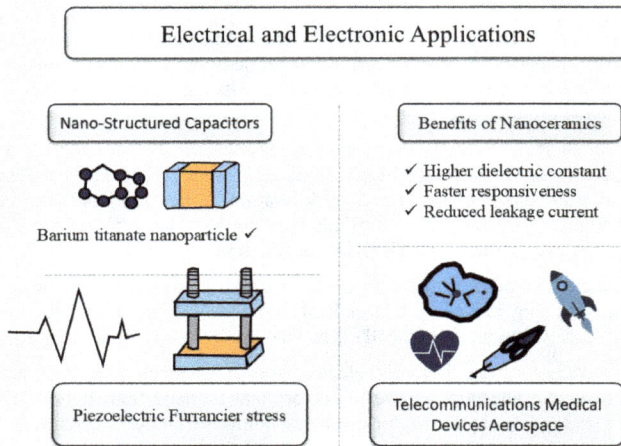

Figure 2. Electrical and electronic applications.

This would be the creation of nano-structured ceramics for use in other capacitors and for insulation purposes. For instance, barium titanate ($BaTiO_3$) nanoparticles are used in producing multilayer ceramic capacitors (MLCCs) with an enhanced dielectric constant that enables the manufacture of smaller and more efficient parts for electronic devices. The small size of the particles improves the packing density and reduces the porosity of the multilayer ceramic capacitors which enables them to store more energy and also reduce the energy that is lost [20–22].

The piezoelectric applications, the use of materials such as zirconate titanate, in their nano-scaled forms, greatly improve their responsiveness to mechanical stress, which makes such materials excellent candidates for use in sensors, actuators, and precision control devices in the fields of microelectronics and robotics. The transverse structure at nano-scale improves the rate at which electric signals are generated as well as the sensitivity and the rate of energy transduction [23–26].

Moreover, nanoceramics offer improved thermal stability and reduced leakage currents, making them reliable for high-frequency and high-voltage electronic systems. These advancements are critical in industries such as telecommunications, medical devices, and aerospace electronics, where performance, size, and energy efficiency are vital [27–29].

By reducing grain boundaries and enhancing structural homogeneity, nanotechnology paves the way for smarter, more compact, and energy-efficient ceramic-based electronic devices. Table 2 shows Advanced nanotechnology-based mechanical enhancements in ceramics.

Materials Research Forum LLC
https://doi.org/21741/9781644903834

Table 2. *Advanced nanotechnology-based mechanical enhancements in ceramics.*

Enhancement Aspect	Nanomaterials Used	Benefits Achieved	Industrial Applications
Hardness	Zirconia (ZrO$_2$) nanoparticles	Improved surface durability and scratch resistance	Cutting tools, wear-resistant tiles
Wear Resistance	Alumina (Al$_2$O$_3$) nanoparticles	Enhanced life cycle under frictional stress	Mechanical seals, valve components
Fracture Toughness	Silicon Carbide (SiC) nanofibers	Reduction in crack propagation under load	Armor ceramics, engine components
Thermal Stability	Titania (TiO$_2$) nanocomposites	Better resistance to thermal shock and phase changes	Heat exchangers, kiln linings
Flexural Strength	Carbon Nanotubes (CNTs)	Higher mechanical strength under flexural load	Aerospace tiles, dental prosthetics

4. Thermal and Environmental Benefits

Nanotechnology significantly improves the thermal and environmental properties of ceramics. One of the thermal properties of interest is the incorporation of nanostructured thermal barrier coatings (TBCs), which are also known for their excellent thermal shock resistance. These coatings are made of materials like nanostructured yttria-stabilized zirconia and help reduce the rate of heat transfer weakening and help sustain structural integrity during severe temperature shifts. This is advantageous in extreme heat situations like in turbines, kilns, and aerospace applications [6,30–33]. As illustrated in Figure 3, Thermal and environmental benefits.

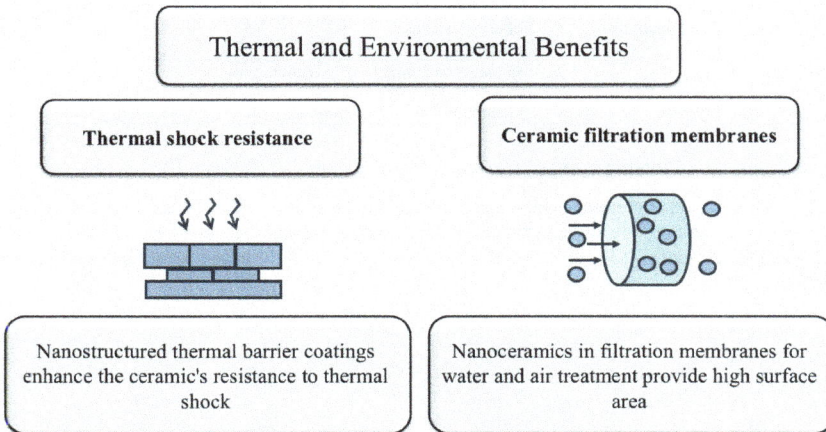

Figure 3. *Thermal and environmental benefits.*

Because of the unique properties of nano-ceramics, having extraordinarily high surface area alongside a highly controlled pore structure, they have proved extremely useful in the field of filtration technologies. Such materials are used in advanced membranes, which can filter heavy metals, bacteria and pollutants from water and air more effectively than ever before. Nanoporous ceramic membranes produced from alumina or titania offer chemical and thermal stability, along with long operational life and high flow rates. Nanoporous membranes help in enabling sustainable solutions for water treatment and air purification in industries [34–37].

By leveraging nanotechnology, the ceramic industry can meet the dual goals of environmental responsibility and thermal efficiency, leading to products that are both functionally superior and environmentally conscious. Table 3 shows Thermal and environmental benefits of nanoceramics.

Table 3. Thermal and environmental benefits of nanoceramics

Application	Nanomaterials Used	Achieved Benefit	Technical Advantage	Potential Challenge
Thermal Shock Resistance	Nano-Zirconia (ZrO_2), Nano-Silica	Prevents cracking from rapid temperature changes	High thermal expansion stability, fine grain structure	High material cost, process complexity
Thermal Insulation	Aerogel-ceramic composites	Reduces heat transfer and conserves energy	Ultra-low thermal conductivity, lightweight	Brittle structure, handling difficulty
Water Filtration	Nanoporous Alumina, TiO_2 membranes	Removes heavy metals, bacteria, and organic compounds	High surface area, chemical resistance	Fouling, cleaning difficulty
Air Purification	TiO_2, ZnO-based nano-ceramics	Photocatalytic degradation of pollutants and gas absorption	UV-activated efficiency, reusable surfaces	Requires light activation, reduced performance in dark conditions
High-Temp Industrial Linings	Yttria-Stabilized ZrO_2 (YSZ)	Maintains structural integrity at extreme temperatures	Excellent phase stability and heat resistance	Coating durability under repeated thermal cycling

5. Antibacterial and Self-Cleaning Ceramics

The use of functional nanomaterials, such as titanium dioxide (TiO_2) and silver nanoparticles (AgNPs), has enabled the development of advanced ceramic surfaces with self-cleaning and antibacterial properties [38–40]. As illustrated in Figure 4, Antibacterial and self-cleaning ceramics

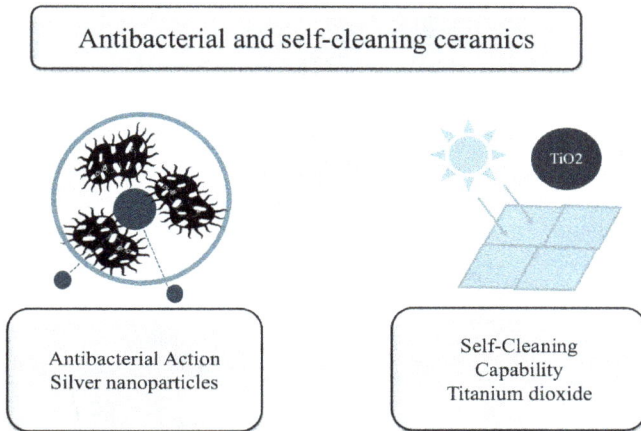

Figure 4. Antibacterial and self-cleaning ceramics.

Antibacterial action: Silver nanoparticles exhibit strong antimicrobial properties by disrupting bacterial cell membranes and interfering with their metabolic functions. When embedded in ceramic glazes or surface coatings, they create an inhospitable environment for microbial growth. This is particularly useful in settings where hygiene is critical, such as hospitals, food preparation areas, and sanitary ware [41–44].

Self-cleaning Capability: Titanium dioxide, especially in its anatase form, demonstrates photocatalytic behavior under UV or visible light. When applied as a nanocoating on ceramic tiles or surfaces, it breaks down organic contaminants, stains, and airborne pollutants, allowing them to be washed away easily by water (a phenomenon known as superhydrophilicity). This reduces the need for chemical cleaners and frequent maintenance [45–49].

These functionalities are now being integrated into "smart ceramic tiles" used in bathrooms, public spaces, and medical facilities, offering long-term cleanliness and hygiene with minimal intervention. The combination of aesthetic appeal and antimicrobial performance enhances both user experience and public health outcomes. Table 4 shows Nanotechnology in antibacterial and self-cleaning ceramics.

Table 4. Nanotechnology in antibacterial and self-cleaning ceramics.

Feature Type	Nanomaterial	Scientific Mechanism	Practical Applications	Technical Advantages	Limitations	Application Method
Antibacterial Function	Silver Nanoparticles (AgNPs)	Releases Ag⁺ ions that bind to bacterial enzymes and DNA, causing cell lysis	Hospital wall tiles, surgical room surfaces, hygiene ceramics	Long-term microbial resistance, works under dark conditions	Expensive; may cause discoloration in light-glazed ceramics	Mixed into glaze or nano-spray coating
Self-Cleaning (Photocatalytic)	Titanium Dioxide (TiO_2 - anatase)	Absorbs UV/visible light, creating electron-hole pairs that oxidize organic pollutants	Smart tiles in bathrooms, façades, public restrooms	Reduces chemical use; continuous surface cleaning with light exposure	Low effectiveness under indoor/dim lighting	Sol-gel dip coating, thermal spraying
Dual Function (Hybrid)	TiO_2 + Ag composite	Combines photocatalysis with direct microbial killing	Touch surfaces in hospitals, subway walls, toilet seats	Enhanced efficacy under various conditions, broad-spectrum action	Complex synthesis; higher production cost	Layered or co-sintered surface films

6. Challenges and Future Trends

Nanoceramics face major challenges in cost, uniformity, and safety. Figure 5 summarizes these key obstacles - cost, dispersion control, and safety issues - alongside emerging solutions. Future progress hinges on developing scalable processes and safer nanoparticle handling protocols.

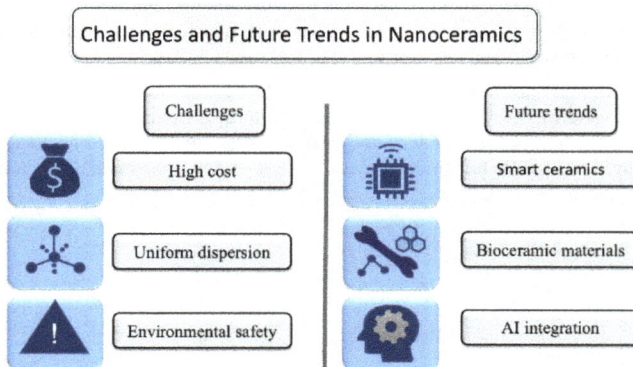

Figure 5. Challenges and future trends in nanoceramics.

Despite existing challenges in manufacturing and cost, the field of nanoceramics is advancing rapidly. Key trends include the emergence of smart ceramics responsive to environmental stimuli, the expansion of bioceramics in medical applications, and the integration of AI for optimized design and quality control. These innovations promise to enhance performance, efficiency, and functionality across various sectors, as summarized in Table 5.

Table 5. Challenges and future trends in nanoceramic applications.

Dimension	Specific Challenge / Trend	Scientific / Technical Context	Strategic Solutions / Emerging Opportunities	Industrial Impact / Outlook
Manufacturing	High production cost	Nanoparticle synthesis involves energy-intensive processes (e.g., sol-gel, hydrothermal methods)	Process optimization, green synthesis methods, scale-up via AI-driven parameter control	Cost reduction is key for mass adoption in consumer and medical markets
Material Dispersion	Non-uniform distribution of nanomaterials	Agglomeration leads to weak zones, affecting mechanical/thermal properties	Use of dispersing agents, surfactants, in-situ functionalization of nanoparticles	Enhances reliability in structural and functional ceramics
Environmental Safety	Health risks from free nanoparticles	Inhalation or leaching concerns, especially in biomedical and construction waste streams	Closed-loop systems, encapsulation methods, and regulation-compliant disposal protocols	Necessary for regulatory approval and environmental certification
Smart Functionality	Responsive ceramic systems (Smart Ceramics)	Nanomaterials that alter properties under external stimuli (pH, temp., electric field)	Integration in sensor devices, building automation, wearable health diagnostics	Expanding field with AI-embedded monitoring and adaptive systems
Bioceramics Evolution	Nano-enhanced medical ceramics	Targeted delivery, osteointegration, antimicrobial surfaces	Bioactive coatings, 3D-printed nano-bioceramics scaffolds	Critical in orthopedics, dental implants, and tissue engineering
AI & Industry 4.0	Digitalization of design & QC processes	AI/ML used to simulate sintering outcomes, optimize compositions, detect defects in real time	Predictive modeling, smart QC sensors, autonomous production lines	Transforms manufacturing speed, precision, and cost-efficiency

7. Conclusion

The application of nanotechnology in the ceramic industry has changed the game entirely by allowing the creation of materials with improved mechanical strength, thermal stability, and multifunctional surface properties. By including engineered nanoparticles like TiO_2, ZrO_2, and Ag, as well as adapting ceramic products, their durability has increased, and their application has become more specialized, like with structural components, biomedical implants, and even self-clearing surfaces. The successful integration of nanotechnology in ceramics stems from innovation in material as well as process control, safety management, and economic feasibility. Although factors like cost of production, dispersion of nanoparticles, and the risk they pose to the environment remain challenges, ongoing research and advancements make streamline processing techniques, digital manufacturing, and AI-directed optimization more commercially viable. In the more distant future, the combination of nanomaterials with biocompatible systems, artificial intelligence, smart infrastructure, and other emerging technologies will enable the creation of multifunctional next-generation ceramics. These encompass responsive materials for environmental sensing, bioactive surfaces for medical applications, and components for clean energy systems.

References

[1]   Ghorbanpour, M. and Shahid, M.A. (2022). *Nano-enabled Agrochemicals in Agriculture*, Academic Press.

[2]   Zinatloo-Ajabshir, S. (2022). *Advanced rare earth-based ceramic nanomaterials at a glance*, Elsevier.

[3]   Visakh, P.M. and Balakrishnan, R. (2022). Nanotechnology in Electronics, Materials Properties, and Devices: State of the Art and Future Challenges. *Nanotechnology in Electronics: Materials, Properties, Devices*. https://doi.org/10.1002/9783527824229.ch1

[4]   Darul, D.R. *et al.* (2024). Industrial applications of nanoceramics: from lab to real-time utilization in the biomedical industry, in *Industrial Applications of Nanoceramics*, Elsevier, pp. 143–158.

[5]   Kiradoo, G. (2021). The transition of traditional pottery-making into advanced ceramics in context to the indian ceramic industry. *Turkish Journal of Physiotherapy and Rehabilitation*.

[6]   Zawrah, M.F. *et al.* (2023). Advanced ceramics: stages of development, in *Advanced Ceramics*, Springer, pp. 1–46.

[7]   Kalendova, A. *et al.* (2024). Applications of Clays in Nanocomposites and Ceramics. *Minerals*. https://doi.org/10.3390/min14010093

[8]   Singh, J. (2006). *Nanomaterials and Nanotechnology*, Springer.

[9]   Jagessar, R. (2021). Nanotechnology and Nanoparticles in Contemporary Sciences. *Journal of Nanosciences Research & Reports*. https://doi.org/10.47363/jnsrr/2021(3)118

[10]   Rani, P. *et al.* (2023). Additive Manufacturing of Ceramic-Based Materials. *Nanotechnology-Based Additive Manufacturing: Product Design, Properties and Applications*.

[11]   Rogenski, E.N. (2021). *The investigation of AM ceramics for the production of a 3D printed high temperature thermocouple*, Youngstown State University.

[12]   Veysel, M.B. and Tuba, B. (2024). Investigation of the effects of precision-casting waste sands on the thermal shock resistance properties of fire clay refractory materials. *Pamukkale University Journal of Engineering Sciences*.

[13]   Eid, N.K. (2025). A Review on the Power of CAD/CAM Technology and the Material Science in Modern Manufacturing. *ERU Research Journal*. https://doi.org/10.21608/erurj.2025.299610.1167

[14]   Lyons, A. (2020). Ceramic products, in *Materials for Architects and Builders*, Springer, pp. 258–264.

[15]   Gaspar, F. *et al.* (2025). Synthetic image generation for effective deep learning model training for ceramic industry applications. *Engineering Applications of Artificial*

Intelligence. https://doi.org/10.1016/j.engappai.2025.110019

[16] Ruivo, L. *et al.* (2021). Energy management in the Portuguese ceramic industry: Analysis of real-world factories. *Energy*. https://doi.org/10.1016/j.energy.2021.121628

[17] Shathi, A.S. *et al.* (2024). Iron removal from red clay using oxalic acid leaching for enhanced ceramic industry applications. *Heliyon*.

[18] Krichen, M. and Ksibi, H. (2025). Sintering behaviour of Tunisian clay–degreaser mixtures: valorizing Weslatiya clay with chamotte as a quartz sand substitute. *Clay Minerals*.

[19] Klunghirun, W. and Serivalsatit, K. (2025). Development of cordierite-mullite refractory castables bonded with magnesium silicate hydrate cement. *Journal of the Thai Ceramic Society*.

[20] Wang, J. *et al.* (2025). Microstructure regulation and dielectric performance analysis of the high-pressure sintered nano-BaTiO₃ ceramics. *Journal of Materials Science: Materials in Electronics*. https://doi.org/10.1007/s10854-025-14619-z

[21] Usukawa, R. *et al.* (2025). Functions of glycerol in the tape-casting process: Revealing with AFM colloid probe measurements. *Journal of the Ceramic Society of Japan*. https://doi.org/10.2109/jcersj2.24112

[22] Wang, Z. *et al.* Formation and Evolution of Oxygen Vacancy Layer in Nano-BaTiO3 Dielectric Ceramics Under Thermal and Electric Field Stimuli. *Available at SSRN 5273441*.

[23] Aboughaly, M. (2025). Emergence of Piezoelectric Materials in Smart Materials. *Green Technology and Smart Materials for Engineering Applications*. https://doi.org/10.1201/9781003516071-4

[24] Xiong, D.C. (2025). Development and Characterization of Flexible PVDF-TrFE-BaTiO3-Ti3C2 MXene-Based Composite Multi-Morphs: Towards the Development of Electro-Active Biomedical-Wearable Devices.

[25] Samrot, A. V. *et al.* (2022). Nanoparticles, a Double-Edged Sword with Oxidant as Well as Antioxidant Properties—A Review. *Oxygen*. https://doi.org/10.3390/oxygen2040039

[26] Karacasulu, L. *et al.* (2025). Sintering Under High Heating Rates. *Annual Review of Materials Research*. https://doi.org/10.1146/annurev-matsci-080323-042441

[27] Ahmed, M.F. *et al.* (2025). Investigation the effect of calcination heating rate on the structural, morphological, thermal and color properties of nano Cobalt Aluminate (CoAl2O4). *Heliyon*. https://doi.org/10.1016/j.heliyon.2025.e42413

[28] Huseien, G.F. *et al.* (2025). Sustainability of Recycling Waste Ceramic Tiles in the Green Concrete Industry: A Comprehensive Review. *Buildings*.

[29] Tang, Q. (2025). Ceramic Automatic Design and Optimization Based on Deep Learning Technology. *Computer-Aided Design and Applications*. https://doi.org/10.14733/cadaps.2025.S1.32-45

[30] Lourenco Alves, C. *et al.* (2025). Challenges and opportunities for increase sustainability and energy efficiency in ceramic tile industry. *International Journal of Applied Ceramic*

Technology. https://doi.org/10.1111/ijac.15097

[31]   Buonomo, B. *et al.* (2025). Life cycle assessment of implementation of an innovative solar thermal technology in Italian ceramic industry. *Thermal Science and Engineering Progress.* https://doi.org/10.1016/j.tsep.2025.103517

[32]   Zhang, J. *et al.* (2025). The environmental sustainability assessment of building ceramic manufacturing based on the LCA-emergy approach. *Environment, Development and Sustainability.* https://doi.org/10.1007/s10668-023-04114-y

[33]   Bhat, C. *et al.* (2025). Critical review of metal-ceramic composites fabricated through additive manufacturing for extreme condition applications. *Mechanics of Advanced Materials and Structures.* https://doi.org/10.1080/15376494.2024.2376337

[34]   Pujiastuti, C. *et al.* (2021). Ceramic Industry Wastewater Treatment by Chemical Coagulation Process. *Journal of Research and Technology.* https://doi.org/10.55732/jrt.v7i2.411

[35]   Shurygin, M. *et al.* (2021). Effective treatment of the wastewater from ceramic industry using ceramic membranes. *Water Science and Technology.* https://doi.org/10.2166/wst.2021.039

[36]   Chauhan, N. *et al.* (2023). Synthesis of Ceramic Membranes and Their Application in Wastewater Treatment and Emerging Contaminants Removal, in *Energy, Environment, and Sustainability*, Springer, pp. 413–446

[37]   Mohamed, R.M.S.R. *et al.* (2021). Greywater Treatment using Pottery Waste Ceramic Filter. *Jurnal Kejuruteraan.* https://doi.org/10.17576/jkukm-2021-33(4)-12

[38]   Mandal, A. et al. (2023). Theoretical Advances in Polariton Chemistry and Molecular Cavity Quantum Electrodynamics. Chemical Reviews. https://doi.org/10.1021/acs.chemrev.2c00855

[39]   Asadi, A. *et al.* (2025). Ethical Constraints of the Application of Nanobiotechnology-Based Meat Products, in *Innovative Technologies for Meat Processing*, CRC Press, pp. 58–77.

[40]   Prasad, R.D. *et al.* (2024). A Review on Nanotechnology from Prehistoric to Modern Age. *ES General.* https://doi.org/10.30919/esg1117

[41]   Mehta, V.N. *et al.* (2021). Bio-functionalized Silver Nanoparticles: A Versatile Candidate for the Ceramic Industry, in *Handbook of Polymer and eramic Nanotechnology: Volume 1,2*, vol. 1, Springer, pp. 83–98.

[42]   Wafy, K.R. *et al.* (2023). Water disinfection using durable ceramic filter coated with silver nanoparticles synthesized using actinomycetes. *Applied Water Science.* https://doi.org/10.1007/s13201-023-01937-y

[43]   Hosseini-Zori, M. *et al.* (2023). Synthesis of Self-Assembled Ag-ZrO2 Plant Shape for an Effective Antibacterial Ceramic Application. *Journal of Nanostructures.* https://doi.org/10.22052/JNS.2023.02.017

[44]   Pokroeva, Y. *et al.* (2023). Biocidal Protective Glass-Ceramic Coatings for Porcelain Stoneware. *Lecture Notes in Networks and Systems.* https://doi.org/10.1007/978-3-031-

46877-3_35.

[45] Wang, Y. *et al.* (2024). A novel self-cleaning ceramic waste-slag geopolymer with nano-SiO2-TiO2 photocatalytic coating. *Ceramics International.* https://doi.org/10.1016/j.ceramint.2024.02.190

[46] Aminian, M.K. and Fatah, S.K. (2025). Multifunctional coatings with decorative, self-cleaning, anti-slip, and cool-coating properties on ceramic tile, in *Multi-scale and Multifunctional Coatings and Interfaces for Tribological Contacts*, CRC Press, pp. 103–120.

[47] Ferreira-Neto, E.P. *et al.* (2021). Thermally stable SiO2@TiO2core@shell nanoparticles for application in photocatalytic self-cleaning ceramic tiles. *Materials Advances.* https://doi.org/10.1039/d0ma00785d

[48] Stanciu, I. (2025). Study of Properties and Applications of Titanium Oxide, in *Titanium Dioxide - Uses, Applications, and Advances*, IntechOpen.

[49] Pauzan, M.A.B. *et al.* (2022). Development of Ceramic (Inorganic) Membranes for Oil/Water Separation, in *ACS Symposium Series*, vol. 1407, ACS Publications, pp. 185–216.

Chapter 9

Glaze Preparation Techniques for Industrial Ceramic Products

Zahide Bayer Öztürk[1*], Betül Yıldız[2*], Elif Eren Gültekin[3*], Kubra Karadeniz[4],
Cumhur Eren Isık[5], Iskender Isık[6], Fatih Sen[4*]

[1]Nevsehir Haci Bektas Veli University, Department of Metallurgy and Materials Engineering,
50300, Nevsehir, Türkiye

[2]Department of Metallurgy and Materials Engineering, Bilecik Şeyh Edebali University, Bilecik,
11100, Türkiye

[3]Airframe and Powerplant Maintenance, School of Civil Aviation, Selçuk University, Konya,
Türkiye

[4]Sen Research Group, Department of Biochemistry, Kutahya Dumlupinar University, Kutahya
43000, Türkiye

[5]Department of Handicrafts, Kutahya Fine Arts Vocational School, Kutahya Dumlupınar
University, Evliya Çelebi Campus, 43000 Kutahya, Türkiye

[6]Department of Materials Science & Engineering, Faculty of Engineering, Kutahya Dumlupınar
University, Evliya Çelebi Campus, 43100 Kutahya, Türkiye

z.ozturk@nevsehir.edu.tr, fatihsen1980@gmail.com

Abstract

This chapter provides a scientific analysis of the composition, production processes, and
functional properties of ceramic glazes. It begins with the definition of ceramic products,
followed by a detailed discussion of the glaze concept, its application purposes, and its
classification. The Seger formula, which is widely used in glaze formulation, is explained in
detail, and the fundamental oxides that constitute glaze compositions are analyzed in terms of
their effects on the physical and chemical properties of the glaze. In addition to primary raw
materials, auxiliary components used in the production process are identified, and the parameters
that must be considered in glaze composition are clarified. Practical information is provided
through sample glaze recipes designed for different ceramic product groups, and the glaze
production process is presented systematically. Furthermore, test methods employed for glaze
quality control are outlined, and finally, the key functional properties expected from ceramic
glazes in today's ceramic industry are briefly summarized. This chapter aims to offer a
comprehensive evaluation of ceramic glazes from structural, functional, and technological
perspectives.

Keywords

Ceramic Glaze, Glaze Production, Materials Used in Ceramic Glazes, Classification of Glazes

1. Introduction

The ceramics sector is included under the heading "manufacture of other non-metallic mineral products", numbered 23 according to the NACE 2 Industrial Branches classification, which is the international coding system. This category comprises production activities that primarily involve materials of mineral origin. It includes the fabrication of end products such as glass (including flat, hollow, fiber, and technical glassware), ceramic goods, bricks, tiles, and fired clay items. The production of shaped and ready-to-use stone and other mineral products is also included in this section. Ceramic products are grouped under seven headings: Refractory ceramic products; Heat-insulating ceramic products of siliceous filtered earth, refractory ceramic retorts, crucibles, spray tubes, pipes, and the like [1,2].

Ceramic tiles; Manufacture of non-refractory ceramic floor or wall tiles and tiles, mosaic cubes, borders and the like, non-refractory ceramic floors and ceramic paving stones. Sanitary products made of ceramics; Washbasins, bathtubs, toilets, and other similar ceramic products. Ceramic household and ornamental articles; Tableware and other household or toilet objects, statuettes, figurines, and other decorative items. Ceramic insulation materials and fittings; Electrical insulation materials and fittings of ceramics. Other technical ceramic products: Ceramic and ferrous magnets, laboratory products of ceramics, chemical and industrial products [1,3,4].

Other ceramic products not classified elsewhere; Ceramic pots, jars, and various similar products used for the transportation or packaging of goods, ceramic products not elsewhere classified [5,6]. Glazes are used on a wide range of materials, including tableware, sanitary ceramics, decorative items, refractories, tiles, electrical insulators, and technical ceramics. They are also occasionally applied to less typical bases like cement and graphite [7,8]. Various ceramic products are shown. The glaze is composed of a combination of vitrified oxides (frits) and various inorganic components that, upon firing, create a thin glassy or semi-glassy coating (ranging from 0.15 to 1.00 mm) over the ceramic body. This layer enhances the surface by improving chemical stability, resistance to liquids and gases, durability against abrasion and scratching, mechanical strength, ease of cleaning, and by imparting decorative and aesthetic qualities[9,10].

2. Purposes of Ceramic Glazing

Glass coatings are used for several purposes, including creating a glaze on porous and rough surfaces to make them easier to clean, increasing the variety of decoration and aesthetic appearance using different decorating techniques, enhancing the strength and surface hardness of the product, and increasing its resistance to chemical influences and external factors. [11,12].

3. Classification of Ceramic Glazes

Ceramic glazes are classified in various ways in the literature. In some sources, the classification is made according to the type of components in the glaze, according to the type of ceramic on which it is applied, and the melting temperature of the glaze. Therefore, the classification of glazes varies from source to source, and the classification is made by considering only a few features [3,13–15]. Tables 1 and 2 show the most commonly used classification methods.

Table 1. Ceramic glaze classification according to chemical composition, suggested by Parmelee[16].

	Criterion	Classification
Raw glazes	Lead-containing	-Glazes containing alumina
		-Glazes lacking alumina
	Non lead-containing	-Those formulated with alkaline-earth elements
		-Glazes incorporating both alkaline and alkaline-earth components
		-Compositions with alkaline, alkaline-earth elements, and zinc oxide
		-Boron-based glazes
		-Salt glazes
Fritted glazes	Lead-containing	-Alumina-based glazes
		-Glazes lacking alumina and boron
	Non lead-containing	-Boron-based glazes
		-Glazes lacking boron
		-Glazes containing high BaO content

Table 2. Classification of glazes based on various criteria [16].

Criterion	Classification
Fusibility	-Fusible
	-Hard or low-fusibility
Presence of an important component	-Lead-based
	-Lead-free
Further application and firing processes	-Single-firing glazing
	-Single-firing pavement
	-Double conventional firing covering
	-Double fast-firing covering
Production application	-Bases
	-Airbrushing (pulverised)
	-Pips
	-Serigraphy
Effect on the finished product	-Shining
	-Matte
	-Semi-matte
	-Satin
	-Transparent
	-Opaque
	-Coloured

3.1 Types of Glazes According to Their Optical Appearance

Although there are various classifications for glazes, ceramic glazes in the industry are mostly categorized according to their optical properties. These properties generally vary depending on factors such as glaze thickness, surface roughness, the refractive index of the glassy phase in the glaze, the refractive index of the crystals and the phases formed as a result of phase separation, the difference in refractive index between the glassy phase and the crystals, as well as the type, size, shape, and quantity of crystals present. Based on these optical characteristics, glazes can be transparent, opaque, matte, or glossy. The optical behavior of glazes is influenced by the reflection and scattering of light caused by small particles or irregular cores within the glaze [16,17]

Transparent, Opaque Glazes: Since transparent glazes have a homogeneous refractive index, the incident light spreads from the air-glaze interface without being reflected or scattered. In opaque glazes, opacity occurs as a result of the diffusion, reflection, and refraction of the incident light due to particles or air bubbles in the glaze [18]. In Figure 1, the interaction of light with transparent glaze (a) and opaque glaze (b) is shown schematically. If the glaze contains crystals or if phase separation has occurred in the glaze, then there are interfaces in the glaze. These interfaces in the glaze itself also cause reflection. The reflection occurring at the inner interfaces is called scattering. Scattering is a concept that forms the basis for opacity[11,18].

Matte, Satin, and Glossy Glazes: The concepts of matte and gloss in glazes are usually explained by relating them to the amount of light reflected uniformly from the glaze surface. Some light is reflected from the surface at the same angle as it hits the surface. These lights are called uniformly reflected lights. Some light is reflected from the surface at different angles. These lights are called diffusely reflected lights. Figure 3 shows schematically the surfaces that show different reflection behaviors. In Figure 3a, the high-gloss surface showed a higher amount of uniform reflection than diffuse reflection. In Figure 3b, the satin surface showed a very small amount of uniform reflection and a high amount of diffuse reflection. In Figure 3c, the matte surface showed no uniform reflection at all; all reflections occurred diffusely [19].

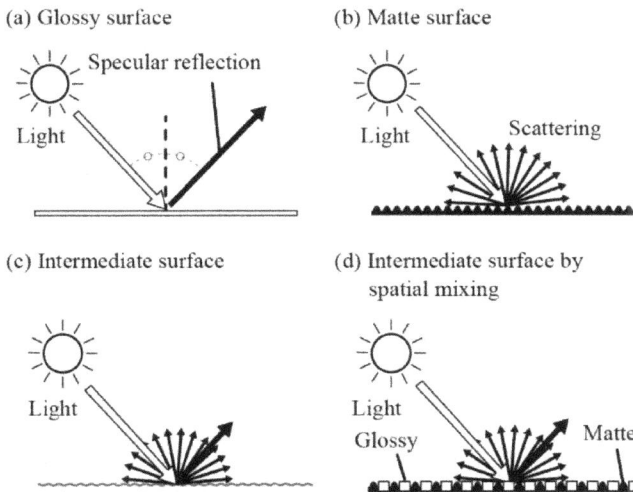

Figure 1. Appearance of material removal by reflected light on (a) glossy, (b) satin, and (c) matte surfaces [20], reprinted with permission from MDPI.

4. Seger Formula in Glaze Design

When planning a glaze study, first of all, it is determined what is expected from the glaze, the product to which the application will be made is clarified and depending on this product, it is determined at which temperature ranges and for how long it will be fired. The raw materials and their ratios to be used in the glaze are adjusted according to the determined criteria. The products to which the glazes are applied and the firing temperatures of the products are different. Both these reasons and the features sought in the conditions of use, and the fact that different raw materials can be used for the same oxide, have necessitated the emergence of very different glaze recipes. In order to easily compare glaze recipes with each other in terms of their compositions and to be able to easily interpret them in terms of melting behavior and usage characteristics, German chemist Hermann Seger developed the glaze formula named after him [21–24].

The Seger formula classifies glazes into three groups according to the type and properties of the oxides. These are basic oxides, amphoteric oxides, and acidic oxides. Basic Oxides: The oxides in this group are alkaline and alkaline earth melting oxides and are symbolized as RO and R_2O (R: Element, O: Oxygen). Basic oxides, when used alone or with other oxides, show differences in terms of their melting points and other properties they add to the glaze. According to Seger, the total amount of basic oxides (PbO, MgO, CaO, BaO, ZnO, K_2O, Na_2O, Li_2O) in the glaze should be one mole. Amphoteric Oxides: The oxides in this group are oxides that provide glaze stability and act as bases against acids and as acids against bases. They are symbolized as R_2O_3. The most important of these oxides is Al_2O_3 [24–26].

Acidic Oxides: These oxides, also called glass-forming oxides in glazes, are symbolized as RO_2. The acidic oxides in the Seger formula are SiO_2, ZrO_2, SnO_2, and B_2O_3 [18]. Although the formula of B_2O_3 is similar to amphoteric oxides, it is included in this group due to its acidic character.

5. The Role of Oxides in Glaze Formation

Silicon Dioxide: It is the oxide used in all glazes and is added into the glaze as clay-kaolin, feldspars, or free quartz. When used in glaze, it forms a glassy structure with an amorphous phase together with basic oxides. While it reduces expansion in glazes, it increases expansion in engobes [27,28].

Aluminum Oxide: The glaze usually contains Al_2O_3 (5-15 wt.%), when it is more than 25 wt.%, it sinks into the glaze structure and provides a matte surface. Aluminum oxide is added to the glaze as clay-kaolin, feldspar, or free alumina. It increases the melt viscosity of the glaze, prevents cracking by reducing glaze expansion, and increases resistance to acids and bases [29,30]. Sodium and Potassium Oxide: Sodium and potassium oxide are the strongest melting oxides used in glazes. However, there are some situations that limit their use, the most important of which is that they create expansion cracks in glazes due to the fact that they increase the expansion too much. They are used very little in glazes produced at high temperatures, and if they are used too much, they cause the glaze to melt too much. Depending on the increase in the sodium oxide ratio in glazes, the softening temperature decreases and the expansion coefficient increases. The most important feature of glazes containing potassium oxide compared to glazes containing sodium oxide is wider the melting range. The need for potassium oxide can be obtained from potassium feldspar or nepheline syenite. In addition to these raw material sources, sodium carbonate can be added to the frit to meet the need for sodium oxide [16,31,32].

Zirconium Dioxide: The most important compound of zirconia in its natural state is zircon ($ZrSiO_4$). Zirconia enhances glaze surfaces with excellent properties due to its high refractive index (1.92–1.97), strong resistance to wear and impact, stability under thermal shock, and outstanding chemical durability [29].

Calcium Oxide: It is an oxide that is commonly used in glaze composition. It has very good melting properties at low temperatures (1100°C). Calcium carbonate ($CaCO_3$), dolomite ($CaCO_3.MgCO_3$), wollastonite ($CaSiO_3$), calcium sulphate ($CaSO_4$) and calcium phosphate ($Ca_3(PO_4)_2$) are used as calcium oxide sources [12].

Magnesium Oxide: Magnesium oxide ensures that the glaze adheres to the body; its decrease or increase in the glaze changes the viscosity of the glaze. Magnesium oxide is also an effective melter and provides stability to the glaze; prevents cracking; provides mechanical resistance to the glaze. Dolomite, talc and magnesite are included in the glaze [12].

Boric Acid: Boric acid, whose chemical formula is $B_2O_3.3H_2O$, is an oxide with a very low melting point and lowers the melting point of other oxides. It is highly soluble in acids and water. Since it is soluble in water, it is used by adding frit into the glaze. Since its melting feature is outstanding, it reduces the melting point of the glaze. It reduces expansion and increases resistance to acids. It increases the surface spread of the glaze while increasing the surface brightness[12,33].

Wollastonite: It has been observed that incorporating 16-22 wt.% wollastonite in raw glaze formulations results in glossy surfaces, while increasing the content to around 30 wt.% produces matte finishes [34]. Zinc Oxide: In glaze formulations, ZnO functions like alkaline earth oxides but has a higher dissolution rate at lower temperatures. As a result, adding a small amount of ZnO to a high-temperature porcelain glaze enhances fluxing action, aiding the melting of other oxides and thereby improving surface smoothness and gloss. When a large amount of ZnO is added, it reacts with the other components of the glaze to form gahnite ($ZnAl_2O_4$) or willemite (Zn_2SiO_4) crystals that create a matte surface [35].

Barium Oxide: Generally used as barium carbonate ($BaCO_4$). BaO additives significantly enhance both the mechanical strength and chemical resistance of materials. Common sources of barium oxide are BaO-containing frits, barium sulfate ($BaSO_4$), and barium carbonate ($BaCO_3$). However, barium sulfate is less frequently used because it releases sulfur dioxide during firing. Similarly, barium carbonate is only occasionally utilized as a raw material, primarily due to its toxicity concerns [36].

6. Raw Materials for Ceramic Glazes

For various reasons, particularly economic efficiency, ceramic coatings are typically not produced by directly mixing individual oxides. Instead, naturally occurring minerals are preferred, as they facilitate rapid melting and reduce production costs, provided they meet the necessary processing requirements. Since many minerals contain multiple oxide components, the choice of raw materials is inherently limited by the coexistence of these oxides within a single source.

Table 3 lists the most commonly used raw materials in ceramic coating formulations. As ceramic coatings are generally applied in the form of aqueous slips, highly water-soluble materials are excluded from the table, as they cannot be used directly unless they are first converted into frits.

When selecting raw materials for glaze formulations, several critical factors must be considered, including the complete chemical composition of the material (with expected impurities), stability and consistency of composition over time, particle size distribution, availability and geographical origin, handling and storage properties, suspension behavior in water, processing performance, environmental impact, and overall cost. [37].

Table 3. Commonly used raw materials [37].

To add	Possible raw materials	Other oxides introduced
Li_2O	Spodumene	Al_2O_3, SiO_2
	Petalite	Al_2O_3, SiO_2
Na_2O	Feldspars	K_2O, Al_2O_3, SiO_2
	Nepheline syenite	K_2O, Al_2O_3, SiO_2
K_2O	Feldspars	Na_2O, Al_2O_3, SiO_2
	Muscovita mica	Al_2O_3, SiO_2
CaO	Wollastonite	SiO_2
	Calcium carbonate	-
	Dolomite	MgO
MgO	Heavy magnesium oxide	-
	Magnesium carbonate	-
	Dolomite	CaO
	Talc	CaO, SiO_2
SrO	Strontium carbonate	-
BaO	Barium carbonate	-
ZnO	Zinc oxide	-
PbO	Lead bisilicate	SiO_2
Al_2O_3	Corundum	-
	Alumina hydrate	-
	Feldspars	Na_2O, K_2O, SiO_2
	Nepheline syenite	Na_2O, K_2O, SiO_2
	Kaolin clay	SiO_2
	Ball clay	SiO_2
	Muscovita mica	K_2O, SiO_2
	Pyrophylitte	SiO_2
SiO_2	Quartz sand, glass sand, flint	-
	Feldspars	Na_2O,K_2O, Al_2O_3
	Nepheline syenite	Na_2O, K_2O, Al_2O_3
	Wollastonite	CaO
	Kaolin clay	Al_2O_3
	Ball clay	Al_2O_3
	Muscovite mica	K_2O, Al_2O_3
	Pyrophylitte	Al_2O_3
	Talc	CaO, Al_2O_3
	Zircon	ZrO_2
ZrO_2	Zircon	SiO_2
TiO_2	Anatase	-
	Rutile (pigmentary)	-
	Rutile (refined ore)	-

7. Other Raw and Auxiliary Materials Used in Ceramic Glazes

Frit: A frit is a type of glass that has been melted, rapidly cooled (quenched), and then ground into a fine powder to be used as a component in glaze formulations. Frits are highly valuable to glaze manufacturers for several reasons. One key advantage is their ability to provide alkaline oxides in an insoluble form, which is important since few insoluble sources of K_2O, Na_2O, and B_2O_3 exist. Although sodium and potassium are present in feldspar, they are bound with significant amounts of alumina and silica, which limits their effectiveness as fluxes, particularly in low-temperature glaze applications. B_2O_3 is available to the pottery in only one naturally occurring mineral, which is insoluble: colemanite.

To obtain these valuable oxides in an insoluble form, they are combined and melted with other oxides through fritting, producing a stable, insoluble material that can be incorporated as a standard ingredient in glaze formulations. Another reason for using fritted material is the poisonous nature of lead oxide. When lead oxide is fritted with sufficient quantities of other oxides, it becomes non-poisonous. Whenever possible, it is advisable to use lead as a frit rather than the raw state [38].

Other reasons for making and using frits include reducing the risk of silicosis caused by crystalline silica dust, as frit dust, although fine-grained, is less sharp-edged; achieving oxide blends in proportions not found in any natural material; lowering the firing temperature of a glaze recipe by pre-fusing part of its constituents; producing colored stains; and serving as a base for glazes to which various additions can be made to create different effects [8].

Frit formulations are typically engineered to deliver a high proportion of the target oxides, most commonly the active fluxing agents such as PbO, K_2O, Na_2O, and B_2O_3, while also incorporating sufficient quantities of additional oxides to ensure that the resulting glass is both chemically stable and water-insoluble. These additional oxides often include SiO_2, Al_2O_3, CaO, ZnO, and MgO.

The process of frit production is relatively straightforward and closely parallels traditional glass manufacturing techniques. Precisely measured raw materials are introduced into a crucible or furnace and subjected to high temperatures until they fuse into a homogeneous molten phase. This molten material is subsequently discharged from the furnace and rapidly quenched in a water tank, where the sudden thermal shock causes it to fracture into fine granules (as illustrated in Figure 4a) [38].

Coloring agents: Coloring oxides and ceramic dyes called pigments are used in coloring glazes (Figure 4b). Coloring oxides dissolve in the glaze and both give color and contribute to the formation of the glassy structure. The dissolution and glassy structure formation properties of these oxides vary depending on the type of oxide, amount, glaze composition, firing temperature, firing atmosphere and amount of usage.

Dyes used in coloring glazes are colorants obtained from various metal oxides and do not dissolve in the glaze. They give color by dispersing in very fine particles in the glaze. For this reason, they are more advantageous in coloring glazes compared to oxides. Because dyes are less affected by oxides against changes in firing temperature, atmosphere, and glaze composition. In addition, the use of dyes in high-grade glazes provides advantages in terms of both color variety and color stability. Table 4 shows the classification of ceramic pigments used in glaze applications [39,40].

Coloring oxides and dyes can be used alone in coloring glazes, or a wide variety of colors can be obtained by using several of them together [13].

Pigments can be classified according to various criteria, including their origin, color characteristics, chemical composition, production methods, and intended applications. The table below presents a classification scheme specifically for ceramic pigments employed in glaze formulations [40].

Table 4. Classification of ceramic pigments used in glaze applications [39,40].

Non oxides	Cd(Sx,Se1-x)
Metallic colloids	Au, Ag, Se, Pt, Cu
Metallic oxides	Cu_2O, CuO, NiO, MnO, MnO_2, Fe_2O_3, Cr_2O_3, Co_3O_4
Naturally colored	Spinel
	Phyrochlore
	Olivine
	Garnet
	Phenacite
	Periclase
Colored by the addition of colored compounds	Zircon
	Baddeleyite
	Corundum
	Rutile
	Cassiterite
	Sphene
White pigments	Sb_2O_3, As_2O_3, CeO_2, SnO_2, TiO_2, ZrO_2, $ZrSiO_4$

Mill additives: A number of auxiliary additives are used to adjust the rheological properties of the glaze. In order to obtain a smooth surface, the glaze must be able to flow well at high liter weights. The main auxiliary additives used are;

Binders: Although certain glaze compositions with high clay content may exhibit sufficient green strength for handling, the majority of dry-applied coatings remain fragile and susceptible to damage during pre-firing processes. To improve the mechanical integrity of these coatings, the incorporation of binders or hardening agents becomes essential. Binders function as temporary adhesives, maintaining the adhesion of glaze particles to the ceramic surface until vitrification occurs during firing. Typically, binder content ranges from 0.5% to 3% by weight, with 0.5% being the most common. Various organic binders are employed for this purpose, among which cellulose ethers, water-soluble cellulose derivatives, such as CMC (Carboxymethyl cellulose), are the most widely used [18].

Deflocculants: Within a slip suspension, solid particles may exist either as individually dispersed entities or as loosely connected clusters known as flocs. According to Stokes' Law, larger particles or aggregated clusters settle more rapidly than smaller, well-dispersed ones. Consequently, achieving proper dispersion of particles within the slip is essential for maintaining suspension stability. This is typically managed through the use of substances known as deflocculating agents or electrolytes. These agents operate by introducing opposing charges similar to magnetic poles within the suspension, thereby preventing particle aggregation. The required amount of deflocculant varies depending on the targeted viscosity for the chosen application technique and the need to optimize the solid loading. Deflocculants are generally categorized into two groups: polyanionic and alkali cationic types. The former includes complex sodium phosphate compounds such as sodium tripolyphosphate, tetrasodium pyrophosphate, and sodium metaphosphate, whereas the latter consists of monovalent alkali salts [18,41,42].

Suspending Agents: When glazes are applied using an aqueous slip, the formulation must contain a certain proportion of colloidal substances characterized by their plate-like morphology and submicron particle sizes to ensure the stable suspension of components denser than water. Among these, clay materials serve as the primary suspending agents. Typically, kaolin, ball clay, or their combinations are incorporated into the mixture, with additions reaching up to 12% by weight [18,43].

8. Design Considerations for Glaze Compositions

Some features need to be taken into consideration when preparing the glaze composition. Melting: During firing, glazes melt and both react with the body, forming an intermediate layer and spreading over the body to create a covering layer. Melting is a desired feature in glazes, and the melting feature depends on many factors. The type of basic oxides in the glaze composition, the acid/base ratio of the glaze, the ratio between basic oxides and quartz, the ratio between Al_2O_3/SiO_2 and the ratio between B_2O_3/SiO_2 are the factors that determine melting in glazes [13]. Designing an effective glaze composition requires more than merely attaining the correct oxide formulation, although this remains a fundamental aspect. It is equally important to select raw materials that will undergo melting or dissolution within the specific time interval during which the glaze is exposed to peak temperatures in the kiln. This melting process must be carefully timed to occur only after gases released from the maturation of both the glaze and the ceramic body have escaped. When these conditions are properly accounted for, the resulting glaze surface is typically free of defects [11,15,44].

Expansion coefficient of glaze: The expansion coefficient is the change in the size (shortening-extension) of an object as a result of heating its unit length by 1°C. This change varies depending on the temperature. In ceramic glazes, the expansion coefficient is an important factor that shows the conformity between the body and the glaze. The expansion coefficients of the ceramic body and the glaze should be close to each other. If there is a large difference between the expansion coefficients of the body and the glaze, glaze errors occur. The expansion coefficient of ceramic glazes can be determined with a dilatometer or by the calculation method based on the Seger formula, chemical composition of the glaze. Each of the oxides used in the glaze composition has its expansion coefficient, and the expansion coefficient of the glaze can be calculated by using these coefficients [13,14].

Surface tension of a glaze: Surface tension is defined as the resistance that liquids exhibit against increasing and expanding their surface area. In other words, surface tension is the energy or work required to increase the surface area of a liquid. When a liquid is dropped onto a solid surface, the liquid partially spreads and then collects as droplets. This is because the surface tension of the liquid is high. As the surface tension of the liquid increases, the collection increases. When ceramic glazes are heated above their softening temperature, they become a liquid with increasing fluidity and spread on the body. In this case, if the surface tension of the glaze is much greater than the body, the glaze collects on the surface as drops. In industrial terms, the collection of the glaze on the body in this way is an undesirable situation and is seen as a glaze defect. Generally, the surface tension of ceramic glazes is 300 dyn/cm, and a glaze with a surface tension above this value collects on the body as drops. Therefore, except for special cases, there should be a harmony between the glaze and the body in terms of surface tension. The surface tension of a glaze can be calculated from the Seger formula or its chemical composition [13,45].

Viscosity of glaze: Viscosity is the resistance of any fluid to movement. In other words, it is a term that expresses the internal friction between the liquid layers during flow. Ceramic glazes melt during firing and exhibit a viscous property. The viscosity property in glazes varies depending on the glaze composition and firing temperature. If the viscosity of a glaze is too low, the glaze flows from the upper areas of the body to the bottom during firing and accumulates in certain areas. For this reason, the glaze layer thickness on the piece varies. In addition, if the viscosity is too high in glazes, it is difficult for the gases in the structure to escape during firing. This causes swelling, crater formation, and pinhole glaze errors on the glaze surface. These are undesirable situations, and the viscosity of the glaze applied to the body must be adjusted [13,46].

9. Glaze Types Applied to Different Ceramic Products

Ceramic tile glazes: Glazes with very different compositions are used in ceramic tile factories for very different decoration features. Different glazing methods for the tile industries are presented. The compositions of the glazes vary according to the optical properties of the desired glaze and the type of tile to be applied (wall tile, floor tile, porcelain tile). In general, in opaque glossy, transparent glossy glazes for wall tiles, over 90% frit by weight and close to 10% kaolin is used. In matte wall tile glazes, frit is generally used at a rate of 30-60% by weight. Compared to wall tile glazes, porcelain and floor tile glaze formulations typically contain a lower proportion of frit and involve more complex recipes that incorporate a broader range of raw materials. Although it varies according to the desired glaze type (opaque, matte, etc.), the oxide ranges that can be used in wall and floor tiles are given in Table 5 [10].

Table 5. Oxidation variation range in glaze composition for floor and wall tiles [47].

Oxide	The percentage by weight of oxides (%)
SiO_2	50-61
Al_2O_3	19-31
B_2O_3	10-13
CaO	4-7
Na_2O	5-8
The remaining portion of the composition typically consists of minor impurities, with each of Fe_2O_3, MgO, K_2O, and TiO_2 present at levels below 1%, and trace amounts—generally less than 0.1%—of ZrO_2, BaO, PbO, P_2O_5, and SrO.	

Ceramic sanitaryware glazes: In the manufacture of vitrified ceramic products, white glazes are predominantly employed. Nonetheless, colored ceramic glazes are also produced, albeit in significantly smaller volumes. Table 6 shows the compositional variations observed in white glazes used for vitrified ceramics [48].

Porcelain tableware glazes: The dipping method, which is extremely cheap, usable and does not require much equipment, has been used in the porcelain tableware industry for many years [12,49]. The dipping and robot glazing method is a more preferred method for glazing porcelain because it ensures that complex-shaped products are glazed well from all sides, and also requires less labor and is a fast method.

Table 6. Range of white glaze formulations applied in vitrified ceramics[48].

Raw material	Wt.%
Kaolin	5-15
High-purity sodium and/or potassium feldspar	20-40
Nepheline syenite	0-15
Quartz	15-30
Marble	0-22
Wollastonite	0-26
Dolomite	0-16
Magnesium carbonate	0-5
Barium carbonate	0-6
Talc	0-6
Zinc oxide	0-4
Zirconium silicate	8-15

Since the products that are biscuit-fired are suitable for dipping in shape, after being glazed in this way, the bottom parts are wiped and subjected to glaze firing. The density, viscosity, grain size distribution of the glaze, porosity of the material, thickness, working temperature and the experience of the worker are very important. The glaze that is taken into the tank in the factories is continuously circulated by passing through the sieve. The glazing process is carried out by dipping the products into the tank by the worker [12,50].

When the dipping method is applied with glazing machines, biscuit products are taken to the belts and automatically transferred to the booms determined in appropriate dimensions according to the product with the help of a vacuum. The arms where the booms are located are immersed in the glaze and rotated through the glaze until the end of the determined time to ensure the ideal glaze thickness, ensuring that the product absorbs the glaze. The reason for the rotation movement here is to ensure the spread of the glaze and faster drying. When the glaze is dry, the glazed products are taken from the vacuum heads and transferred to the kiln loading belt [2,6].

In the spray glazing method, spray guns are used. Biscuit products are placed on wire stands and passed through a path with many spray guns from various angles. When the path is completed, the product is taken from the machine and placed on the kiln cars after the feet are wiped [51].

10. Glaze Manufacturing Process

The conventional processing steps for preparing ceramic coating slips typically include the mixing of raw materials, particle size reduction, dispersion in water, elimination of undesired components, and the incorporation of minor additives to tailor the slip's properties.

Ball milling is widely regarded as the primary industrial technique for producing ceramic coating slips and is illustrated in Figure 2. These ball mills consist of sealed cylindrical containers rotating around a horizontal axis. Their sizes vary from small laboratory-scale units to large-scale production mills capable of processing several tons of slip. Industrial-scale mills are generally made from metal and internally lined with wear-resistant materials. To facilitate particle size reduction, the mills are charged with spherical or cylindrical grinding media. The interaction between the lining and the grinding media is optimized to minimize contamination of the slip during milling [1,11,14].

The reduction in particle size is achieved through the cascading motion of the grinding media, which impacts larger particles or agglomerates within the rotating mill. Although effective, ball milling is inherently energy-inefficient, making it essential to optimize variables that influence grinding performance. These include the material and design of the mill lining, the type and size of the grinding media, the proportions of media, ceramic batch, and water, the rotational speed of the mill, the initial particle size of the feed material, and the physical characteristics such as hardness and consistency of the input [52,53].

In wet milling operations, accurate control of the material load is vital. Empirical data suggest that optimal grinding conditions occur when the grinding media and associated voids occupy approximately 55% of the total mill volume. This occupancy should remain consistent across different mill sizes and types of grinding media. Meanwhile, the dry batch typically comprises 11% to 18% of the mill's volume, depending on the density and type of media employed. Higher-density grinding media can accommodate larger batch volumes due to their greater momentum during operation. Water content significantly affects grinding outcomes and usually constitutes 50% to 55% of the total slip volume [6,54].

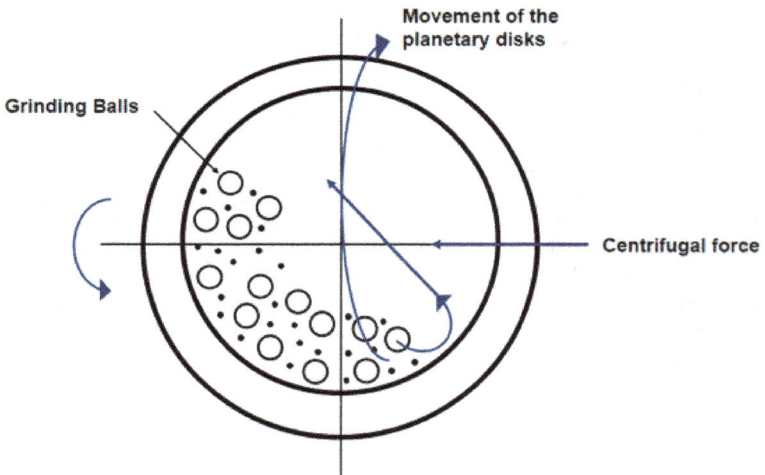

Figure 2. Remote Section of a commonly used ball milling bowl.

If excess water is added, the slurry becomes too fluid, and the dry particles may avoid proper contact with the media, reducing grinding effectiveness. Conversely, if water is insufficient, the slip becomes too viscous, impeding media motion and reducing the efficiency of particle breakage. Therefore, managing the specific gravity of the slip is a critical parameter.

The optimal specific gravity of the slip depends on the density of the dry materials. For instance, in a lead-free system with an average specific gravity of 2.5, a solid-to-water ratio of approximately 67:33 is suitable. In contrast, a high-lead oxide system with an average specific gravity of 4.5 typically requires a 79:21 solid-to-water ratio.

Materials Research Forum LLC
https://doi.org/21741/9781644903834

Mill speed is another key operational variable. As the mill rotates, the contents are subjected to gravitational, centrifugal, and frictional forces, which cause relative movement between the grinding media and the mill lining. At low rotational speeds, gravity dominates, causing the materials to remain at the mill's bottom. Excessively high speeds, however, result in the contents adhering to the mill wall due to centrifugal force, eliminating the relative motion necessary for grinding. The threshold at which this condition occurs is referred to as the critical speed of the mill, which can be theoretically calculated [3,55,56].

Critical speed (RPM): 21 to $30/(D)^{1/2}$ where D is the mill diameter in meters.

Empirical observations suggest that operating a ball mill at approximately 60% of its critical speed yields optimal performance. At this rotational speed, frictional forces elevate the dry batch materials and grinding media to a height from which they cascade effectively, promoting efficient particle size reduction through attrition as the slip-coated media tumble over one another [56–58].

In most cases, all raw materials are loaded into the mill prior to the commencement of grinding. For industrial-scale operations, milling durations typically range from 6 to 18 hours. In contrast, laboratory-scale milling is generally completed within one to two hours.

Particle size analysis serves as the primary control mechanism for evaluating milling effectiveness. Once the desired particle size is achieved, the slip is discharged from the mill. This may occur either by gravity or by pumping the slip into storage tanks for subsequent application. To ensure the removal of oversized particles and contaminants, it is standard practice to screen the slip through a 40- or 60-mesh sieve and to pass it through magnetic separators to eliminate residual iron oxide impurities [11].

11. Quality Control Tests for Ceramic Glazes

Liter weight: The liter weight of the glaze is measured using a 100 cm³ metal pycnometer with a tare of 200 g. The glaze is filled into the dry and clean pycnometer, and the perforated lid of the pycnometer is closed. After the glaze overflowing from the hole in the lid is cleaned, the pycnometer is weighed. The tare of the pycnometer is subtracted from the weighing result, and the remaining result is multiplied by 10 to be expressed in g/l [4].

The sieve residue (percentage): 100 g of glaze is poured onto a 45 μm vibrating sieve, and the glaze on the sieve is thoroughly washed with water. The material on the sieve is transferred onto aluminum foil or plates with the help of a pipette and dried in an oven. The dried material is weighed, and the obtained value is multiplied by 10 to find the sieve residue value in g/l [4,53,59].

Viscosity: Viscosity is the resistance of liquids to flow. For the ground glaze to be glazed homogeneously, the glaze must show a certain fluidity. Fluidity is inversely proportional to viscosity. If fluidity increases, viscosity decreases. If fluidity decreases, viscosity increases [4]. The most practical way to determine fluidity is with a flow viscometer. The process consists of determining the time required for the mud filled into a known volume chamber with a conical bottom and a known diameter outlet to be emptied. The viscosity, in other words, the flow time measurement of the prepared glazes is performed using a Ford-Cup with a 4 mm diameter hole. A thoroughly mixed glaze is filled into the Ford cup, the hole of which is kept closed. When the hole of the Ford Cup is opened, the chronometer (timer) is started simultaneously. The chronometer is stopped as soon as all the glaze in the container is emptied. The value read from the stopwatch gives the flow time of the glaze in seconds, or in everyday language, the viscosity value [4,24].

12. Recent Functional Trends in Ceramic Glaze Technology

Guided by technological advancements and evolving societal expectations, recent research has increasingly focused on developing innovative and functional glazed ceramic products that improve the quality of life while supporting environmental sustainability goals. In the tile industry, particular emphasis is placed on creating hygienic surfaces that reduce water consumption and on producing materials with superior mechanical strength, chemical durability, and inherent antibacterial and antifungal properties. In the tableware and sanitaryware sectors, self-cleaning surface technologies have become increasingly important, while photoluminescent glazes have emerged as a significant area of interest. These advancements encompass antibacterial, antimicrobial, and antifungal properties; self-cleaning capability and photocatalytic performance; enhanced mechanical and chemical stability; and photoluminescent behavior. [55,60–63].

13. Conclusion

The systematic investigation of ceramic glazes presented in this chapter has led to several key findings:

Compositional Understanding: A clear relationship has been established between glaze composition and the resulting physical and chemical properties. The roles of various oxides, particularly fluxes, glass formers, and stabilizers, have been identified as critical in tailoring glaze behavior.

Seger Formulation Efficiency: The Seger formula proves to be an effective tool for predicting and adjusting glaze properties, allowing for more controlled experimentation and optimization during formulation.

Raw Materials and Additives: The selection of high-purity raw materials and the strategic use of auxiliary additives are crucial for achieving desired glaze qualities such as gloss, opacity, or texture.

Process-Property Link: The production process, including firing conditions and application techniques, significantly influences the final glaze characteristics. Process control is therefore essential for consistent product quality.

Functional Properties in Modern Applications: In contemporary ceramic manufacturing, glazes are expected not only to serve decorative purposes but also to provide functional benefits such as chemical resistance, anti-slip surfaces, photocatalytic activity, and antibacterial effects.

Based on these findings, the following recommendations are proposed for researchers and industry practitioners:

Holistic Formulation Approach: Glaze design should integrate both compositional and process-related parameters to meet specific performance targets.

Functional Glaze Development: Greater emphasis should be placed on developing glazes with multifunctional properties to meet the demands of emerging applications, particularly in architectural, sanitary, and technical ceramics.

Standardization of Testing Methods: The implementation of standardized testing protocols for glaze performance will enhance quality control and facilitate cross-comparison among studies.

References

[1]   L. Koroglu, C. Peksen, M. Ince, E. Ayas, Cost-effective and eco-friendly transparent ceramic glazes for porcelain bodies, J. Aust. Ceram. Soc. 59 (2023) 1263–1269.

[2]   T. Cao, W. Dong, Z. Chen, B. Li, Q. Bao, Z. Hu, P. Li, X. Gu, In-situ preparation of titanic-based superhydrophilic self-cleaning glaze via a simple one-time firing route, Ceram. Int. 50 (2024) 32026–32038.

[3]   Z. Bayer Ozturk, Y. Karaca, E. Ubay, Enhancing thermal properties and surface quality of lappato glazed porcelain tiles through milling time optimization, J. Therm. Anal. Calorim. 149 (2024) 7279–7287.

[4]   H. Ergin, Y. Yıldırım, A. Çırpın, H. Turgut, K. Kayacı, Technological and economical investigation of glaze preparation using dry stirred media mill, Physicochem. Probl. Miner. Process. (2023). https://doi.org/10.37190/ppmp/166261

[5]   V. Dotsenko, I. Ukrainets, D. Tebiev, V. Strelkov, L. Dyadyuchenko, G. Krapivin, Synthesis, plant growth regulating activity and herbicide antidote activity of new pyrano[2,3-d]pyrimidines, in: Proc. 20th Int. Electron. Conf. Synth. Org. Chem., MDPI, Basel, Switzerland, Switzerland, 2016: p. a048. https://doi.org/10.3390/ecsoc-20-a048

[6]   M. Ozidede, J.O. Ogbah, B. Erabo, A. Tejiri, Ceramic Tools, Equipment, And Material Handling: Safety Guide For Educational Studios And Workshops, (n.d.).

[7]   J. García-Ten, M. Dondi, J.V.M.B. Vieira Lisboa, M. Vicent Cabedo, L. Pérez-Villarejo, E. Rambaldi, C. Zanelli, Critical raw materials in the global high-throughput ceramic industry, Sustain. Mater. Technol. 39 (2024). https://doi.org/10.1016/j.susmat.2024.e00832

[8]   B. Almamari, N. Alsaadi, Q. Alsalhi, Developing Low Temperature Glazes for Omani Clay Artworks: Laboratory Investigation to Reduce Costs by Using Frits, New J. Glas. Ceram. 13 (2023) 17–34. https://doi.org/10.4236/njgc.2023.132002

[9]   S. Ghosh, K.S. Pal, N. Dandapat, J. Ghosh, S. Datta, Glass-ceramic glazes for future generation floor tiles, J. Eur. Ceram. Soc. 33 (2013) 935–942. https://doi.org/10.1016/j.jeurceramsoc.2012.11.008

[10] M. Bustillo Revuelta, Ceramic Products, in: Constr. Mater. Geol. Prod. Appl., Springer, 2021: pp. 339–374.

[11] P. Colomban, Glazes and Enamels, in: Encycl. Glas. Sci. Technol. Hist. Cult., Wiley, 2021: pp. 1309–1325. https://doi.org/10.1002/9781118801017.ch10.6

[12] Ö. Yıldırım, F. Tazeoğlu Filiz, Çağdaş Seramik Sanatında Yeni Bir İfade Biçimi Olarak Yoğun Sırlama Ve Sırla Şekillendirme, Adıyaman Üniversitesi Sos. Bilim. Enstitüsü Derg. (2023) 272–292. https://doi.org/10.14520/adyusbd.1232894

[13] Y.S. Kaplan, C. Tasarımı, A. Dalı, G.S. Enstitüsü, M. Sinan, G. Sanatlar, Sanat

Seramiğinde Sır Kullanımına Farklı Bir Bakış 1, (2023) 14–27.

[14] L. Huasheng, S. Suhaily S, B. Qifu, S. Xiaoli, S. Lihua, Firing Process and Factors Affecting the Copper-Based Green Glaze of Jun Porcelain in Industrial Ceramic Production, Glas. Ceram. 81 (2025) 421–427.

[15] Q. Du, Application of Ceramic Glaze Based on Big Data, in: J. Phys. Conf. Ser., IOP Publishing, 2021: p. 42062.

[16] R. Casasola, J.M. Rincón, M. Romero, Glass–ceramic glazes for ceramic tiles: a review, J. Mater. Sci. 47 (2012) 553–582. https://doi.org/10.1007/s10853-011-5981-y

[17] E. Willhauk, R. Harikantha, Glass Ceramics for Household Appliances, in Low Thermal Expansion Glass Ceramics, Springer International Publishing, Cham, (2005): pp. 51–119. https://doi.org/10.1007/3-540- 28245-9_3

[18] T. Pradell, J. Molera, Ceramic technology. How to characterise ceramic glazes, Archaeol. Anthropol. Sci. 12 (2020). https://doi.org/10.1007/s12520-020-01136-9

[19] R. Sfez, S. De-Botton, D. Avnir, R. Wakshlak, Sol–gel glazes - a safe glass and ceramics coloring approach, J. Sol-Gel Sci. Technol. 102 (2022) 562–573. https://doi.org/10.1007/s10971-021-05699-4

[20] Y. Yoshimizu, H. Yasuga, E. Iwase, Quantification of Visual Texture and Presentation of Intermediate Visual Texture by Spatial Mixing, Micromachines 13 (2022) 255. https://doi.org/10.3390/mi13020255

[21] H. Norsker, J. Danisch, Glaze Formula Calculations. in: Glazes -for the Self-Reliant Potter, Vieweg+Teubner Verlag, Wiesbaden, (1993): pp. 135-148. https://doi.org/10.1007/978-3-663-06865-5_16

[22] W. Tasang, A. Artitkawin, S. Lawanwadeekul, S. Mulinta, SciTech Res. J. 7 (2024) 111–126.

[23] Y. Morito, A holistic approach to ceramic sculpture: Its history, theory, and materiality, Cambridge Scholars Publishing, 2022.

[24] E. Youssef, N. Mostafa, J.E. Khoury, T. Merhej, R. Lteif, Glaze surface defects causes and prevention controls, J. Ceram. Sci. Technol. 14 (2023) 1–10.

[25] H. Norsker, Glazes—for the Self-Reliant Potter: A Publication of Deutsches Zentrum für Entwicklungstechnologien—GATE. A division of the Deutsche Gesellschaft für Technische Zusammenarbeit (GTZ) GmbH, Springer Science & Business Media, 2013.

[26] K. Pekkan, E. Taşçı, Y. Gün, Development of temmoku glazes and their applications onto different tiles under industrial fast firing conditions at 1180 C, J. Aust. Ceram. Soc. 56 (2020) 489–497.

[27] G. Eramo, Ceramic technology: how to recognize clay processing, Archaeol. Anthropol. Sci. 12 (2020) 164.

[28] K. Pekkan, E. Taşçı, V. Uz, Production of metallic glazes and their industrial applications, J. Aust. Ceram. Soc. 51 (2015) 110–115.

[29] G. Topateş, B. Alıcı, B. Tarhan, M. Tarhan, The effect of zircon particle size on the surface properties of sanitaryware glaze, Mater. Res. Express 7 (2020) 015203. https://doi.org/10.1088/2053-1591/ab657d

[30] A. Çırpın, E. Yakar, Production and Characterization of ZrO2–Al2O3–K2O Based

Ceramic Whitening Agents, Glas. Ceram. 79 (2022) 194–201.

[31] P. Saha, D. Sarkar, Whiteware and Glazes, in: Ceram. Process., CRC Press, 2019: pp. 215–245.

[32] Z.B. Ozturk, B. Yildiz, Effect of alumina characteristics and concentration on thermal behavior, phase evolution, and aesthetic properties of tile glaze coatings, Glas. Phys. Chem. 42 (2016) 257–262.

[33] A.C. Kılıç, İngirgen (Redüksiyon) Ortamda Sırsız Seramik Pişirim Teknikleri Ve Uygulama Yöntemleri, Ankara Üniversitesi Güzel Sanatlar Fakültesi Derg. 5 (2023) 360–378. https://doi.org/10.58608/augsfd.1224148

[34] T. Kronberg, L. Hupa, The impact of wollastonite and dolomite on chemical durability of matte fast-fired raw glazes, J. Eur. Ceram. Soc. 40 (2020) 3327–3337. https://doi.org/10.1016/j.jeurceramsoc.2020.03.033

[35] J. Kim, D. Jeong, J. Choi, U. Kim, Changes in the glaze characteristics and moderate antibacterial activity of ceramic tile glazes with the addition of ZnO, J. Asian Ceram. Soc. 10 (2022) 241–252. https://doi.org/10.1080/21870764.2022.2038044

[36] J. Partyka, K. Gasek, K. Pasiut, M. Gajek, Effect of addition of BaO on sintering of glass–ceramic materials from SiO2–Al2O3–Na2O–K2O–CaO/MgO system, J. Therm. Anal. Calorim. 125 (2016) 1095–1103. https://doi.org/10.1007/s10973-016-5462-2

[37] H. Moradi, L.N. Nguyen, Q.-A.V. Nguyen, C. Torres, Glaze Epochs: Understanding Lifelong Material Relationships within Ceramics Studios, in: Sixt. Int. Conf. Tangible, Embed. Embodied Interact., ACM, New York, NY, USA, 2022: pp. 1–13. https://doi.org/10.1145/3490149.3501310

[38] Academic E-Books, Acad. E-Books (2017). https://doi.org/10.2307/j.ctt1wf4ds0.

[39] F. Andreola, L. Barbieri, F. Bondioli, M. Cannio, A.M. Ferrari, I. Lancellotti, Synthesis of chromium containing pigments from chromium galvanic sludges, J. Hazard. Mater. 156 (2008) 466–471. https://doi.org/10.1016/j.jhazmat.2007.12.075

[40] G. Costa, M.J. Ribeiro, J.A. Labrincha, M. Dondi, F. Matteucci, G. Cruciani, Malayaite ceramic pigments prepared with galvanic sludge, Dye. Pigment. 78 (2008) 157–164. https://doi.org/10.1016/j.dyepig.2007.11.004

[41] M.F. Gazulla, A. Barba, A.M. Orduña, Y. Bautista, Stability of ceramic glaze compositions. Correlation between partial dissolution and rheological properties. Part 1, Glas. Technol. J. Glas. Sci. Technol. Part A 53 (2012) 101–108.

[42] Q.-A. Dar, E.M. Schott, S.E. Catheline, R.D. Maynard, Z. Liu, F. Kamal, C.W. Farnsworth, J.P. Ketz, R.A. Mooney, M.J. Hilton, Daily oral consumption of hydrolyzed type 1 collagen is chondroprotective and anti-inflammatory in murine posttraumatic osteoarthritis, PLoS One 12 (2017) e0174705

[43] M. Kavanová, A. Kloužková, J. Kloužek, Characterization of the interaction between glazes and ceramic bodies, Ceram. - Silikaty 61 (2017) 267–275. https://doi.org/10.13168/cs.2017.0025

[44] P. Liu, F. Wang, H. Luo, J. Zhu, P. Shi, Y. Hao, B. Feng, Analysis of the influence of iron source and its occurrence state on the color of celadon glaze, Ceram. Int. 48 (2022) 18425–18432.

[45]   K.N. Sundari, D.C. Birawidha, H. Prasetia, I.W. Mudra, Y. Hendronursito, The high-hardness ceramic glazes based on basalt from Bali Province for ceramic body coatings, Physicochem. Probl. Miner. Process. 60 (2024).

[46]   B. Yıldız, Effect of particle size distribution on the properties of celsian based glazes, J. Aust. Ceram. Soc. 60 (2024) 1495–1504.

[47]   S.S. Anufrik, N.N. Kurian, I.I. Zhukova, K.F. Znosko, M. V. Belkov, Chemical Composition of Ceramic Tile Glazes, J. Appl. Spectrosc. 83 (2016) 764–770. https://doi.org/10.1007/s10812-016-0360-8

[48]   M.P. Desole, L. Fedele, A. Gisario, M. Barletta, Life Cycle Assessment (LCA) of ceramic sanitaryware: focus on the production process and analysis of scenario, Int. J. Environ. Sci. Technol. 21 (2024) 1649–1670. https://doi.org/10.1007/s13762-023-05074-6

[49]   E.I. Cedillo-González, P. Chierici, M. Buttazzo, C. Siligardi, E. Blasi, A. Ardizzoni, Correlating the physico-chemical properties of two conventional glazed porcelain stoneware tiles in relation to cleanability and sanitization, Mater. Today Commun. 34 (2023) 105191.

[50]   J.V. Owen, E. Adlakha, D. Carter, Crawling Glazes on Mid-Century Modern Maritime Canadian Studio Pottery: Shared or Re-Created?, Mater. Cult. Rev. 96 (2023) 63–82.

[51]   S. Chitwaree, J. Tiansuwan, N. Thavarungkul, L. Punsukumtana, Energy saving in sintering of porcelain stoneware tile manufacturing by using recycled glass and pottery stone as substitute materials, Case Stud. Therm. Eng. 11 (2018) 81–88. https://doi.org/10.1016/j.csite.2018.01.002

[52]   I. Celades, V. Sanfelix, A. López-Lilao, S. Gomar, A. Escrig, E. Monfort, X. Querol, Channeled PM10, PM2. 5 and PM1 emission factors associated with the ceramic process and abatement technologies, Int. J. Environ. Res. Public Health 19 (2022) 9652.

[53]   G. Boschi, Sustainability and environmental impact of the Italian ceramic tile industry, (2023).

[54]   M. Sisti, D. Guidetti, F. Altimari, F. Andreola, L. Barbieri, I. Lancellotti, L. Casini, F. Colombo, R. Arletti, R. Fantini, Sustainable glazes for ceramic tiles: Exploiting inertized man-made vitreous fibres waste as a resource, Ceram. Int. 51 (2025) 4195–4205.

[55]   O. Savvova, O. Yefimov, O. Tur, I. Zaitseva, O. Fesenko, Modern trends in the creation of luminescent silicate ceramic and glass materials for industry and technology, in: IOP Conf. Ser. Earth Environ. Sci., IOP Publishing, 2024: p. 12007.

[56]   C. Cuviella-Suárez, D. Borge-Diez, A. Colmenar-Santos, Production Line: Process and Energy Modeling, in: Water Energy Use Sanit. Manuf. Using Model. Process. Water Energy Account. Decarbonisation, Springer, 2021: pp. 43–95.

[57]   C.A.C. Vela, Islamic Ceramics: Trade and Technology in the Southern Garb Al-Andalus (X-XI Century), (2022).

[58]   C. Cuviella-Suárez, D. Borge-Diez, A. Colmenar-Santos, Planning for Energy and Water Management, in: Water Energy Use Sanit. Manuf. Using Model. Process. Water Energy Account. Decarbonisation, Springer, 2021: pp. 21–42.

[59]   G. Rubino, F. Trovalusci, S. Vesco, Development and characterization of a sustainable process for the implementation of lightweight ceramics, Ceram. Int. 49 (2023) 11474–

11484.

[60] E. Sánchez, J. García-Ten, V. Sanz, A. Moreno, Porcelain tile: Almost 30 years of steady scientific-technological evolution, Ceram. Int. 36 (2010) 831–845. https://doi.org/10.1016/j.ceramint.2009.11.016

[61] G. Han, N. Zheng, W.U. Junming, Ceramic art reality in the context of technology history, Cult. Reli. Stud. Engl. Version 9 (2021) 17.

[62] G.A. Khater, B.S. Nabawy, A.A. El-Kheshen, M. Abdel-Baki, M.M. Farag, A.G. Abd Elsatar, Preparation and characterization of low-cost wollastonite and gehlenite ceramics based on industrial wastes, Constr. Build. Mater. 310 (2021) 125214.

[63] B.A. Türkmen, Ş.K. Özbilen, T.B. Duhbacı, Improving the sustainability of ceramic tile production in Turkey, Sustain. Prod. Consum. 27 (2021) 2193–2207.

Materials Research Forum LLC

https://doi.org/21741/9781644903834

Chapter 10

Glazing Methods in Industrial Ceramics

Didem Oge[1], Ceren Peksen[2*], Hussein Elaibi[3], Farah Mutlag[3,4], Hiranur Taner[3], Cumhur Eren Isık[5], Fatih Sen[3*]

[1]Ondokuz Mayıs University, Institute of Graduate Education, Department of Art and Design, 55000 Samsun, Türkiye

[2] Ondokuz Mayıs University, Faculty of Fine Arts, Department of Ceramic and Glass, 55000 Samsun, Türkiye

[3]Sen Research Group, Department of Biochemistry, Kutahya Dumlupinar University, Kutahya 43000, Türkiye

[4]Ministry of Education, Karbala Education Directorate, Karbala, 56001, Iraq

[5]Department of Handicrafts, Kutahya Fine Arts Vocational School, Kutahya Dumlupınar University, Evliya Çelebi Campus, 43000 Kutahya, Türkiye

cpeksen@omu.edu.tr, fatihsen1980@gmail.com

Abstract

Ceramic glazes cover the surface of the ceramic body with various oxide mixtures that make up their composition, forming a layer similar to a glassy structure as a result of the firing process. Glaze protects the ceramic surface to which it is applied from environmental conditions such as moisture, dirt, and heat, and enhances the value of the ceramic product to which it is applied with its different colors and textures. The two primary purposes of glazing in ceramic products are to provide a hygienic and cleanable structure for porous and micro-rough surfaces and to give the structure an aesthetically pleasing appearance. Ceramic glazes are classified according to their composition, surface properties, and optical properties. Additionally, ceramic glazes can be classified based on their firing characteristics, such as single firing, double firing, or different firing atmospheres. Another classification is based on whether the glazes are used in industrial applications or in artistic ceramics. Industrial glazes are glazes used for various products manufactured in the ceramic industry. The most important properties sought in ceramic glazes used in the ceramic industry are compatibility with the substrate and resistance to cracking. The glazing process involves applying one or more layers of glaze with a total thickness of 75-500 µm to ceramic surfaces using various methods. The glazing process, which is applied to create a glassy surface during firing, varies depending on the properties of the products, the tools and equipment used, and the type of glaze. Glazing can be done manually, semi-automatically, or automatically, and may also vary depending on the type of ceramic product. In this study, the methods used in the glazing process of industrial ceramic products were examined.

Keywords

Industrial Ceramics, Ceramic Glazes, Industrial Glazes, Glazing, Glazing Techniques

1. Introduction

Ceramic glazes are mixtures obtained from ceramic raw materials of a specific particle size and composition, which form a glass-like structure on the ceramic surface as a result of the firing process, and the layer formed on the ceramic body [1–3]. The fundamental difference between glaze and glass is that glaze has a large surface area to volume ratio and is formed through interaction with the ceramic body and the atmosphere. Glass is essentially a closed system, and its chemical composition does not change significantly during firing. Glaze, on the other hand, is an open system, and its chemical composition changes during firing due to interaction with the ceramic and the evaporation of certain compounds. The interaction between glaze and ceramic body results in the diffusion of elements from the glaze into the body and from the body into the glaze. As a result, glazes are not homogeneous and exhibit composition gradients on the glaze surface due to diffusion [3–5]. Secret: A mixture of various oxides that melt as a result of thermal reaction and cover the surface of ceramic products, protecting the ceramic surface from moisture, dirt, temperature, and other environmental conditions, and enhancing the value of the ceramic product with which it is applied with its different colors and textures [1–5].

To achieve smoothness on ceramic surfaces, a glass base layer or a glazed ceramic base layer must be used. The formation of the necessary physical and chemical bonds between the glaze and the substrate, and the strength or weakness of these bonds for various reasons, are factors that determine the success of the glaze. Surface compression can also be achieved with glaze. The glaze composition should be selected to have a lower coefficient of expansion than the matrix material. For the glaze layer to be flawless, it generally needs to remain free of cracking and crazing in the ceramic body. However, in glazes produced for artistic purposes, such glaze defects or other variations are intentionally created [6,7].

There are two main purposes of glazing ceramic products. These are to obtain a more hygienic and easier to clean structure by covering surfaces that are generally porous and micro-rough with glaze, and to give the product an aesthetically pleasing appearance. The use of glaze in ceramic structures provides a smooth and glossy surface, protects against liquids and gases, enhances resistance to mechanical stress, improves electrical, chemical, and optical properties, preventing the leakage of harmful and toxic substances from the body during functional use, and inhibiting the growth of microorganisms on the surface, thereby facilitating easy cleaning. Adding aesthetic value to glazed bodies with different colors and textures, protecting decorations applied beneath the glaze, and creating a protective layer on bodies that exhibit colored firing are also among the functions of ceramic glazes [6,8–11].

In the classification of ceramic glazes, it is observed that different characteristics are taken into consideration in the literature. In these classifications, glazes known by the names of ceramic products are classified according to their composition, surface characteristics, and optical properties. It is also possible to classify ceramic glazes according to their firing characteristics, such as single firing, double firing, or different firing atmospheres. Another classification is based on whether the glazes are used in industry or in artistic ceramics. Industrial glazes are those used for the various products produced in the ceramic industry. Examples include porcelain glazes, wall and floor tile glazes, medical device glazes, and insulator glazes. Artistic glazes, on the other hand, are classified based on the artistic effects they produce on ceramic surfaces and the firing atmosphere. Artistic glazes include matte glazes, crackle glazes, gathering glazes, flowing glazes, crystal glazes, and aventurine glazes in an oxidation firing atmosphere, and lustrous glazes, Chinese red glazes, and celadon glazes in a reduction firing atmosphere [2,6,8,10].

The most important characteristics sought in ceramic glazes used in industry are compatibility with the body to which they are applied and resistance to cracking after firing. The firing temperature of the body and glaze must be compatible. In artistic ceramics, however, it may be desirable for the glaze to crack for artistic purposes. Such characteristics play a significant role in the classification of glazes. Before making a classification, it is important to consider various factors such as the raw materials used in the composition of the glazes, the product on which they will be applied, firing temperatures, firing atmospheres, and whether the glaze is opaque or transparent [8,12].

Glazes can be applied to ceramic surfaces as raw glazes or as compositions that have undergone fritting, either in dry form or as liquids with a specific water content. The raw materials and compounds that make up the glaze are very diverse, and some of the elements contained in these raw materials are acidic in nature and form glass when heated. Others are glass-forming elements and are generally basic in nature. Some compounds partially vitrify and vitrify. These are intermediate group elements that exhibit amphoter properties [3,10].

2. Secret Raw Materials

The raw materials that make up the glaze determine its behavior and properties. Glazes are composed of inorganic raw materials. Silica is the main component (e.g., glass-forming), but other elements such as fluxes (e.g., alkali, alkaline earth, boron, or zinc), opacifiers (e.g., zirconium or titanium), or pigments (e.g., iron, chromium, cobalt, or manganese) are also present. The raw materials used in glaze production can be categorized into three main groups [8,13]. Figure 1 illustrates raw materials in ceramic glaze production.

Raw materials that produce basic oxides are used as fluxes. They contain the chemical formulas RO and RO_2 (R is the name of the element, O is oxygen). The ratio, type, and relationship of basic oxides with other oxides directly affect the melting point of the glaze. To lower the melting point of the glaze, oxides with a high fluxing coefficient must be used. For example: PbO, CaO, ZnO, BaO, MgO, K_2O, Na_2O, Li_2O [14–16].

Raw materials that yield amphoteric oxides exhibit both acidic and basic properties. They contain R_2O_3 chemical formulas. Al_2O_3 is a general representative [17,18].

Raw materials that produce acidic oxides; they enable glass formation and provide strength. They contain RO_2 chemical formulas. SiO_2 and B_2O_3 are general representatives [19–21].

Silica is the main glass former and the only network former used in some frits and glazes. The higher the silica content in the glaze, the higher the firing temperature. However, a glaze with a high silica content has high resistance to mechanical stress and chemical attack. In low-temperature glazes, the ratio of silica to flow is approximately 2:1. For glazes at the highest temperatures, the ratio increases to 10:1 (mol). An increase in silica reduces the thermal expansion coefficient of the glaze and its fluidity at the glaze's maturation point. Some silica sands contain high amounts of feldspar minerals and metal oxides (alumina and alkali oxides) [15,16,22].

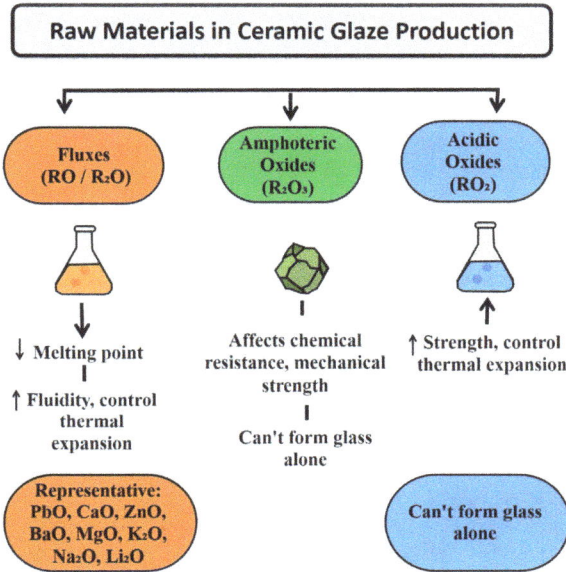

Figure 1. Raw materials in ceramic glaze production.

Alumina cannot form glass on its own, but it can be included in the network as an intermediate oxide. In glazes that do not contain silica, it is common for the composition to contain some aluminum. The presence of aluminum as a pure chemical or mineral component in a glaze or frit affects the chemical resistance, mechanical strength, thermal expansion, viscosity, and surface tension of the molten glaze, as well as crystallization and the behavior of the glaze surface [3,6,8].

Alkali metals generally lower the melting points of frits and glazes and increase their fluidity at firing temperatures, thereby enhancing the development of luster. The refractive indices of some frits can be increased by adding alkali oxides. The thermal expansion of glassy compositions is significantly affected by the inclusion of alkalis and is widely used to control the crack resistance of glazes. An increase in the amount of alkali oxides used in glaze compositions can increase chemical resistance and extend the vitrification range [3,23,24].

In the coloring of ceramic glazes, coloring oxides and various dyes are generally used. The coloring of ceramic glazes is influenced by the type and amount of colorant, the composition of the glaze to be colored, the kiln atmosphere, and the firing temperature. If any one of these factors changes, the desired color effect cannot be achieved in the coloring of the glaze [6,8,10,25,26].

Copper oxide (CuO, Cu_2O), iron oxide (FeO, Fe_2O_3, Fe_3O_4), cobalt oxide (CoO, Co_2O_3, Co_3O_4), chromium oxide (Cr_2O_3), manganese dioxide (MnO_2), tin dioxide (SnO_2), zirconium dioxide (ZrO_2), titanium dioxide (TiO_2), nickel oxide (NiO, Ni_2O_3), carbonates, phosphates, and selenium compounds are used [27–29]. Depending on the composition of the copper oxide glaze, it produces all shades of green and blue. In alkaline lead-free glazes, it produces Egyptian blue, and in boron glazes, especially boron-tin glazes, it produces turquoise. When a normal glossy glaze is saturated with copper compounds (8-25%), black matte metallic glazes are obtained [6].

Iron oxide is one of the most important coloring oxides found in varying amounts in all raw materials. It affects the glaze and body color. Depending on the kiln atmosphere and glaze composition, it produces colors ranging from yellow and red to brown, gray, and black. It is also used in the production of special artistic glazes such as celadon, aventurine, and temmoku [21,30].

In ceramic glazes, cobalt oxide, which can produce all shades of blue from light blue to dark blue, is more difficult to dissolve in the glaze if it is not ground very well because it is harder than other color-producing oxides. Using $CoCO_3$ instead of CoO makes dissolution in the glaze easier. In the black coloring of a transparent glaze, cobalt oxide is used in specific ratios with iron, chromium, and manganese oxides [3,6].

Chromium oxide, which generally produces a green color, turns gray and brown when used in increasing amounts in zinc glazes. Opaque surfaces can be obtained with chromium oxide, which has high refractory properties. The well-known chrome red is obtained in basic glazes with abundant lead in a neutral and oxidizing kiln atmosphere with chromium oxide [6,30].

Manganese compounds are used directly in glaze compositions in the form of oxides and carbonates, and are also found in pink, brown, and gray-black pigment recipes. They offer a variety of colors in ceramic glazes, depending on the composition. In lead-based glazes, they create a color scale ranging from beige to various shades of brown, while in alkaline and boron-based glazes, they form a color spectrum from pink and purple to brown [21].

The most distinctive feature of tin oxide is its opacity in glazes. In addition to its opacity, it is a very expensive oxide. Tin oxide also forms the basis of many ceramic paints. When used with vanadium, it produces green colors, and when used with copper, it produces pink colors [6].

The most commonly used form of zirconium oxide for opacification is zirconium silicate. Zirconium-containing glazes vary from completely transparent to highly opaque, from completely glossy to completely matte, and can be used on all types of ceramic substrates. Zirconium silicate is an efficient and economical opacifier primarily used in glazes. Glazes prepared with zirconium oxide produce a blue color when combined with copper. To achieve hard, mechanically resistant matte white glazes, glazes containing zirconium are supplemented with over 20% talc [3,6].

The primary role of titanium in frits and glazes is to increase acid resistance. The most notable feature of titanium oxide is that it mattifies and crystallizes glazes. Due to these properties, it is generally preferred in artistic glazes. In cobalt glazes, colors range from gray-blue to green; in copper glazes, colors range from yellow to blue; and in chromium glazes, colors are dirty gray. In manganese glazes, titanium oxide causes the color to turn gray [6].

Nickel oxide or nickel carbonate is added to glazes. The addition of $NiCO_3$ to matte glazes is more advantageous in terms of its easy solubility. In low-lead glazes containing zinc oxide, the presence of alkalis at temperatures above 1060°C, combined with the addition of nickel compounds, produces colors ranging from pink to moss green and blue. Nickel, which has a high surface tension, causes the same phenomenon in glazes when added in excessive amounts and tends to accumulate in the glaze [6].

Table 1 is a summary of Section 2 and illustrates Raw Materials in Ceramic Glaze Production: Types, Properties, and Effects, providing an overview of the main types of raw materials used in ceramic glaze production, their properties and effects, and showing the relationship between different chemical categories and their role in determining the final glaze characteristics.

Table 1. Raw materials in ceramic glaze production: types, properties, and effects.

Category	Representative Oxides & Compounds	Role / Properties	Special Notes
Fluxes (Basic Oxides, RO / R_2O)	PbO, CaO, ZnO, BaO, MgO, K_2O, Na_2O, Li_2O	Lower melting point, increase fluidity, control thermal expansion, can increase chemical resistance and refractive index	High fluxing coefficient oxides used for low melting point
Amphoteric Oxides (R_2O_3)	Al_2O_3	Affects chemical resistance, mechanical strength, viscosity, thermal expansion, surface tension, and crystallization	Cannot form glass alone; enters network as intermediate oxide
Acidic Oxides (RO_2)	SiO_2, B_2O_3	Main glass formers, increase strength, control thermal expansion, affect firing temperature	High SiO_2 = high firing temp, low expansion, high resistance
Silica (SiO_2)	Quartz, silica sands	Glass former, high mechanical & chemical resistance, controls thermal expansion & fluidity	Low-temp glazes: ~2:1 silica:flux (mol), high-temp: ~10:1
Alumina (Al_2O_3)	Pure oxide or minerals	Improves resistance, viscosity, surface tension, crystal formation	Common even without silica
Alkali Oxides	Na_2O, K_2O, Li_2O	Lower melting point, increase luster, adjust thermal expansion, improve crack resistance	Alkalis can extend vitrification range
Colorants & Pigments	CuO, Cu_2O, FeO, Fe_2O_3, Fe_3O_4, CoO, Co_2O_3, Co_3O_4, Cr_2O_3, MnO_2, SnO_2, ZrO_2, TiO_2, NiO, Ni_2O_3, carbonates, phosphates, selenium compounds	Provide glaze colors, opacity, and special effects	Color depends on type, quantity, kiln atmosphere, firing temp
Copper Oxides	CuO, Cu_2O	Greens, blues, turquoise; in saturated form gives black matte metallic	Egyptian blue in alkaline lead-free; turquoise in boron-tin glazes
Iron Oxides	FeO, Fe_2O_3, Fe_3O_4	Yellows to blacks; special effects like celadon, temmoku	Atmosphere & composition dependent
Cobalt Oxides/Carbonates	CoO, $CoCO_3$	Blues; mixes with Fe, Cr, Mn for blacks	$CoCO_3$ dissolves more easily than CoO
Chromium Oxide	Cr_2O_3	Greens; gray/brown in zinc glazes; chrome red in leaded basic glazes	High refractory; opaque surfaces possible

Manganese Compounds	MnO₂, MnCO₃	Beige to browns (lead glazes); pink to purple/brown (alkaline/boron glazes)	Found in pigments as well
Tin Oxide	SnO₂	Opacifier; green with V, pink with Cu	Expensive; also base for ceramic paints
Zirconium Compounds	ZrO₂, ZrSiO₄	Opacifier, can be glossy/matte; blue with Cu; hard matte whites with talc	Economical; wide range of effects
Titanium Oxide	TiO₂	Acid resistance, mattifying, crystallization	Color shifts with Co, Cu, Cr, Mn
Nickel Oxide/Carbonate	NiO, Ni₂O₃, NiCO₃	Pink to moss green/blue in Zn-alkali glazes	High surface tension; excessive amounts cause pooling

3. Preparation of Secrets

The glaze to be prepared must be weighed according to the proportions of the raw materials in the recipe, and the mill in which this mixture will be ground must be selected to be large enough to accommodate the raw materials, water, and mill balls. In commonly used glaze production methods, it is necessary to mix the components, reduce the particle size, achieve homogeneous distribution in water, remove unwanted materials, and add chemical additives to change the physical properties of the glaze [3,8,31]. Figure 2 illustrates the Preparation of ceramic glaze.

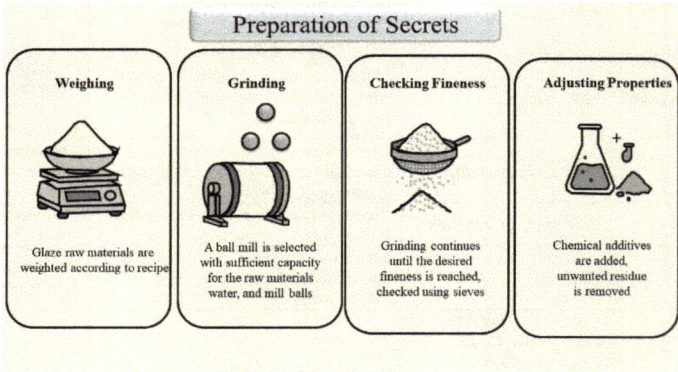

Figure 2. Preparation of ceramic glaze.

Ball mills are generally used in the grinding process, and grinding is carried out with water. The grinding process in mills continues until the glaze reaches a certain fineness. The desired fineness of the glaze is checked using sieves, and under normal conditions, the glaze should leave no residue on a sieve with 9,000-10,000 holes per cm², or the residue ratio should be no more than 0.1%. Over-grinding the glaze, or grinding it too finely, can cause certain important glaze and glazing defects. These defects include: glaze accumulation on the piece, glaze flaking off the piece in leaf-like patterns, and contamination of the glaze with other materials from the mill's abrasive components (mill inner walls and balls) in proportion to the length of the grinding time. In ball mills, the type of mill lining, the type and size of the grinding medium, the weight and ratio of each grinding medium, the glaze components in the mill, the mill speed, the particle size of the feed material, and the consistency and hardness of the feed content are characteristics that affect efficiency. The types of balls used for grinding glazes are: flint, porcelain, steatite, and high-density aluminum [3,6,32–34].

The specific gravity of the glaze plays an important role in the glazing process and firing. How the glaze moves on the ceramic surface, its adhesion, and the effects of adhesion affect the thickness and smoothness of the glaze. If the specific gravity of the glaze is high, it settles more quickly and becomes difficult to apply evenly. This can result in the formation of thicker or thinner glaze layers on the ceramic surface. On the other hand, if the glaze has a low specific gravity, it may not settle evenly and may tend to flow or drip during firing. This can affect the appearance and functionality of the glazed surface after processing. The grinding material ratios for glazes with different specific gravities are provided in Table 2. Glaze manufacturers typically formulate glazes within a specific weight range, and ceramic manufacturers can adjust the weight of the glaze by adding or removing water or adding another material to adjust the viscosity of the glaze. In general, understanding and controlling the specific gravity of ceramic glazes is important for ensuring that the finished glazed product has the desired consistent appearance and function [3,15,24,35–37].

Table 2 *Water and grinding material ratios for clays with specific gravities of 2.5 and 4.5 [3].*

Specific gravity	Water %	Grinding material %
2,5	32,8	67,2
4,5	21,4	78,6

One of the most significant errors that may occur during grinding in glaze mills is the collapse of the glaze. This error, which is most commonly observed in glazes containing alumina, can be explained by the "hydrolysis" of alkalis. As a result of this hydrolysis, the glaze transforms into an alkaline-reactive liquid, leading to its inevitable precipitation and hardening. As a preventive measure, it is recommended to grind the glaze with less water and in a shorter time during the initial stage. Additionally, the glaze must contain a high proportion of kaolin and clay additives. However, the ratios of these additives must not exceed the limits specified in the Seger formula. Some organic surfactants, such as dextrin, cellulose, or other organic adhesives, can also help prevent the problem when used at a maximum ratio of 2-3%. Weak acids like vinegar or substances like ammonium chloride and ammonium oxalate can dilute the glaze while also helping it maintain a better suspension. A more practical solution is often applied to prevent glaze settling, and by adding a small amount of bentonite, settling can be prevented both during grinding and during storage before use [6,38–40]. At the same time, it is very important to control both the viscosity of the glaze at the time of application and the viscosity formed by the heat effect inside the kiln. These two parameters can be controlled by the chemical composition of the glaze, particle size, and auxiliary chemicals used. To ensure that the prepared glazes consistently maintain the same properties, there are rules that must be followed at every stage of preparation. First and foremost, the glaze recipe must be weighed accurately and completely. However, the properties of the weighed glaze raw materials must be continuously monitored to ensure that they are of consistent type, quality, and particle size.

The selection and characteristics of the mill used to grind the glaze also affect the success of glaze preparation. The mill used to grind the glaze should not be changed unless absolutely necessary, and the weights of the raw material, water, and mill balls, which are defined as mill fillings, must always be kept the same. Considering the importance of the relationship between water and the ground raw material, the density of the prepared glaze must be kept under strict control. An areometer, a device used to measure density, is also used to measure glazes and determines the density of glazes in terms of Baumé degrees. Controlling the specific gravity of the glaze is also the most commonly used method for controlling and determining the density of the glaze. Prepared glazes may need to be passed through magnetic separators before use. This precaution is most commonly taken for white, transparent, or light-colored glazes. Free iron particles that mix into the glaze during all preparation stages are separated from the glaze and retained in special electromagnets called magnetic separators, aided by the glaze's suitable flowability [6,41–43]. Before applying glazes, that is, before starting the process known as "glazing," certain precautions must be taken. Most of these precautions begin with checking the glaze while it is still in the preparation stage. In order to maintain a successful glazing process, the water content of the glaze must be carefully adjusted. Pieces glazed with an excessively watery glaze become thin, while those glazed with a less watery, darker glaze become thick, and these differences can lead to visible defects. The pieces to be glazed must be prepared and free of defects before the glazing process. In general, pieces with thick walls can be glazed without biscuit firing. However, thin pieces must be biscuit-fired to the appropriate temperature according to the type of ceramic technology in order to gain resistance before glazing. As a result of frequent changes in biscuit firing temperature, underfired pieces absorb more glaze than normally fired pieces. The most significant error observed in overfired pieces is the thin glazing and prolonged wetness of the piece due to the glaze not being absorbed. If the surfaces of the pieces to be glazed are not kept dust-free, grease-free, and dry, glazing defects such as clumping and failure to absorb glaze may occur. Another defect that occurs during glazing is that pieces with very thin walls quickly become wet and expel water during glazing. To prevent this defect, which is often only noticeable after firing, the first surface of thin pieces should be glazed first, and the second surface should be glazed only after ensuring that the first surface is completely dry. Other precautions include not keeping the glaze's specific gravity too low and, if necessary, lightly wetting the glaze or the piece [6,8,15,44].

4. Sorting Techniques

Glazing is the process of applying one or more layers of glaze with a total thickness of 75-500 μm to ceramic surfaces. This method, which is used to create a glassy surface during firing, varies depending on the tools used, the type of glaze, and the methods employed. During firing, the glaze slowly melts at every point, spreading over the body to form a glassy structure. The hardening of the glaze occurs as the temperature is lowered, causing it to solidify. In ceramic technology, glazing methods vary not only due to these differences but also depending on the characteristics of the products. This process is performed to impart a range of technical and aesthetic properties to the fired ceramic product, including water resistance, cleanability, luster, color, texture, and chemical/mechanical durability [45–47].

Glazing methods in ceramic technology vary depending on the characteristics of the products. These include the thickness of the product, whether biscuit firing has been carried out, the size and shape of the piece, and whether it is a single piece or mass-produced. Glazing can be done manually, semi-automatically, or automatically, and may also vary depending on the type of ceramic product. Glazing methods according to ceramic product types are provided in Table 3.

Table 3. *Glazing methods according to ceramic product type [6].*

Ceramic Product	Product Condition	Glazing Method
Porcelain tableware	Biscuit (900- 950°C)	Dipping, Spraying, Pouring
Porcelain insulator	Raw	Dipping, Spraying, Electrostatic
Vitreous china tableware	Sintered, Biscuit, Raw	Dip, Spray
Vitreous china sanitary ware	Raw	Spray, Electrostatic
Floor and wall tiles	Biscuit (900-1000°C)	Slip, Spray
Glazed ceramic	Raw	Dipping, Pouring, Spraying
Akçini	Biscuit (Max. 1150°C)	Dipping, Spraying, Pouring

4.1 Immersion method

The dipping method is one of the simplest, most efficient, fastest, and most effective methods used in glazing, producing a smooth layer on the product. In this method, ceramic products are dipped into the glaze, moved in a controlled manner, and then removed and drained so that the glaze forms a film layer on the product. The duration the product remains in the glaze and the glaze's density directly affect whether the glaze is thick or thin. When measured with a boumetre, the glaze's density should be between 45-50 boumes; when measured by liter weight, it should be between 1450-1500 g/l. The time the biscuit remains in the glaze varies between 5-15 seconds, depending on the density of the glaze [3,6,48].

In the dipping process, the surface texture of the fired glaze is affected by many interrelated factors. These include liquid density, glaze viscosity and thixotropy, glaze grinding fineness, the porosity and surface texture of the product to be glazed, the thickness of the product, the dipping time, and user experience [3,37,49].

Application of the method: The surface of the product is cleaned, lightly moistened if necessary, and prepared for glazing. Areas on the product surface where glazing is not desired can be covered with hot paraffin or water-based paraffin emulsion using a brush to prevent glazing. The product is dipped into the glaze with an adjustable density using tongs or by hand. Application speed is an important variable in the dipping process, and the product is left in the glaze for 10-15 seconds before being removed. If a thick glaze is applied, it may flow due to the kiln's firing regime, causing buildup at the bottom and posing a risk of adhesion to the kiln's refractory plates. Finally, the product is prepared for firing by applying a glaze touch-up [6,46,50].

Depending on the thickness of the ceramic product, this method can be applied after biscuit firing in thin structures and after drying in thick products. Glazing by dipping can be done automatically with machines in industry (tableware production) as well as by hand. A method called automatic dipping is also used in porous products. In this application, the product is placed evenly along an inclined support and dipped into the glaze tank using a moving belt. In the first stage, the back surface of the product is glazed, and after cleaning the feet, the other side of the product is transferred to another set and subjected to the glazing process. The drying and placement processes are carried out in a specific order to complete the glazing process [3,6,10].

4.2 Spray method

It is the most economical glazing method. With this method, the glazing process can be applied by spraying a liquid glaze mixture onto dry or biscuit products, forming thin or thick coatings on the surface. Among the advantages of spraying are the simplicity and portability of the equipment, high working speed, ease of process control, and the ability to fully automate the process. Surfaces with variable porosity and large, complex products can be glazed evenly [6,10,51,52].

The glaze is sprayed onto the product using special spray guns called pistols. The pistol allows the amount of air and pressure at the tip to be adjusted to optimize the glaze output and thickness on the body. The density of the glaze should be measured with a boumetre and adjusted to 60-65 boume, while the air pressure of the compressor should be adjusted to 7-8 bar. The nozzle opening of the spray gun is adjusted according to the desired spray density, the glaze is filled into the spray gun reservoir, and the product placed in the glazing booth is subjected to the glazing process so that it is evenly glazed on all sides. The glazed product is then retouched and prepared for firing [6,10,53,54].

For this purpose, both human hands and machines can be used. Machines spray glaze onto ceramic products according to a program prepared in advance by a glazing expert and can glaze all sides of the ceramic product flawlessly. Glaze loss due to excessive spraying and backfire are factors that must be taken into consideration. Some of the factors affecting glazing include: the nozzle opening of the spray gun, the width of the glaze cone, the amount and pressure of the spray air, the filtration of the glaze through appropriate screens, the density of the glaze, and the spray distance. The glaze quality of the product subjected to the firing process is related to the surface properties of the product. This situation is related to factors such as the density and viscosity of the glaze, the thixotropy of the glaze, the fineness of the glaze grinding, air pressure, air flow, glaze feed rate, spray gun nozzle opening, user expertise, vitrification degrees, product temperature, position and number of spray guns. In this method, the powdered glaze particles formed due to the spraying pressure pose a health hazard to humans. Therefore, glazing should be performed in glazing booths equipped with exhaust fans and water curtains [3,6,19].

4.3 Pouring method

Glazing is a classic method used for specific shapes or decorative purposes, such as large ceramic products. When there is insufficient glaze volume for the dipping method, glazing is preferred. In industry, glazing by pouring involves the continuous flow of glaze from the lower opening of a continuously fed reservoir or from a point near the center of a circular surface, forming a continuous film to create a thin glaze layer on horizontal products. Two pouring systems are used in industrial glazing for pouring. In the first system, the products to be glazed pass under a glaze reservoir that can flow a glaze curtain at a certain flow rate. In the second system, the glaze is poured onto a bell-shaped mold, creating a circular glaze curtain that flows evenly from all sides. The products to be glazed also pass under the circular glaze curtain here. To achieve successful glazing using the glazing method, the cross-section of the glaze curtain (glaze film) must flow with the same thickness everywhere [6,45,55,56].

Another application of the glazing method is centrifugal glazing. Biscuit-fired or unfired floor and wall tiles, raw bricks, and coarse ceramics such as tiles move on a belt at a steep or slight angle, and the glaze is sprayed onto the surfaces of the pieces by centrifugal force from a rotating disc system. The success of glazing depends on the glaze reservoir being well supplied, the glaze flowing evenly everywhere, the speed of the glaze belt being carefully selected, and the glaze density being at the desired levels [3,8,19,44,57,58].

4.4 Glazing method using brushes and similar materials

This method is preferred for inaccessible areas of ceramic surfaces where glazing methods such as spraying, dipping, or pouring cannot be applied. It involves applying a thin layer of glaze to the surface of biscuits or dry semi-finished products using a brush or similar material. The prepared glaze is placed in a container, and the brush is dipped into the container and applied to the ceramic surface in the desired amount. Important considerations during application include ensuring the surface is free of dust and foreign particles, and that the glaze thickness applied to the surface is uniform throughout. Glazing with a brush or similar tools is used for products that are too small to be glazed with other methods, test plates, and retouching areas that remained unglazed after previous firing processes. Since it is a manual application, the experience in controlling the brush, the fact that some glazes dry quickly, and determining the appropriate glaze thickness constitute the challenges of this method. This method is also applied to achieve an artistic appearance on ceramic surfaces [3,8,19,45,51,59].

4.5 Glazing in the rapid single firing method

This method, also known as monoporoza, is a technology in which tiles are shaped in a single pattern and continuously by dry pressing, glazing tape is used, and glazing is carried out in kilns with fast firing capabilities in the shortest possible time. In this method, tiles are pressed and dried to achieve a mechanical resistance of 30 kg/cm2. Then, a special primer is applied to the tiles, and glazing is performed using the spray method on the glazing band. Since the tiles are not subjected to firing, various decorative applications or screen printing can be applied to the tiles in certain quantities. Subsequently, the ceramics are fired at a temperature of 1250°C for 35-45 minutes in computer-controlled passage-type (narrow-section) monoporoza kilns. In this glazing method, glazes with high viscosity and softening points are preferred. This allows for the release of gases that may form during the firing process from the structure of the ceramics that are glazed without biscuit firing [6,60,61].

4.6 Glazing using the dry pressing method

First used in Italy, this method is known as "mono pressatura" and has been used most widely in the industry for ceramic floor and wall tiles. It is a single glazing-pressing method using a specially designed hydraulic press for the glaze and ceramic surface. The pressing of the tile and glaze takes place simultaneously. The ceramic/glaze unit is single-fired. Initially applied as an alternative to decoration for enhancing color and texture, it later became a unique surface provider in tiles with specially formulated and sized particle mixtures, offering both aesthetic design and wear resistance properties [3,19,62,63].

Dry powdered secrets brought to a certain particle size are pressed onto dry tiles in a second stage. The tiles are fired in narrow-section roller passage kilns using a single firing method [6]. The main advantage of this technique is the elimination of a separate sealing line. The seals produced using this technique have a strong interface and a reduced micro-bubble structure. This normally provides improved hardness and wear resistance. Seals are usually textured or have a granite effect and are 1 to 2 mm thick. The dry pressing method has been applied in the tile industry in recent years and has shown technological advancement in terms of energy savings, pollution control, health, and safety [19,20,24,64].

4.7 Salt brining method

The salt glazing technique is much more unique than other glazing techniques in terms of application. It is a process of coating the product to be glazed without the need for the long and meticulous stages required in the preparation of glazes. In this technique, known as "salt glazing," only ceramic products fired at high temperatures of 1200–1300°C can be glazed. Since the decomposition of salt in unfired ceramic products does not produce enough glass phase to form a glaze, this method cannot be used for glazing. In this method, glazing is performed in open-fire kilns, whose firing and operating principles have remained unchanged since the Middle Ages. These kilns use wood, coal, fuel oil, or gas as fuel. The glazing is created by the salt vapor formed at high temperatures inside the kiln. Salt (NaCl) is added to the kiln in a moist state when the kiln atmosphere begins to reach the sintering temperature of the ceramics. The salt is either added directly to the kiln or poured onto the fuel material used in the combustion chamber to ensure its entry into the kiln. This process cannot be performed at random intervals or quantities. The salt quantity is calculated as 1.5/2 kg per 1 m³ of kiln capacity. The salt addition temperature is set at 1250/1320°C for light-colored sintered ceramics (glazed tiles, gre, stoneware) and 1140/1200°C for red-fired ceramics. The salt added to the kiln at the specified temperatures reacts with the combustion gases and water vapor in the kiln atmosphere under the influence of heat, coating the surface of all ceramics in the kiln with a thin glaze layer. To ensure that the salt vapor is evenly distributed throughout the kiln, the kiln flue damper is closed to maintain a slight pressure inside the kiln. After creating a reducing atmosphere in the furnace atmosphere with an excess of combustible gases, the furnace is tightly sealed to prevent "reoxidation" (re-oxidation) [6,65,66].

4.8 Dusting method

Although not very practical, this technique used in coating metals with enamel, is a glazing method applied to ceramic products. In cases where the recipe contains water-soluble substances and glazing is not possible, ceramic glazes can be applied to ceramic products by sprinkling them dry. In this method, the materials in the glaze recipe are ground dry or, if the grinding process has been carried out, they are thoroughly mixed in their dry state and sieved, then sprinkled onto flat-surfaced products at the desired thickness. On sloped surfaces, organic adhesives with a fine structure, such as gum arabic or Arabic gum solutions, can be used to ensure that the powdered glazes adhere to the surface [6,18].

4.9 Electrostatic coating

This method, which has high start-up costs, has been used since the 1960s specifically for glazing medical devices and insulators. This technique was first developed for organic paints. It is a method preferred by modern businesses that produce in series and in large quantities. This system is created by connecting multiple spray guns and mounting them with a bracket that moves vertically with the structure. Ceramic bodies are rotated on semi-finished product molds to ensure that the glaze flow is applied to the entire surface. Technically, it involves using low air pressure to ensure that the glaze particles adhere to the product under an electric charge. For this reason, 100,000 volts of static electricity is applied to the ceramic glaze in the tank, causing the glaze molecules with opposite charges (+, -) to align with the same charges (+, +, etc.). Therefore, by applying 100,000 volts of static electricity to the ceramic glaze inside the tank, the glaze molecules with opposite charges (+, -) are brought to the same charges (+,+). The ceramic, which is made conductive by being wetted and creates an attractive field for electrically charged glaze particles, is glazed by the glaze particles being sprayed onto it using a spray gun in the glazing booth [2,6,67,68]. With this method, the sprayed particles cover the ceramic surface homogeneously. Due to the limited movement of the spray gun, glazes may not reach products with geometric depths, and this method is not suitable for such products. New adjustments are required for different models and each new Project [20,69,70].

5. Conclusion

Industrial glazes are used for various products manufactured in the ceramics industry and are classified as porcelain glazes, wall and floor tile glazes, sanitary ware glazes, and insulator glazes. Glazing methods vary depending on the characteristics of the substrate to which they are applied. The thickness of the product, whether biscuit firing has been performed, the size and shape of the piece, and whether it is a single piece or mass-produced are important factors in determining the glazing technique. Glazing can be done manually, semi-automatically, or automatically. In the dip glazing method, the most commonly used glazing technique, ceramic products are dipped into the glaze, moved in a controlled manner, and then removed and drained to form a film layer of glaze on the product. This method, preferred for glazing porcelain tableware, porcelain insulators, vitreous china tableware, fired ceramic, and enameled products, is a simple, efficient, and fast glazing technique. In the spray glazing method, which is the most economical glazing method, a low-viscosity glaze mixture is sprayed onto dry or biscuit products, allowing for the application of thin or thick coatings on the surface. This method, preferred for glazing porcelain products, floor and wall tiles, sanitary ware ceramics, vitrified tiles, and earthenware products, is widely used due to the simplicity and portability of its equipment, working speed, ease of process control, and the ability to fully automate the process. The pouring method is a classic technique used for specific forms or decorative purposes, such as large ceramic products. When the glaze volume is insufficient for the dipping method, the pouring method is preferred for glazing.

The brush glazing method is a manual technique used for areas of ceramic surfaces that are inaccessible to other glazing methods, such as spraying, dipping, or pouring. It is also used for products that are too small to be glazed by other methods, test plates, or areas that remain unglazed after previous firing processes. Glazing using the fast single-firing method, also known as mono-porosa, is a process where ceramic tiles are shaped in a single pattern and continuously using dry pressing, a glazing band is used, and glazing is performed in furnaces with the fastest firing capability in the shortest time. Glazing using the dry pressing method is a simultaneous glazing-pressing method where the glaze and ceramic surface are processed together in a specially designed hydraulic press. The main advantage of this technique is the elimination of a separate glazing line.

The salt glazing technique has a much more unique quality than other glazing techniques in terms of application. It is the process of coating the product to be glazed with glaze without the need for the long and meticulous stages required in the preparation of glazes. Although it is not very suitable for application, the powder coating method used in the enamel coating of metals is also a method that finds application in ceramic products. In cases where the recipe contains water-soluble raw materials and glazing is not possible, ceramic glazes can be applied to ceramic products by sprinkling them dry using this method. In electrostatic glazing, a method preferred by industrial establishments that produce in series and large quantities, there is a system connected to multiple spray guns and mounted with vertical movements of the structure. Ceramic bodies are rotated on semi-finished product molds to ensure that the glaze flow is applied to the entire surface. Technically, this method, which uses low air pressure to enable glaze particles to adhere to the product under an electric charge, is widely used in the glazing of industrial ceramic products.

Referances

[1]   E. Cuce, S.B. Riffat, A state-of-the-art review on innovative glazing technologies, Renew. Sustain. Energy Rev. 41 (2015) 695–714. https://doi.org/10.1016/j.rser.2014.08.084

[2]   R.A. Eppler, D.R. Eppler, Which Colors Can and Cannot Be Produced in Ceramic Glazes, in: 2008: pp. 281–288. https://doi.org/10.1002/9780470314340.ch30

[3]   T. Pradell, J. Molera, Ceramic technology. How to characterise ceramic glazes, Archaeol. Anthropol. Sci. 12 (2020) 189. https://doi.org/10.1007/s12520-020-01136-9

[4]   J. Molera, T. Pradell, N. Salvadó, M. Vendrell-Saz, Interactions between Clay Bodies and Lead Glazes, J. Am. Ceram. Soc. 84 (2001) 1120–1128. https://doi.org/10.1111/j.1151-2916.2001.tb00799.x

[5]   T. Kronberg, As a result, glazes are not homogeneous and show composition gradients over the glaze surface due to diffusion, 2020.

[6]   E. Gliozzo, Ceramic technology. How to reconstruct the firing process, Archaeol. Anthropol. Sci. 12 (2020) 260. https://doi.org/10.1007/s12520-020-01133-y

[7]   D.W. Richerson, W.E. Lee, Modern Ceramic Engineering, CRC Press, 2018. https://doi.org/10.1201/9780429488245

[8]   T. Pradell, J. Molera, Ceramic technology. How to characterise ceramic glazes, Archaeol. Anthropol. Sci. 12 (2020). https://doi.org/10.1007/s12520-020-01136-9

[9]   D.W. Richerson, The Magic of Ceramics, Magic Ceram. (2012). https://doi.org/10.1002/9781118392317

[10]   E. Gliozzo, Ceramic technology. How to reconstruct the firing process, American Ceramic Society, Westerville, Ohio SE - ix, 332 pages : illustrations (some color) ; 24 cm, 2020. https://doi.org/10.1007/s12520-020-01133-y

[11]   J.F. Shackelford, R.H. Doremus, Ceramic and glass materials: Structure, properties and processing, Springer US, Boston, MA, 2008. https://doi.org/10.1007/978-0-387-73362-3

[12]   E. Şölenay, S. Aygün, M. Sıratça, Seramik sanat eğitiminde sırlama ve pişirme yöntemleri el kitabı, Murat Kitabevi Yayınları, Ankara SE - viii, 115 sayfa : resim (bazıları renkli) ; 24 cm, 2009. https://doi.org/LK - https://worldcat.org/title/1037985096

[13]   P. Colomban, B. Kırmızı, G.S. Franci, Cobalt and associated impurities in blue (And green) glass, glaze and enamel: Relationships between raw materials, processing, composition, phases and international trade, Minerals 11 (2021) 633. https://doi.org/10.3390/min11060633

[14]   G. Simsek, O. Unsalan, K. Bayraktar, P. Colomban, On-site pXRF analysis of glaze composition and colouring agents of "Iznik" tiles at Edirne mosques (15th and 16th-centuries), Ceram. Int. 45 (2019) 595–605. https://doi.org/10.1016/j.ceramint.2018.09.213

[15]   T. Kronberg, L. Hupa, The impact of wollastonite and dolomite on chemical durability of matte fast-fired raw glazes, J. Eur. Ceram. Soc. 40 (2020) 3327–3337.

https://doi.org/10.1016/j.jeurceramsoc.2020.03.033

[16] J. Partyka, K. Pasiut, P. Jeleń, M. Leśniak, M. Sitarz, Comparison of the impact of the addition of three alkaline earth metal oxides BaO, SrO and ZnO on sintering of glass–ceramic glazes from the SiO2–Al2O2–CaO–MgO–Na2O–K2O system, J. Therm. Anal. Calorim. 138 (2019) 4341–4347. https://doi.org/10.1007/s10973-019-09022-w

[17] G. Monrós, M. Llusar, J. Badenes, R. Galindo, Sol-Gel ceramic glazes with photocatalytic activity, J. Sol-Gel Sci. Technol. 102 (2022) 535–549. https://doi.org/10.1007/s10971-022-05787-z

[18] L. Bloomfield, Colour in glazes, Bloomsbury Publishing, 2012.

[19] R. Eppler, Glazes and glass coatings, American Ceramic Society, Westerville, Ohio SE - ix, 332 pages : illustrations (some color) ; 24 cm, 1999. https://worldcat.org/title/43076811

[20] B.A. Jacoby, Sized Dry Glazes: A Decorating Alternative, in: 2008: pp. 166–172. https://doi.org/10.1002/9780470313916.ch17

[21] R. Hopper, The ceramic spectrum : a simplified approach to glaze and color development, 2nd ed., n, Krause, Iola, WI SE - 256 pages : color illustrations ; 28 cm., 2008. https://doi.org/LK - https://worldcat.org/title/1306353462

[22] P. Colomban, Glazes and Enamels, in: Encycl. Glas. Sci. Technol. Hist. Cult., Wiley, 2021: pp. 1309–1325. https://doi.org/10.1002/9781118801017.ch10.6

[23] M.B. Almendro-Candel, M.M. Jordán Vidal, Glasses, Frits and Glass-Ceramics: Processes and Uses in the Context of Circular Economy and Waste Vitrification, Coatings 14 (2024) 346. https://doi.org/10.3390/coatings14030346

[24] P. Saha, D. Sarkar, Whiteware and glazes, in: Ceram. Process. Ind. Pract., CRC Press, 2019: pp. 215–245. https://doi.org/10.1201/9781315145808-7

[25] R. Sfez, S. De-Botton, D. Avnir, R. Wakshlak, Sol–gel glazes - a safe glass and ceramics coloring approach, J. Sol-Gel Sci. Technol. 102 (2022) 562–573. https://doi.org/10.1007/s10971-021-05699-4

[26] B. Almamari, N. Alsaadi, Q. Alsalhi, Developing Low Temperature Glazes for Omani Clay Artworks: Laboratory Investigation to Reduce Costs by Using Frits, New J. Glas. Ceram. 13 (2023) 17–34. https://doi.org/10.4236/njgc.2023.132002

[27] C. Altundal, Characterization of Glazes of the 11-14th-Centurt Potsherds and Tiles from Comana Pontica in Tokat, Turkey, (2023).

[28] H. Ettoumi, M. Dammak, M. Toumi, Exploring the temperature-induced structural evolution and optical properties of an acid-based geopolymer incorporating CuO and Fe, Polym. Bull. 82 (2025) 2891–2909. https://doi.org/10.1007/s00289-025-05649-y

[29] D. Mahapatra, S. Panda, S.K. Singh, P. Singh, A. Kumar, M.R. Majhi, Optical, Mechanical, and Physicochemical Properties of Na2O-P2O5-SiO2 Based Tint Glass for Building Construction Applications, Silicon 15 (2023) 6851–6865. https://doi.org/10.1007/s12633-023-02542-z

[30] A. Ducksworth, A. Ducksworth, Pottery : From The Source Pottery : From The Source, (2024).

[31] HAN Guangda, ZHENG Naizhang, WU Junming, Ceramic Art Reality in the Context of Technology History, J. Cult. Relig. Stud. 9 (2021) 17. https://doi.org/10.17265/2328-

2177/2021.07.002

[32] J. Lv, Z. Wang, S. Ma, Calculation method and its application for energy consumption of ball mills in ceramic industry based on power feature deployment, Adv. Appl. Ceram. 119 (2020) 183–194. https://doi.org/10.1080/17436753.2020.1732621

[33] K. Kishore, M.K. Sinha, A. Singh, Archana, M.K. Gupta, M.E. Korkmaz, A comprehensive review on the grinding process: Advancements, applications and challenges, Proc. Inst. Mech. Eng. Part C J. Mech. Eng. Sci. 236 (2022) 10923–10952. https://doi.org/10.1177/09544062221110782

[34] Z. Bayer Ozturk, Y. Karaca, E. Ubay, Enhancing thermal properties and surface quality of lappato glazed porcelain tiles through milling time optimization, J. Therm. Anal. Calorim. 149 (2024) 7279–7287. https://doi.org/10.1007/s10973-024-13297-z

[35] M. Bailey, Glazes cone 6 : 12400C, A & C Black ; University of Pennsylvania Press, London, Philadelphia, Pa. SE - 128 pages : color illustrations ; 24 cm., 2001. https://doi.org/LK - https://worldcat.org/title/48417574

[36] B. Yıldız, Effect of particle size distribution on the properties of celsian based glazes, J. Aust. Ceram. Soc. 60 (2024) 1495–1504. https://doi.org/10.1007/s41779-024-01058-y

[37] H. Moradi, L.N. Nguyen, Q.A.V. Nguyen, C. Torres, Glaze Epochs: Understanding Lifelong Material Relationships within Ceramics Studios, in: ACM Int. Conf. Proceeding Ser., ACM, New York, NY, USA, 2022: pp. 1–13. https://doi.org/10.1145/3490149.3501310

[38] L.A. Dobrzański, L.B. Dobrzański, A.D. Dobrzańska-Danikiewicz, Additive and hybrid technologies for products manufacturing using powders of metals, their alloys and ceramics, Arch. Mater. Sci. Eng. 102 (2020) 59–85. https://doi.org/10.5604/01.3001.0014.1525

[39] L.C. Giannossa, I.D. van der Werf, Overview of materials in cultural heritage: Pottery, in: Chem. Anal. Cult. Herit., Walter de Gruyter GmbH, Berlin/Boston, 2020: pp. 3–6. https://doi.org/10.1515/9783110457537-001

[40] B. Jing Han, Development of Fire-Protective Water-Borne Intumescent Coating Incorporated With Rubberwood Ash for Steel, (2022).

[41] M. Sisti, D. Guidetti, F. Altimari, F. Andreola, L. Barbieri, I. Lancellotti, L. Casini, F. Colombo, R. Arletti, R. Fantini, A.F. Gualtieri, Sustainable glazes for ceramic tiles: Exploiting inertized man-made vitreous fibres waste as a resource, Ceram. Int. 51 (2025) 4195–4205. https://doi.org/10.1016/j.ceramint.2024.11.396

[42] M. Ajadi, S. Lawal, S. Babalola, Physicochemical Characterization of Cullet to Glaze for Environmental Sustainability and Entrepreneurial Development, Int. J. Environ. Sci. 6 (2023) 1–12. https://doi.org/10.47604/ijes.2113

[43] V. Lappalainen, Recycled Glass in Coloured Ceramic Glazes: Case Roska Ceramics, (2025).

[44] M.F. Islam, Characterization of Pin-Holing Defect and Its Preventive Measures in Tableware Ceramic Industry, (2019).

[45] K.J. Anderson, Ceramic Glazes, Thames & Hudson, SE -, 1993. https://doi.org/10.1557/s0883769400039130

[46] P. ÇALIŞKAN GÜNEŞ, Uzak Doğu Seramik Sırlarından Temmoku, Güzel Sanatlar Enstitüsü Derg. (2019) 214–221. https://doi.org/10.32547/ataunigsed.604759

[47] A.J. Bell, Ferroelectrics: The role of ceramic science and engineering, J. Eur. Ceram. Soc. 28 (2008) 1307–1317. https://doi.org/10.1016/j.jeurceramsoc.2007.12.014

[48] M.G.A. Brito, F.L.B. Amaral, C.P. Turssi, R.T.B. Hofling, F.M.G. França, Effect of chemical or mechanical finishing/polishing and immersion in staining solutions on the roughness, microhardness, and color stability of CAD-CAM monolithic ceramics, Acta Odontol. Latinoam. 36 (2023) 86–95. https://doi.org/10.54589/aol.36/2/86

[49] H.A. Adawi, M.M. Al Moaleem, N.M. Al Ahmari, M. Shariff, M.A. Qahhar, S.M.H. Muharraq, N.A. Alghazali, Assessment of color stainability of computer-aided design and computer-aided manufacturing (cad/cam) ceramic materials after hot and cold coffee immersion at different time intervals, Med. Sci. Monit. 27 (2021) e932745-1. https://doi.org/10.12659/MSM.932745

[50] A. Kelimeler, the Investigation of the Use of the Colemanite in Vitrification Instead of the, 27 (2019) 1–8.

[51] M. Kurt, M. Bankoğlu Güngör, S. Karakoca Nemli, B. Turhan Bal, Effects of glazing methods on the optical and surface properties of silicate ceramics, J. Prosthodont. Res. 64 (2020) 202–209. https://doi.org/10.1016/j.jpor.2019.07.005

[52] Z. Qian, M. QingLong, X. YongQian, G. Lin, The robot intelligent spraying glazing system for sanitary ceramics industry, in: J. Phys. Conf. Ser., IOP Publishing, 2020: p. 12028. https://doi.org/10.1088/1742-6596/1653/1/012028

[53] G. Acikbas, N. Calis Acikbas, E. Ubay, H. Karaer, The influence of varying Cu doping concentrations on the microstructure, phase evolution and surface wettability of ceramic glazes modified with nano Cu-ZnO, Appl. Phys. A Mater. Sci. Process. 130 (2024) 354. https://doi.org/10.1007/s00339-024-07546-z

[54] F. Güngör, B. Altun, Effect of Glaze Composition and Sintering Process on the Pyroplastic Deformation Behaviour of Bone China, Orig. Res. Artic. Seramik-Journal Turkish Ceram. Soc. 1 (2021) 6–13. https://orcid.org/0000-0002-2405-0358

[55] E. Türkel, F.R. Yayman, In-Mold Glaze and Slip Applications in Ceramics, Ceram. - Art Percept. 124 (2025) 142–149.

[56] A. Çamoğlu, A. Çelik, M. Alaf, Development of Glaze Compositions to Increase the Durability of Decorative Paints Used in Glazed Porcelain, Bilecik Şeyh Edebali Üniversitesi Fen Bilim. Derg. 11 (2024) 37–46. https://doi.org/10.35193/bseufbd.1218432

[57] M.J. Ribeiro, D. Tulyaganov, Traditional Ceramics Manufacturing, in: PoliTO Springer Ser., Springer, 2021: pp. 75–118. https://doi.org/10.1007/978-3-030-85776-9_3

[58] M.B. Er, an Archaeometrical Investigation on Provenance and Technological Properties of Seljuk Period Pottery From Komana (Tokat), (2020) 276.

[59] R. de Castro Albuquerque, B. Ometto Sahadi, C. Bosso André, F. Allen Rueggeberg, M. Giannini, Resistance of glaze application on indirect restorative CAD/CAM materials against abrasive toothbrushing wear, Int. J. Prosthodont. 37 (2023). https://doi.org/10.11607/ijp.8604

[60] K. da C. Lima, R.G. Vivanco, P.R.B. Rodrigues, A.L.P. Caetano, P. de S. Fernanda de

      Carvalho Panzeri, Long-term effect of firing protocols on surface roughness and flexural strength of lithium disilicate glass-ceramic, Braz. Dent. J. 34 (2023) 79–86. https://doi.org/10.1590/0103-6440202305496

[61] N.S.M.N. Izam, Z. Itam, S. Beddu, N.A. Razeman, A. Syamsir, M.H. Zawawi, N.M. Zahari, N.L.M. Kamal, D. Mohamad, The characteristics and production method between glazed and unglazed of ceramic and porcelain tiles by firing process: A review, in: IOP Conf. Ser. Earth Environ. Sci., IOP Publishing, 2023: p. 12056. https://doi.org/10.1088/1755-1315/1135/1/012056

[62] B. Ji, A.A. Alrayes, J. Zhao, Y. Feng, Z. Shen, Grinding and polishing efficiency of a novel self-glazed zirconia versus the conventional dry-pressed and sintered zirconia ceramics, Adv. Appl. Ceram. 118 (2019) 46–55. https://doi.org/10.1080/17436753.2018.1472904

[63] R.A. Al-Dabbagh, A.Y. Imam, K.M. Baik, M. Eldemellawy, A. El-Etreby, The effect of re-pressing and glazing on the optical properties of high-density micronized heat-pressed glass ceramic, Ceram. Int. 50 (2024) 37217–37224. https://doi.org/10.1016/j.ceramint.2024.07.113

[64] L. Taylor, The ceramics bible : the complete guide to materials and techniques, Chronicle books, 2011. http://ccl.worldcat.org/oclc/706025438

[65] V. Occari, H. Möller, C. Fenwick, P. Quinn, I.C. Freestone, M. Chaouali, P. von Rummel, The technology of polychrome glazed ceramics in Ifriqiya: new data from the site of Chimtou, Archaeol. Anthropol. Sci. 16 (2024) 61. https://doi.org/10.1007/s12520-024-01974-x

[66] D.E. Southan, The Science And Art Of Dental Ceramics. Vol. Ii. Bridge Design And Laboratory Procedures In Dental Ceramics, by J. W. McLean, Quintessenz Verlag, 1981. https://doi.org/10.1111/j.1834-7819.1981.tb04008.x

[67] S. Boxu, Development of Advanced Spectrally Selective Coating for The Application of Energy-efficient Windows, (2022).

[68] F. Russo, Innovative composite enamel coatings with improved abrasion resistance and mechanical properties, (2023). https://iris.unitn.it/retrieve/handle/11572/381529/654817/phd_unitn_Russo_Francesca.pdf

[69] A.S. Bhatt, Photocatalytic ceramic coatings, in: Adv. Ceram. Coatings Fundam. Manuf. Classif., Elsevier, 2023: pp. 187–195. https://doi.org/10.1016/B978-0-323-99659-4.00001-2

[70] S. Kreve, A.C. dos Reis, Effect of surface properties of ceramic materials on bacterial adhesion: A systematic review, J. Esthet. Restor. Dent. 34 (2022) 461–472. https://doi.org/10.1111/jerd.12799

Materials Research Forum LLC
https://doi.org/21741/9781644903834

Chapter 11

Glaze and Body Defects in Industrial Ceramics

Rayane Mahious[1,2*], Abdullah Seyrankaya[3], Ebru Halvaci[2], Selcuk Erdogan[2], Fatih Sen[2*]

[1]Department of Biology, Faculty of Art and Science, University of Dumlupinar, Kutahya, Türkiye

[2]Sen Research Group, Department of Biochemistry, Kutahya Dumlupinar University, Kutahya 43000, Türkiye

[3]Faculty of Engineering, Department of Mining Engineering, Dokuz Eylul University, Izmir, Türkiye

rayane.mahious@ogr.dpu.edu.tr, fatihsen1980@gmail.com

Abstract

Glaze and body defects create an important challenge in the industrial ceramics industry, as they directly impact on the quality of the product, both functionally and aesthetically. Many of these defects become apparent after the glazing process, although their causes are often linked to the former stages of preparation, shape or glazing application. Several defects are related to problems in the glaze layer, resulting in a mismatch in thermal expansion coefficients between glaze and ceramic body. Cracking and peeling, which occur because of poor adhesion between layers; pinholes and bubbles caused by gases, as well as crawling, which causes the glazed layer to retreat from parts of the ceramic surface. Cracks are among the most common defects that occur in ceramic bodies during the drying or firing stages, resulting from rapid drying or uneven shrinkage. The delamination, blisters, and black core negatively affect the surface's resistance to liquids and pressure. The severity of these defects lies in the fact that they reduce the life of the product and give rise to industrial rejection. Therefore, to reduce these defects, an intensive understanding of factors that affect each production phase, from the choice of raw materials, through formation and drought conditions, and from firing conditions and glass.

Keywords

Glaze, Defect, Cracking, Peeling, Pinholes, Delamination, Blisters, Black Core

1. Introduction

Ceramics consist of inorganic substances made from a blend of metallic and non-metallic elements; with their characteristics defined these elements combined. Ceramics are the most adaptable among all categories of materials [1–4]. Their adaptability stems from the chemical makeup of their bonds, which predominantly features strong ionic and covalent bonds in different ratios. These connections define several characteristics of ceramic materials, such as relatively elevated melting points, a significant modulus of elasticity, strong wear resistance, low thermal conductivity, great hardness, and brittleness associated with cohesion and limited ductility. They serve as effective electrical insulators owing to their absence of conduction electrons and the nature of their mixing, which leads to the formation of chemical bonds [2,3,5–7]. Ceramics find applications across various end users in the construction sector, including clay blocks and bricks, wall and floor tiles and sanitary fixtures; in consumer products, like cookware (glass ceramics), tableware (porcelain), or pottery; and in industrial manufacturing and processing, encompassing refractory materials, filters, and cutting tools, as along with advanced technological applications as elements in solid oxide fuel cells, nuclear ceramics for nuclear fuel, coating reactors and ceramic-based batteries. Their electrical characteristics enable them to serve as capacitors, high-temperature superconductors, piezoelectric and ferroelectric substances, insulators, variable resistors, and bases for integrated circuits. They also serve as infrared windows and are employed in the fabrication of lasers and high-pressure sodium lamps because of their optical characteristics. Ceramic armor, known for its great hardness and compressive strength, is utilized in armored vehicles and body armor to withstand bullet penetration. Alumina, boron carbide, silicon, and titanium diboride rank among the most commonly utilized materials [8–11]. Industrial ceramics are categorized into carbides, oxides, borides, nitrides, silicates, and glass-ceramics. Compared to metals and polymers, ceramics' capacity to endure elevated operating temperatures positions them as a superior engineering material [3,12]. Their rigidity and durability are like those of metals and are typically chemically inert. Many ceramics act as insulators for heat and electricity and are durable in harsh conditions. Many ceramics have qualities including high electrical resistance, excellent thermal resistance, increased chemical stability, strong corrosive resistance, exceptional wear resistance, temperature strength, and considerable hardness [3].

2. The Concept of Glazing

A glaze is a final coating layer that melts and permanently bonds with the clay surface through a crystalline structure during the firing process. Aluminum is essential in the glaze composition, serving as a stabiliser that ensures the glaze adheres strongly to the ceramic base [13–16]. Engobe is a layer that is occasionally applied initially on a substrate beneath the glaze layer in many ceramic tiles, to enhance the substrate's opacity and minimize physical–chemical differences, improve compatibility between the substrate and the glaze, and decrease surface defects in the ceramic material [14,17]. Depending on the glaze composition and the preferred outcome, the glazing surface finish can be shiny or dull, and it may be clear, solid, white, or tinted. By altering the ratios of components such as alumina and silica, various types of glazes and engobes can be created. For instance, increasing the alumina while decreasing the silica results in a matte glaze [14].

The main characteristics that characterize a glaze can be categorized into microstructures, such as bubbles and cracks and compounds. Within the former, microstructures may be primary (during firing/cooling) or secondary (alterations after burial), whereas the latter can be crystalline or amorphous. Non-crystalline state phase: denotes the varied melt formed during firing that incorporates specific elements from the ceramic material. The composition of the melt depends on the firing temperature and the existence of certain particles that stay undissolved [18].

Undissolved substances: these are the original glaze elements that have partially reacted, usually displaying rounded forms with surfaces that are dissolving. Coarse sand particles are common and can add to the opacity of glazes. Typical crack patterns are often observed around quartz grains due to the alpha-to-beta phase change at 573 °C [18,19].

Crystalline substances: these substances are created from elements that primarily turn insoluble as they cool down. Slow cooling encourages the formation of the crystallites. Some compounds have a higher tendency to crystallize compared to others. Crystalline materials may also form due to transformations taking place during burials. Bubbles: They are inherent to the glazes and originate from multiple sources. In the beginning phase of glaze melting, bubbles form from gases emitted by the decomposition of organic matter and carbonates, sulfates, and the removal of hydroxyl groups from clay minerals, or from air and moisture trapped in the glaze [18,20,21].

Bubbles are beneficial as their mixing and agitation promote the uniformity of the melt. The total removal of these bubbles can be accomplished depending on the glaze thickness, viscosity of the melt, the surface tension of bubbles, temperature and time of firing. At higher temperatures, fresh bubbles form from boiling (gases that change their composition and become insoluble or those whose solubility decreases with increased temperature), though distinguishing between the two types of bubbles is difficult. Crazing: tiny hairline cracks on the glazed surfaces resulting from the tensions that develop between the ceramic and the glaze as they cool. Crazing happens when the glaze contracts more extensively than the ceramic it is strongly attached to. A craze pattern can emerge immediately after removal from the kiln or even years later. Cracks encourage the breakdown of the coatings [18,22].

Crazing: fine hairline fissures on the glazed surfaces caused by the stresses that arise between the ceramic and the glaze during the cooling process. Crazing occurs when the glaze shrinks more significantly than the ceramic it is firmly bonded. A craze pattern may appear right after being taken out of the kiln or even years afterwards. Fissures promote the deterioration of the coatings [18].

Like glass, colors of the glazes are created by presence of transition metals, which are subjected to firing in either oxidizing or reducing environments. The color associated with the glazes results from elements (manganese, iron, cobalt, copper,). Iron oxidation states, coordination numbers (four, five, six), and spatial arrangement, of nearby atoms. The color arises from the metal ions absorbing light at specific wavelengths. Iron serves as the main colorant because it is the most common impurity present in rocks and can produce an array of colors. Fe^{2+} exhibits considerable absorption in the infrared, with a broad tail extending into the red and yellow, resulting in a blue tint for the glaze. Fe^{3+} shows multiple diminished absorption bands in the ultraviolet and blue areas, imparting a yellow tint to the glaze. Both are generally present in the glaze, with the Fe^{2+}/Fe^{3+} ratio affected by the firing atmosphere and temperature; elevated temperatures lead to a greater ratio. Totally eliminating iron from the glaze is generally impractical as it can be part of various clay mineral compositions; selecting materials that are naturally low in iron serves as the main approach to avoid iron's coloring effect. To achieve the desired coloring effect, reducing firing will create a blue tint, while an oxidizing firing will result in a yellowish/brown tone. For the same amounts of iron, the coloring effect of Fe^{2+} is far more important than that of Fe^{3+}; therefore, it was common to incorporate small quantities of other elements like manganese, arsenic, or antimony that engage in a redox reaction in which iron is oxidized [18,23].

3. Historical Development of Glazing

Glazes are stable glass-like coatings used on ceramic earthenware, which were historically created by cooling melted oxides or minerals applied to the surfaces of ceramic items. Initially, glazes represented a significant advancement for earthenware as they not only sealed the porous ceramic surface to prevent liquid evaporation but also allowed for a wide range of decorative options [18,24–26]. The initial glazes appeared around 3500 BC in Eastern Mediterranean regions, created by potters aiming to replicate the valuable blue stone lapis lazuli [13,18,26]. To achieve this, tiny beads were fashioned from steatite ($Mg_3Si_4O_{10}(OH)_2$) and subsequently covered with azurite or malachite powders, which are natural copper ores displaying blue and green hues, respectively. Upon firing, the coating reacted with the steatite to produce a thin film of colored glass. Subsequently, potters began to test various mixtures of crushed and ground stone blended with water to cover the surfaces of pottery. They thus found combinations that entirely coated the surfaces of their ceramic items with a waterproof glass-like layer. As time went on, potters discovered how to create glazes in various colors and textures, utilizing multiple firing cycles at different temperatures to achieve layered effects [13,26]. In Babylon during the second millennium BC, lead glazes were created. Lead served as a flux, enabling the glaze to develop at reduced temperatures. Pigments that fade at elevated temperatures can be subsequently utilised, leading to more vibrant and diverse hues. Lead glazes can be applied to pre-fired ceramic surfaces or over another glaze fired at a higher temperature, permitting extensive artistic expression [13,26,27]. In the 8th century BC, the artisans in Mesopotamia and Persia began using a different glaze additive: tin oxide. This additive produces a white opaque coating that can entirely conceal the brown or reddish hues of clay earthenware. Glazed earthenware was created to address the need for storing and transporting liquids and food. Subsequently, because of the aesthetic options provided by glazes, they began to be utilized as ornamental coverings for walls in the shape of tiles. A prominent and early instance of glazed tile usage in ancient Mesopotamia is the Ishtar Gate, located on inner wall of Babylon, constructed during rule of King Nebuchadnezzar II (in the 6th century BC). Tiles in blue, gold-plated, and reddish hues were utilized to create forms of both actual and mythical creatures [13,26].

4. The Functional and Artistic Aspects of Glazing

Glazes influence the look (hue and shine) of ceramics, but they also serve a purpose; they render ceramics waterproof, allowing them to hold liquids and food (kitchen and dining items) [28]. Glazes enhanced the worth of ceramic items and, by selecting colors and patterns, rendered them trendy and, in certain instances, transformed them into luxury products, thus becoming items for commerce [29]. As demand rose, they were replicated, sold for less, and eventually became widely used [30]. The analyses revealed the artefacts, originating from Anatolia, Caucasus, Iran, and Central Asia, had black outlines to separate colored areas and date back from 12th to the 19th centuries. Three categories of glaze were identified according to their silicate network nanostructure: a lead-heavy glaze like Byzantine, Zirid, Hafsid, and Andalusian artefacts, an alkali-lead glaze typical of Ottoman Iznik-Kutahya pieces, and a mixed alkali glaze prevalent in Safavid creations. The micro composition of glaze was determined by color [31]. Moreover, every area and governing dynasty created a distinctive style or hallmark of glazed pottery, which, alongside their prevalence in archaeological sites relative to other materials, renders them especially useful for dating .Research on glazes yields details regarding their chemical makeup, microstructure, firing methods, and the form and design of decoration. The properties of the glazes are directly linked to recent technical and scientific developments as well as trends. Additionally, examining glazes provides indirect insights into the development of technical skills, the exchange of specific materials unavailable locally, the presence or lack of connections between regions, and emergence and acceptance of new trends. Acquiring new technical skills and ensuring access to raw materials are essential for effectively developing a product. The acceptance of a new glazed ceramic that resembles an imported glazed item is not inherently linked to utilizing the same production techniques or materials. The acceptance and reproduction of new patterns and shades frequently lack precision, making them distinguishable from the original [18,32].

5. Transparent Glazing: Characteristics and Industrial Challenges

A transparent glaze is among the most essential glazes in the realm of ceramic tiles. It is commonly used on the decorative layer to provide an outstanding decorative appearance while guaranteeing a smooth and easy-to-clean surface. Consequently, effectiveness of transparent glaze is essential for ceramic tiles [16,33–35]. A transparent glaze primarily consists of an amorphous glass phase, that exhibits weak mechanical properties and low chemical stability, which restricts longevity and use of the tiles (e.g., inadequate for extremely high traffic locations like airports or squares). Therefore, enhancing the characteristics of transparent glaze, particularly its mechanical properties, is an essential priority for the tile industry. Glass-ceramics are polycrystalline substances produced by managing the nucleation and crystallization of the original glass, resulting in outstanding mechanical characteristics, optical features, and chemical resistance. A significant use of glass-ceramics has been creating glass-ceramic glazes featuring high hardness and wear resistance to enhance the functionality of conventional glazes [33,36–38]. Yekta et al. enhanced the hardness of the glaze by forming gahnite ($ZnO-Al_2O_3$) and a β-quartz solid solution within a $ZnO-Al_2O_3-SiO_2-ZrO_2$ framework [39]. Ghosh et al. created a glass-ceramic glaze based on mullite-cordierite within the $CaO-MgO-Al_2O_3-SiO_2$ framework, achieving a maximum hardness of 8.38 ± 0.07 GPa for the glaze [40]. Cai et al. created a spinel glass-ceramic glaze featuring high opacity and hardness, serving as a substitute for the conventional zirconium opaque glaze. Nonetheless, the majority of studies have concentrated on mechanical characteristics instead of optical features, leading to the scarcity of transparent glass-ceramic glazes [41,42].

6. Financial Factors and Their Influence on Glazing Standards

Pre-prepared glasses are readily available, but they are notoriously costly. Many ceramic producers and artists, across various educational and industrial stages, depend on imported materials instead of local resources, leading to elevated production expenses. A significant ongoing discussion in ceramics and pottery research is the formulation of glaze recipes in artists' studios to create innovative pottery [43–45]. Nevertheless, minimal debate has emerged concerning the inclusion of clay in ceramic glaze formulations. To create glazes in a separate lab at a reduced cost, it is essential to minimize variety of materials incorporated in ceramic glaze formulations [46,47]. For these glazes to be suitable for ceramic arts and crafts, utilizing professional glaze materials is essential. The Oxford English Dictionary (OED) states that the word "frit" was initially used in 1662 to describe a calcined blend of sand and additives prepared for melting in a crucible to produce glass [48]. Frits are, in fact, elements in glaze formulations that are melted and regrinded before being added to the glaze mixture [47–50].

6.1 The Definition of Frits

Frits are fundamentally raw materials that are combined, heated, liquefied, pulverized, and ground into fine powder. Employing frits can decrease the harmfulness of raw materials like lead, barium, and zinc; reduce glaze's melting point; and prevent evaporation of unstable substances [51]. Frittes are produced through this process: they are combined with silica, a type of glass, to guarantee that the producer provides the essential component for creating glazes. The raw material is subsequently melted into a glassy form, which is insoluble, and in the end, the fruit is milled to create a powder for coating. Due to the varying use of ferrite among different ceramic manufacturers, ferrites are classified into four types: lead ferrite, lead-boron ferrite, clear-coated borosilicate borax ferrite (without lead), and white-solid borosilicate ferrite (without lead). Ceramic artisans utilize ferrites as replacements in various fundamental glaze formulations. In other terms, every ferrite denotes the basic standard components of glaze, comprising silica, alumina, and flux [49,52]. While ferrites can present difficulties for ceramic producers relying heavily on raw glazes, they have become crucial due to their ability to address numerous challenges and enhance glazed ceramic items. Although ferrites are undeniably more costly and premium compared to raw glazes, their benefits frequently surpass the production expenses. Numerous reasons for employing ferrites as an alternative to raw materials align closely with the motivations for opting for commercial stains rather than raw metal oxides. Typically, greater amounts of frit lead to improved firing quality and accuracy, reduced melting temperatures, fewer glaze imperfections, enhanced clarity, a smoother finish, more vibrant colors, quicker firing, and decreased thermal expansion [48]. Generally, glazes made up of 85% frit and 15% clay are costly for mass-production ceramic manufacturers, and the expenses will increase further for independent artists and small to medium pottery businesses [48,53].

7. Categories of Glaze Surface Defects

In industry, the quality of a production piece is determined by the glazing surface, which impacts both the product's functionality and aesthetics. Defects in the finished glaze surface negatively affect product quality. Most of these defects are only noticeable upon exiting the kiln, even if their primary source lies elsewhere in the production process. Identifying the causes of glaze defects is challenging, so thorough inspection throughout the production process is crucial to avoid all defects and maintain the integrity of the piece [54–56].

7.1 Glaze Pinhole Defects

Every glaze includes some bubbles. Minor bubbles have negligible effects on the quality of the glaze and can be overlooked, particularly if they are separate from the exterior surface of glaze. Yet, as bubbles enlarge or rise nearer to surface, they disrupt it and create a lack of conformity. D. Fortuna and A. Angeli state that the creation of pinholes occurs due to nearly spherical gas bubbles exceeding 80 μm in diameter trapped between the body and glaze. Two scenarios define the pinhole in a microscope, reliant on various production factors that affect the creation of pits: The initial scenario involves a spherical gas bubble situated beneath a highly thin glazing layer with its thickness ranging from 80 to 800 microns [57,58]. The bubble's form is influenced by two differing pressures: the internal pressure within the bubble and the external pressure, both resulting from the cooling process. The second instance: A semi-spherical void on the glaze surface with a size ranging from 80 to 800 microns and a hollow in the glaze layer is formed by a gas bubble that has ascended to the top and exploded [57–59].

Numerous studies have shown that the decomposition of raw materials causes pinholes, resulting in a reaction within the glaze as it transitions from a liquid state to a solid state in the cooling zone. Most gases are emitted prior to the glaze melting over, without causing any flaws in the glaze surface. While firing, the gas bubbles need to break through the surface tension of the melted glaze to break free from the glossy surface, where they subsequently explode. If the glaze can repair the surface, the bubbles exit glaze without creating pinholes. Insufficient fluidity of the glaze drastically impacts outgassing of glaze or body when heated, influencing the breakdown of raw materials, particularly calcium carbonate, by emitting CO_2 [18,60]. At this point, the burst gas bubbles rupture the surface, forming pinholes. Typically, melted glaze fills the craters, but this could be more efficient if the soaking duration is too brief or if the pinholes appear after the firing cycle. This issue can be addressed by selecting suitable raw materials and ideal firing conditions to create a glassy surface that coats the body. Another flaw is pitting, which occurs when the body possesses excessive thickness and dual-glazed layers. Air gets entangled, particularly during the firing phase when gases are unable to escape through the glaze until the melting phase of the glaze has finished [58,60].

The gases that occupy bubbles originate from at least one of several sources:

• Air captured among the glaze particles, due to the evaporation of water, constitutes roughly 40% of the glaze volume 10. Water evaporation creates tiny gaps among the glaze particles. If the glaze particles are small, the glaze will have few pores.

• Air is trapped within the pores of the body.

• Stage of glaze spraying that creates bubbles from too many layers. Bubbles are likely to form if the initial layer dries too quickly, as this layer will show surface porosity that leads to air leakage and bubble creation when applying the second glaze layer. It is essential to guarantee the saturation of the initial layer.

• Unsuitable drying process for the cast items.

• Gases released from the breakdown of glazing materials. Substances like calcium and barium carbonates, porcelain clay, talc, zirconium silicate, and organic compounds undergo gas generation when these substances break down while burning. Additionally, quartz helps enhance the retention of this gas.

• Decomposition of organic binders or glaze contaminants that can result in gas comprises of rust from mixing tools and SiC, which is utilised in certain kiln furniture.

• Gases released from the breakdown of glaze impurities that may enter the glaze during the mixing process, such in the form of silicon carbide, calcium carbonate, and gypsum particles. Additionally, the presence of sodium and magnesium impurities sulphate and the sulfur levels from the kiln environment [58,61].

7.2 Crawling

A frequent issue in manufacturing is glaze crawling. From a morphological standpoint, crawling is a typical deformation in ceramic production. The defect is an exposed region in the glazed surface, defined as "the break in the glaze layer." The extent of unglazed region may vary from small to large based on the underlying reason for the defect. The edge of the glaze surrounding the fault is even, curved, and enlarged. The shrinkage of glaze is immediately identifiable, as it unpredictably creates small islands of glaze, leaving parts of the fired body visible. When crawling occurs, unevenly shaped areas that are either unglazed or partially covered disrupt the uniformity of the fired glaze. This non-conformity may vary from a single small, exposed area of the ceramic body to glaze beading or could be affected by the surface tension of the glaze. Another sign of this flaw arises from the outer edges of the retracted glaze, which are modified following the effect of the surface tension. Usually, glaze shrinkage occurs in the curved region of the surface being glazed, as the glaze layer tends to be thicker in this area. Typically, the formation of this defect manifests as a crack, appearing as a disruption in the glaze that coats the body. This issue may arise from mechanical factors, such as uneven thickness of the glaze on the body, resulting in the emergence of this flaw, which is grounded in surface tension [58].

• Rapid contraction of the glaze while drying. The drying conditions need to be altered to prevent cracks from developing in the fragile coating. Unfortunately, numerous drying cracks are so tiny that they remain invisible before firing. When firing, the material retracts due to surface tension, causing crawling. Shrinkage problems can also occur if a second wet coating is applied before the first layer has dried. Magnesia can lead to glaze crawling due to its effect of inducing glaze dryness. Unwanted shrinkage of the glaze post-application can occur due to the excessive fineness of its particle composition; a thicker glaze, especially in the concave areas of the item, tends to absorb more glaze; and dust or granules on the surface of the body may lead to glaze shrinkage [58].

• Inadequate adhesion between glaze, and body due to dust on body hindering proper binding; insufficient glazing of the body; improper glaze application; the accumulation of soluble salts on the body's surface interfering with the glaze, potentially causing lifting during firing, and soluble salts from the body may build up on the surface. These salts can raise the glaze from the body, causing crawling. A rise in firing temperature may result in an enhancement of the glaze fluidity to allow the glaze to refill the crater [58,62].

7.3 Cracking and Peeling

Crazing and peeling are frequent issues in ceramics. The two defects arise from considerable variations between the body's expansion coefficient and that of glaze. When expansion coefficient of glaze surpasses that of the body, it creates tension on the glaze surface, resulting in fine cracks called crazing. Conversely, if the body's expansion coefficient exceeds that of the glaze, it will create compressive stress on glaze surface, resulting in the breakage of several glaze chips, a defect referred to as peeling. Ceramic materials are fragile substances, exhibiting significant resistance to compression. However, they easily break under pressure. Consequently, crazing is the most prevalent defect as it happens when ceramic materials experience excessive tensile stress resulting from the disparity in thermal expansion coefficients between the body and the glaze. The coefficient of thermal expansion isn't the sole reason for crazing; numerous other factors also play a role, such as the distribution of particle sizes in glaze. Crazing and peeling typically happen when the forces created within the glaze (tension and compression) surpass the glaze's mechanical strength. While firing, a transitional layer develops between glaze and body, which directly influences the manifestation of crazing and peeling. The makeup of the intermediate layer relies on the composition of the body and glaze, as well as the glaze characteristics during firing, such as the viscosity of the molten glaze and its surface tension. Furthermore, it relies on the firing temperature and firing curve [58].

8. Ceramic Body Defects

The ceramic body is the basic component of the manufacture of ceramic products. It contains a mixture of inorganic raw materials (soil, field spat, silica sand, and other mineral oxides), which are formed at high temperatures to achieve their final hardness, dried, and removed. The ceramic body plays a fundamental role in determining the mechanical and physical properties of the product, such as strength, soundness, thermal shock resistance, and shrinkage during firing. Ceramic bodies are classified according to the type of soil or the required temperature for glass (such as stone vessels, porcelain, and ceramics). Mineral composition and thermal properties are the most important factors that determine behaviour of the ceramic body during making, drought, and firing processes. Studies suggest that the structural properties of a ceramic body depend on the type of binding, shape, pores, and bonds formed during shooting. Ceramic body compatibility with glazed layers also requires the thermal expansion and coefficient of shrinkage properties to avoid the formation of cracks or surface defects [63,64].

8.1 The relation between glaze and body defects in industrial ceramics

Matching the thermal expansion coefficients of the ceramic body and surface coating is essential. Glazes and ceramic materials are closely connected and chemically and physically interact during the firing process. Chemical reactions take place among the distinct elements, altering the chemical makeup and, if the firing duration is sufficient, possibly resulting in the development of interfacial layers on the surfaces. Describing the relationship between glazes and ceramic bodies yields insights into application characteristics, firing procedures, and the spectrum and severity of stresses within the glaze layer. Ceramic products with glaze frequently display leftover stress as a result of the production method. Residual stresses may arise from inhomogeneities in the microstructure and the existence of tiny defects. All these flaws can result in a loss of functional characteristics or in the degradation of the ceramic item. A consistent glazed surface finish on a ceramic body serves both an aesthetic and protective role, while also needing to create a strong bond with its substrate. The glaze should be devoid of imperfections to fulfill its function. To suggest a suitable surface finish for a particular ceramic body, it is essential to understand the fundamental properties, the characteristics of glaze, the ceramic body, and the stress interactions between glaze and ceramic body. The primary element influencing stress relationships between the glaze and the ceramic body is the disparity in their thermal expansion coefficients [13,42].

8.2 Categories of Ceramic Body Defects

8.2.1 Cracks

Cracks are one of the most common defects that occur in ceramic objects during the drying or firing stages. These cracks occur due to different object components shrinkage during drying or heating, as the material loses moisture and shrinks. However, if this shrinkage occurs unevenly between different regions of the object, it leads to an accumulation of internal loads exceeding the mechanical strength of ceramics, resulting in cracks [65] .In addition, rapid temperature changes during firing can lead to uneven expansion and contraction between different regions of the object, especially if there is a difference in chemical or grain composition in a ceramic object. It produces thermal stress leading to cracks. These cracks can be superficial, appearing on the surface of the piece, or inner and invisible, but they reduce durability and stiffness, making the product useless. The difference in the coefficient of thermal expansion between the components of objects also plays an important role in presence of cracks. For example, if there is a difference in expansion and contraction of various mineral particles or stages in a body, it causes mechanical stress, leading to the presence of cracks [65,66].

8.2.2 Delamination

Delamination is a common structural defect in ceramic objects. It refers to the formation of different layers in an object, resulting in poor compaction, air entrapment during forming or insufficient air removal during preparation processes, such as casting or extrusion. These unnatural layers occur when particles are not properly bonded or when small air bubbles get stuck between layers of materials, preventing complete conditions during formation or pressing [67]. Although this defect cannot be displayed before firing, it becomes clear during the final drying or firing processes, causing internal cracks or sudden breakage in the final product, reducing its mechanical resistance or affecting its functional properties, especially in design or architectural applications requiring high strength. This defect is often seen in products manufactured by dry pressing or extrusion, where there is inadequate removal of air. In order to reduce the occurrence of this defect, it is necessary to improve the ventilation processes, ensure a homogeneous distribution of materials during casting, and use the correct pressure and compression density. The use of techniques such as pressure gradients or vacuum ventilation is also effective in reducing this problem [67,68].

8.2.3 Blisters

Blisters are a common ceramic body defect characterized by the presence of bubbles that form either on or just below the surface. This defect is mainly produced in the firing phase when gases are trapped due to the incomplete combustion of organic materials such as binders or lubricants. When the surface of the ceramic seals under premature sintering, there is no escape route for the trapped gases, resulting in blisters. These blisters not only affect the visual appeal of the final product but also weaken its mechanical integrity by creating localized stress points and reducing density [69,70].

8.2.4 Black Core

The black core typically results from errors in the firing procedure. Insufficient air circulation during firing results in a deficiency of oxygen inside the tile, leading to the buildup of unburned carbon deposits. This phenomenon has been particularly noted in ceramics subjected to high-temperature firing. Moreover, incorrect firing duration or temperature can lead to the clay not fully sintering, potentially resulting in the "Black Core" defect. Another key element influencing defect occurrence is the quality of the raw materials utilized. Inadequate mixing of raw materials can cause carbon-rich regions to develop in certain areas. In this instance, adequate combustion does not take place in these areas during firing, leading to the formation of "Black Core". This risk notably rises in clays that have high organic matter levels. Research indicates that a 2% variation from uniformity in the clay blend can raise the occurrence of the "Black Core" by as much as 12%. The impact of these defects on ceramic tiles extends beyond visual decline; it also adversely influences the mechanical strength of the tiles [71].

9. Conclusion

Defects in glaze and ceramic bodies present major obstacles in creating high-quality ceramic items. These flaws, such as delamination, blisters, black core, crawling, and cracking, result from intricate interactions between raw materials, manufacturing techniques, and firing parameters.

Their presence not only reduces the visual charm of ceramic pieces but also negatively impacts their structural integrity and operational efficiency. Grasping the underlying reasons for these defects like inadequate material adhesion, trapped air, partial combustion of organic materials, and incompatible thermal expansions is crucial for enhancing manufacturing procedures. By managing factors such as material composition, processing methods, and firing environments, manufacturers can decrease defect rates, thus improving product longevity and lower industrial waste. Ongoing research and technological progress are essential for enhancing ceramic glaze formulations and body compatibility, ultimately resulting in more dependable and high-quality ceramic products.

References

[1]   I. Baker, Lead Zirconate Titanate, in: Fifty Mater. That Make World, Springer International Publishing, Cham, 2018: pp. 111–115. https://doi.org/10.1007/978-3-319-78766-4_21.

[2]   F. Cardarelli, Ceramics and Glasses, in: Mater. Handb., Springer London, London, 2000: pp. 337–369. https://doi.org/10.1007/978-1-4471-3648-4_9

[3]   R. Rakshit, A.K. Das, A review on cutting of industrial ceramic materials, Precis. Eng. 59 (2019) 90–109. https://doi.org/10.1016/j.precisioneng.2019.05.009

[4]   P. Wang, The General Properties and Applications of Ceramic Materials, Appl. Mech. Mater. 174–177 (2012) 215–218. https://doi.org/10.4028/www.scientific.net/AMM.174-177.215

[5]   R. Giordano II, Ceramics overview, Br. Dent. J. 232 (2022) 658–663.

[6]   J.D.. Selvam, I. Dinaharan, R.S. Rai, Matrix and Reinforcement Materials for Metal Matrix Composites, in: Encycl. Mater. Compos., Elsevier, 2021: pp. 615–639. https://doi.org/10.1016/B978-0-12-803581-8.11890-9

[7]   L. Vaiani, A. Boccaccio, A.E. Uva, G. Palumbo, A. Piccininni, P. Guglielmi, S. Cantore, L. Santacroce, I.A. Charitos, A. Ballini, Ceramic Materials for Biomedical Applications: An Overview on Properties and Fabrication Processes, J. Funct. Biomater. 14 (2023) 146. https://doi.org/10.3390/jfb14030146

[8]   O. Adigun Afolabi, O. Akanni Olanrewaju, Processing and Applications of Composite Ceramic Materials for Emerging Technologies, in: 2024. https://doi.org/10.5772/intechopen.1007296

[9]   N. Kordani, Materials, design, and technology of body armor, in: Adv. Healthc. Prot. Text., Elsevier, 2023: pp. 259–301. https://doi.org/10.1016/B978-0-323-91188-7.00006-6

[10]   A.R. BOCCACCINI, Ceramics, in: Biomater. Artif. Organs Tissue Eng., Elsevier, 2005: pp. 26–36. https://doi.org/10.1533/9781845690861.1.26

[11]   C.B. Carter, M.G. Norton, Ceramic Materials, Springer New York, New York, NY, 2013. https://doi.org/10.1007/978-1-4614-3523-5

[12]   W.G. Fahrenholtz, G.E. Hilmas, Ultra-high temperature ceramics: Materials for extreme environments, Scr. Mater. 129 (2017) 94–99. https://doi.org/10.1016/j.scriptamat.2016.10.018

[13]   R. Casasola, J.M. Rincón, M. Romero, Glass–ceramic glazes for ceramic tiles: a review, J.

Mater. Sci. 47 (2012) 553–582. https://doi.org/10.1007/s10853-011-5981-y

[14] R.K. Dhir, J. de Brito, G.S. Ghataora, C.Q. Lye, Use of Glass Cullet in Ceramics and Other Applications, in: Sustain. Constr. Mater., Elsevier, 2018: pp. 327–387. https://doi.org/10.1016/B978-0-08-100984-0.00009-6

[15] S.S. Anufrik, N.N. Kurian, I.I. Zhukova, K.F. Znosko, M. V. Belkov, Chemical Composition of Ceramic Tile Glazes, J. Appl. Spectrosc. 83 (2016) 764–770. https://doi.org/10.1007/s10812-016-0360-8

[16] C. Peng, M. Lv, C. Jing-Pei, C. Peng, M. Ding, J. Wu, Cordierite glass-ceramic used for transparent tile glazes, J. Am. Ceram. Soc. 100 (2017) 4402–4406. https://doi.org/10.1111/jace.15031

[17] S.Ö. Varışli, F. Taşkıran, B. Öztürk, B. Çiçek, Effect of SiO2/Al2O3 ratio on the whiteness of ceramic tile engobes with low zircon content, Cerâmica 69 (2023) 254–260. https://doi.org/10.1590/0366-69132023693913465

[18] T. Pradell, J. Molera, Ceramic technology. How to characterise ceramic glazes, Archaeol. Anthropol. Sci. 12 (2020) 189. https://doi.org/10.1007/s12520-020-01136-9

[19] H. Takahashi, S. Yonezawa, M. Takashima, Glaze-Slurry Particle-Size Distribution Influences on Physical Properties of Some Kutani-ware Glaze, J. Ceram. Soc. Japan 115 (2007) 460–465. https://doi.org/10.2109/jcersj2.115.460

[20] T. Kronberg, L. Hupa, K. Fröberg, Melting Behaviour and Surface Structure of Glazes Containing Wollastonite and Dolomite, in: 2006: pp. 590–595. https://doi.org/10.4028/www.scientific.net/AST.45.590

[21] W. Jiang, Q. Liao, Effect of compositions on the microstructure and surface properties of phase separation-crystallization opacified glazes, Journal-Chinese Ceram. Soc. 35 (2007) 230.

[22] Z.-C. Ma, Eliminating of the Glaze-Bubble Phenomenon on Grey-Glaze Products, Dianci Bileiqi(Insulators and Surge Arresters) (2012) 26–28.

[23] L. Maritan, Ceramic abandonment. How to recognise post-depositional transformations, Archaeol. Anthropol. Sci. 12 (2020) 199. https://doi.org/10.1007/s12520-020-01141-y

[24] D.C. Birawidha, D. Asmi, S. Sembiring, S. Sumardi, F. Bahfie, D. Susanti, Characterization of the Glass Structure of East Lampung's Scoria Basalt (Indonesia) Applied to the Ceramic Body, Powder Metall. Met. Ceram. 61 (2023) 699–707. https://doi.org/10.1007/s11106-023-00357-6

[25] G. Pfaff, 11 Ceramic colors, in: Antraquinonoid Pigment. - Color Fundam., De Gruyter, 2022: pp. 95–102. https://doi.org/10.1515/9783110588071-011

[26] D.W. Richerson, The magic of ceramics, John Wiley & Sons, 2012.

[27] D. V. Hill, R.J. Speakman, M.D. Glascock, H. Neff, The Technology of Mesopotamian Ceramic Glazes, in: 2007: pp. 422–446. https://doi.org/10.1021/bk-2007-0968.ch023

[28] F.G. Melchiades, B.T. Rego, S.M. Higa, H.J. Alves, A.O. Boschi, Factors affecting glaze transparency of ceramic tiles manufactured by the single firing technique, J. Eur. Ceram. Soc. 30 (2010) 2443–2449. https://doi.org/10.1016/j.jeurceramsoc.2010.04.030

[29] J. Wu, H. Ma, N. Wood, M. Zhang, W. Qian, J. Wu, N. Zheng, Early development of Jingdezhen ceramic glazes, Archaeometry 62 (2020) 550–562.

https://doi.org/10.1111/arcm.12539

[30] E. Holmqvist, T. Heinonen, R. Väisänen, A. Pihlman, A. Koivisto, E. Russow, Ceramic fabrics and lead glazes of late medieval redware pots in the Helsinki, Turku and Tallinn regions (ED-XRF, SEM-EDS), J. Archaeol. Sci. Reports 34 (2020) 102627. https://doi.org/10.1016/j.jasrep.2020.102627

[31] P. Colomban, G. Simsek Franci, Timurid, Ottoman, Safavid and Qajar Ceramics: Raman and Composition Classification of the Different Types of Glaze and Pigments, Minerals 13 (2023) 977. https://doi.org/10.3390/min13070977

[32] P. Colomban, Glazes and Enamels, in: Encycl. Glas. Sci. Technol. Hist. Cult., Wiley, 2021: pp. 1309–1325. https://doi.org/10.1002/9781118801017.ch10.6

[33] S. Wang, X. Li, C. Wang, M. Bai, X. Zhou, X. Zhang, Y. Wang, Anorthite-based transparent glass-ceramic glaze for ceramic tiles: Preparation and crystallization mechanism, J. Eur. Ceram. Soc. 42 (2022) 1132–1140. https://doi.org/10.1016/j.jeurceramsoc.2021.11.036

[34] L. Koroglu, C. Peksen, M. Ince, E. Ayas, Cost-effective and eco-friendly transparent ceramic glazes for porcelain bodies, J. Aust. Ceram. Soc. 59 (2023) 1263–1269. https://doi.org/10.1007/s41779-023-00906-7

[35] R.F.A. Abd-Elrahim, H.M. Abdelmotalib, M.A. Hassan, E.D.A.-E. Soliman, Possibility of obtaining Transparent Glaze from the Local clays of Qena Governorate, 0–0 (2023). https://doi.org/10.21608/jfea.2023.252277.1112

[36] S. Wang, X. Li, C. Wang, M. Bai, X. Zhou, X. Zhang, Y. Wang, https://doi.org/10.1016/j.jeurceramsoc.2021.11.036, J. Eur. Ceram. Soc. 42 (2022) 1132–1140. https://doi.org/10.1016/j.jeurceramsoc.2021.11.036

[37] M. Gajek, J. Lis, J. Partyka, M. Wójczyk, Floor tile glass-ceramic glaze for improvement of the resistance to surface abrasion, IOP Conf. Ser. Mater. Sci. Eng. 18 (2011) 112016. https://doi.org/10.1088/1757-899X/18/11/112016

[38] A.R. Molla, A.M. Rodrigues, S.P. Singh, R.F. Lancelotti, E.D. Zanotto, A.C.M. Rodrigues, M. Reza Dousti, A.S.S. de Camargo, C.J. Magon, I.D.A. Silva, Crystallization, mechanical, and optical properties of transparent, nanocrystalline gahnite glass-ceramics, J. Am. Ceram. Soc. 100 (2017) 1963–1975. https://doi.org/10.1111/jace.14753

[39] B.E. Yekta, P. Alizadeh, L. Rezazadeh, Synthesis of glass-ceramic glazes in the ZnO–Al2O3–SiO2–ZrO2 system, J. Eur. Ceram. Soc. 27 (2007) 2311–2315.

[40] T.K. Ghosh, S.K. Das, Bioactive glass-ceramic synthesis using nano and normal silica: a comparative study, J. Aust. Ceram. Soc. 57 (2021) 933–946. https://doi.org/10.1007/s41779-021-00596-z

[41] J. Cai, M. Lu, K. Guan, W. Li, F. He, P. Chen, C. Peng, P. Rao, J. Wu, Effect of ZnO/MgO ratio on the crystallization and optical properties of spinel opaque glazes, J. Am. Ceram. Soc. 101 (2018) 1754–1764. https://doi.org/10.1111/jace.15321

[42] M. Kavanova, Characterization Of The Interaction Between Glazes And Ceramic Bodies, Ceram. - Silikaty (2017) 267–275. https://doi.org/10.13168/cs.2017.0025

[43] O.S. Adelabu, Advancing Ceramics Glaze Formulation with Existing Software Technology Using Locally Available Raw Materials in Nigeria, Adv. Mater. Res. 463–

464 (2012) 266–270. https://doi.org/10.4028/www.scientific.net/AMR.463-464.266

[44] A.O. Samuel, K.I. Bolaji, A.T. Lawrence, Developing Computer Aided Ceramic Glaze Recipes Using Local Raw Materials, Adv. Ceram. Sci. Eng. 2 (2013) 95–105.

[45] B. ALMAMARI, Developing Ceramic Textured Matt Glazes Using Omani Plant Ash: The Contributions of Art Education Teachers, J. Arts Humanit. 5 (2016) 12. https://doi.org/10.18533/journal.v5i12.1047

[46] B. Yıldız, The Effect Of Firing Temperature On The Properties Of Glazes Prepared With Barium-Based Frits, Eskişehir Tech. Univ. J. Sci. Technol. A - Appl. Sci. Eng. 25 (2024) 617–626. https://doi.org/10.18038/estubtda.1581744

[47] K. Dana, S.K. Das, Formulation of Some Frit Compositions for Rapid Once-Fired Wall Tile Glaze, Trans. Indian Ceram. Soc. 62 (2003) 158–161. https://doi.org/10.1080/0371750X.2003.11012099

[48] B. Almamari, N. Alsaadi, Q. Alsalhi, Developing Low Temperature Glazes for Omani Clay Artworks: Laboratory Investigation to Reduce Costs by Using Frits, New J. Glas. Ceram. 13 (2023) 17–34.

[49] B. Almamari, F. Al-Thamari, Z. Al-Zadjali, Q. Alsalhi, Use of Ferro Frit 3110 in Developing Glazes for Omani Earthenware Clays: Evaluation of Color Quality for Ceramic Decoration in High Schools, OALib 07 (2020) 1–8. https://doi.org/10.4236/oalib.1106991

[50] G. Monrós, M. Llusar, J. Badenes, R. Galindo, Sol-Gel ceramic glazes with photocatalytic activity, J. Sol-Gel Sci. Technol. 102 (2022) 535–549. https://doi.org/10.1007/s10971-022-05787-z

[51] H. Huang, J. Yu, F. Liu, H. Zeng, Preparation of A High-Performance Frit Glaze Using High-Potassium Feldspar, IOP Conf. Ser. Earth Environ. Sci. 943 (2021) 012018. https://doi.org/10.1088/1755-1315/943/1/012018

[52] K.Y. Frolenkov, S.I. Matyukhin, L.Y. Frolenkova, Infrared Spectra and Differential-Thermal Analysis of Ceramic Glaze Frits, Glas. Ceram. 75 (2019) 372–374. https://doi.org/10.1007/s10717-019-00088-4

[53] K. Pekkan, B. Karasu, Production of opaque frits with low ZrO2 and ZnO contents and their industrial uses for fast single-fired wall tile glazes, J. Mater. Sci. 44 (2009) 2533–2540. https://doi.org/10.1007/s10853-009-3329-7

[54] E. Çaşin, N. Derin Coşkun, C.E. Işik, Detection And Characterization Of Some Glaze Faults Encountered In Sanitarywares, J. Sci. Reports-A (2023) 108–124. https://doi.org/10.59313/jsr-a.1252415

[55] N. DERİN COŞKUN, E. ÇAŞIN, C.E. IŞIK, Investigation of Defects Occurring in the Casting Shop Process of Ceramic Sanitaryware on the Final Product, Black Sea J. Eng. Sci. 6 (2023) 527–534. https://doi.org/10.34248/bsengineering.1347161

[56] N. Bao, X. Ran, Z. Wu, Y. Xue, K. Wang, Design of inspection system of glaze defect on the surface of ceramic pot based on machine vision, in: 2017 IEEE 2nd Inf. Technol. Networking, Electron. Autom. Control Conf., IEEE, 2017: pp. 1486–1492. https://doi.org/10.1109/ITNEC.2017.8285043

[57] D. Fortuna, A. Angeli, Pinholing, Part 2, Interceram 54 (2005) 80–85.

[58]   E. Youssef, N. Mostafa, J.E. Khoury, T. Merhej, R. Lteif, Glaze surface Defects causes and Prevention Controls, J. Ceram. Sci. Technol. 14 (2023) 1–10.

[59]   C. Liu, J. Zhou, J. Lu, Sinterability, structural evolution and pinhole elimination of high strength self-glazed glass-ceramics sintered from granite sludge by instant glaze firing, Mater. Technol. Reports 3 (2025) 2851. https://doi.org/10.59400/mtr2851

[60]   J. Zamek, Diagnosing glaze blisters, Ceram. Tech. (2008) 23–29.

[61]   G. Huang, Z. Yang, The detachment of an air bubble from glass plates featured with pore structures filled with gas or water, Phys. Fluids 37 (2025). https://doi.org/10.1063/5.0251462

[62]   D. Wattanasiriwech, S. Wattanasiriwech, Effects of particle size of glaze powder and thickness of glaze layer on glaze crawling for Dolomite Wares, Chiang Mai J, Sci 33 (2006) 35–44.

[63]   K. Wiśniewska, W. Pichór, E. Kłosek-Wawrzyn, Influence of Firing Temperature on Phase Composition and Color Properties of Ceramic Tile Bodies, Materials (Basel). 14 (2021) 6380. https://doi.org/10.3390/ma14216380

[64]   M. Syed Fofanah, Characterization of Ceramic Tile Bodies Prepared From Clays Collected from Four, Int. J. Innov. Sci. Res. Technol. (2024) 525–534. https://doi.org/10.38124/ijisrt/IJISRT24OCT307

[65]   B.A. CHAKRAVARTHI, Study on different types of cracks in plain and reinforced concrete, (2014).

[66]   M. Safiuddin, A.B.M.A. Kaish, C.-O. Woon, S.N. Raman, Early-Age Cracking in Concrete: Causes, Consequences, Remedial Measures, and Recommendations, Appl. Sci. 8 (2018) 1730. https://doi.org/10.3390/app8101730

[67]   F.-Z. Semlali Aouragh Hassani, R. Bouhfid, A. Qaiss, Recent advances in the fabrication of hybrid natural fiber composites, in: Hybrid Nat. Fiber Compos., Elsevier, 2021: pp. 113–131. https://doi.org/10.1016/B978-0-12-819900-8.00010-6

[68]   M. Romagnoli, M. Burani, G. Tari, J.M.F. Ferreira, A non-destructive method to assess delamination of ceramic tiles, J. Eur. Ceram. Soc. 27 (2007) 1631–1636. https://doi.org/10.1016/j.jeurceramsoc.2006.05.069

[69]   M. Barsoum, Fundamentals of ceramics, CRC press, 2019.

[70]   M. Engels, S. Link, Bubble control in ceramic glazes*, Part 1, Interceram 55 (2006) 80–84.

[71]   M.U. Taşkıran, K. Kayacı, A.A. Sirkeci, Y. Yıldırım, B. Darcan, Prevention of black core formation in ceramic bodies., Physicochem. Probl. Miner. Process. 61 (2025).

Materials Research Forum LLC

https://doi.org/21741/9781644903834

Chapter 12

Screen Printing Methods and Technical Decoration Applications in Industrial Ceramics

Nihal Derin Coşkun[1*], Cumhur Eren Isık[2], Fatmanur Maran[3], Selcuk Erdogan[3], Iskender Isık[4], Fatih Sen[3*]

[1]Ceramic and Glass Department, Ordu University, 52200 Ordu, Türkiye

[2]Department of Handicrafts, Kutahya Fine Arts Vocational School, Kutahya Dumlupınar University, Evliya Çelebi Campus, 43000 Kutahya, Türkiye

[3]Sen Research Group, Department of Biochemistry, Kutahya Dumlupınar University, 43000 Kutahya, Türkiye

[4]Department of Materials Science & Engineering, Faculty of Engineering, Kutahya Dumlupınar University, Evliya Çelebi Campus, 43100 Kutahya, Türkiye

nihalderincoskun@odu.edu.tr, fatihsen1980@gmail.com

Abstract

Industrial ceramics and design have undergone significant changes in recent years in light of digital developments. In this study, screen printing techniques, which are important production methods obtained through mass production of ceramic materials and are significant in product design, were examined. By reviewing studies on the past, present, and future of these methods, the impact of technology on the design process in the field of industrial ceramics was observed. Screen printing techniques utilize a wide range of materials, including toner, suitable solvents, mesh screens, appropriate inks, lacquer, and more. Ceramic screen-printing techniques encompass methods such as laser printing, injection or digital printing, serigraphy, indirect printing, and others. The oldest of these techniques is serigraphy printing. With the advancement of technology, other printing techniques have emerged, ranging from designs made with computer programs to artificial intelligence designs, and this development is progressing at an increasing rate every day. In the future, it is anticipated that new revisions will emerge as technology advances in the materials used and in the fields of design and technology, particularly in artificial intelligence and computer technology. In this process of transformation in art, ceramic artists and industry are embracing this transformation by keeping up with technology, opening a multidisciplinary field of work. This is achieved through screen printing and various printing methods, which offer designers the opportunity to transfer the entire world in their minds onto ceramic surfaces.

Keywords

Screen Printing Methods, Seramik Applications, Technical Decoration

1. Introduction

Due to its widespread use in the arts, the term "printing" is defined as the transfer of any pattern, shape, image, or text commonly used in various art forms such as graphics, design, and painting to another surface using various printing techniques and reproducing it in the desired number of copies [1,2]. Almost all these printing techniques used in different art forms can be applied to ceramic surfaces. However, in order to apply printing techniques to ceramic surfaces, certain auxiliary materials and ceramic paints are required [3–5].

Ceramics, one of the oldest industrial products that shed light on human history, undergoes stages such as shaping, drying, and firing due to its structure. Due to the high temperatures required for this production process, decorative design techniques using printing methods require technical processes and different materials. While printing in textiles or painting yields immediate results, these processes in ceramics only allow for the final product to be obtained after thermal treatment. Differences may arise between the design and product colors due to temperature, and these issues are addressed through analyses and experience gained during the production process.

The process known as printing has been used in the ceramics industry for centuries due to its suitability for mass production, as it enables the transfer of an image onto another surface (Figure 1). The process of transferring a design to another surface has been around for a long time. However, these methods have been renewed and improved every day considering technological and scientific developments until today. In line with the growing population and changing customer demands, needs in the field of design have also changed and continue to change. Printing methods, which are among the main factors enabling a quick response to this change, have also contributed to the advancement of different printing methods used in ceramic art within the changing definitions of art. At the present stage, ceramic products obtained in terms of production and design are at an advanced level [6,7].

Figure 1. Examples of products manufactured using printing techniques from the past to the present day [8,9].

Printing techniques offer the possibility of transferring a designed, drawn, or photographed pattern, decoration, image, or visual to another surface as many times as desired, making them widely used in industry. Today, it is almost impossible to come across a commercial object without any printing on it. With the increasing use of these printing techniques due to technological developments, they are also frequently used and preferred by ceramic artists [10].

Since its first production, ceramics have not only been objects we use in our daily lives, but also an artistic medium that reflects the emotions, imagination, and inner world of the person who shapes, designs, and forms them. Regardless of the reason for using ceramics in our daily lives, the emotions evoked by the visuals on ceramics have always been of great importance throughout the ages. Creating the three-dimensional structure and visually appealing liveliness of ceramics requires different techniques. From the past to the present, traditional methods such as stamping, engraving, carving, ajour, brush, sponge, paraffin, spraying, and luster have been used on ceramics, but with the development of industry, methods such as lithography, laser printing and transfer, serigraphy (screen printing), pad printing, and digital (inkjet) printing have been discovered and are still in use today. Among these, one of the most important methods is screen printing [11].

This method is applied to both artistic and industrial ceramic surfaces using methods such as laser transfer, mono printing, and image transfer, which are designed in advance in a digital environment using photocopying or inkjet printing [12]. The development of these techniques has been influenced by technological advances and mechanization.

The history of printing techniques used on ceramic objects dates to ancient times. Throughout history, ceramic artifacts have been unearthed in archaeological excavations across all continents of the world. Dating back to approximately 29,000 BCE, ceramic art has undergone similar developmental processes in every region of the world, from early humans to Egyptian art, Greek art to China, Central Asia to Africa. The development of this unique material began with the discovery of fire and has evolved to automated production systems, continuing its progression unabated [13].

When examining ceramic production today, it is evident that the shapes that were once created using wooden tools and other materials have evolved over time, with ceramic artists utilizing various materials to create different patterns and images on a platform. and creating different patterns and images on clay. Approximately 30 years ago, ceramics and porcelain, which were once kept in the most special places in our homes, became easily accessible through mass production, and today, they have given way to a period where the most valuable designs are admired and desired. When we look at the evolution from the oldest works to the present day, we observe that while the artist has expressed themselves in every era, the materials have changed. Ceramics now cover every part of our homes, from bathrooms to kitchens, offering the possibility of living in harmony with them. Today, the use of digital technologies in the ceramic industry has provided ceramic artists with the opportunity to express their emotions and imagination on ceramic objects in a more comfortable and higher quality manner. Because it is wrong to look at ceramics only from an industrial perspective; reflecting people's imagination and visuality on ceramic objects is a beautiful aspect of art. Today, this influence is also reflected in advanced technology products, including nanotechnology, which refers to the 1-100 nm range, and the high produced from it [14,15]. Nanotechnological ceramic designs clearly demonstrate the interdisciplinary collaboration between technology and art and the impact of technology on every field [16,17]. Numerous studies on nanomaterials conducted by researchers demonstrate

Materials Research Forum LLC
https://doi.org/21741/9781644903834

the development and change in this field [18,19]. At the same time, it has been studied in the field of nanotechnology in advanced technology and artistic ceramics [20,21]. Countless studies on ceramics have been published in the literature [22,23].

2. Seal Printing

With the growth and development of the ceramics industry and art, there has been an opportunity for advancement in the methods used in ceramics. These developments encompass many stages, such as design, firing, and shaping. The printing process has also benefited from these advances. In the early days, seals were used to print on ceramic surfaces, but today, many applications are made on glazed and unglazed surfaces. Figures 1 and 2 show historical examples of seal printing. Over time, color has also been added to these works.

Figure 2. Examples of seals and stamps used throughout history [8].

Looking at ancient sources, it is estimated that seal impressions are the oldest printing model used in ceramics. Seal impressions have been used by sculptors and artists working with ceramics to add visual liveliness to their works. Cylindrical seals, which have a very long history, still exist today. In Anatolia, seal examples in the shape of a cylinder feature different types of seals. These are generally found in the Colonial Age seal types discovered at Acem Höyük, Boğazköy, Kültepe, and Alişar (Figure 3).

Figure 3. Seals and impressions obtained from excavations in Anatolia [24].

3. Template Printing

Another type of printing, stencil printing, forms the basis of today's digital screen-printing technique. The first examples of stencil printing were used in Greece and Egypt around 2500 BC, and it was also used in China and Japan in the first and second centuries AD. Referred to as seals or stamps in various historical sources, experiments on ceramics were mostly used in architectural structures made of brick and tile, particularly in temples.

It was part of the architectural structure. The difficulty of drawing on hard surfaces led people at that time to discover printing techniques and different paints (pigments) (Figure 3) [25].

Figure 4. Seal stamp unearthed during excavations at Çatalhöyük [8].

4. Digital Printing

Unlike other art forms, ceramics involves more technological processes due to its dependence on thermal treatments and the calculation of appropriate ratios of raw material components, and its artistic history is categorized according to the development of techniques and materials. Conversely, its industrial dimension also requires design and visual appeal, making it a discipline that combines technical and design aspects to a much greater extent. This process was made more pronounced by the Industrial Revolution, and this development, particularly in France and England, began to be used in various fields for design purposes during the French Revolution and the Industrial Revolution (Figure 5). Today, templates of various types offer artists a wide range of options in decorative applications [26,27].

Figure 5. Examples of ceramic products obtained by digital printing [25].

5. Ceramic Printing Techniques and Materials

5.1 Monochrome printing

In the ceramics industry, various monochrome methods other than screen printing have been used for decoration. In these transfer techniques, the process was carried out using molds obtained by peeling materials such as steel, plaster boards, and copper. After the mold was prepared, a monochrome print was taken on transfer paper, and then these designs were transferred onto ceramic objects. It is not possible to provide precise information about the history of printing on ceramic tiles. However, examples by John Sadler and Guy Freen, who created the first examples, are shown in Figure 6. These examples are among the valuable works in which the developing techniques were applied [28].

Figure 6. Examples of products made using the monochrome printing method [29,30].

Today, among ceramic production items, the tile sector is at the forefront of areas where printing techniques are most widely used, as it is the sector where mechanization and technology are most widely used, the human factor is minimal, mass production and design demands are highest, and it is suitable for production due to its flat surface. Printing techniques, which have been used since the earliest days of ceramic production, have been employed to create various patterns. With the advancement of technology, these techniques have evolved alongside digitalization, offering a wide range of options today. However, both in artistic productions and industrial ceramic productions, all methods from the most primitive to the most advanced continue to find their place in use.

5.2 High pressure

This method is one of the simplest and most primitive methods used in ceramic mass production. This technique, which aims to produce fast, mass-produced, and similar ceramics, is also preferred in fields other than ceramics, such as metal, glass, and textiles. The design is carved into a mold, and the desired parts are painted and pressed onto a flat surface to decorate the ceramic. This printing process can be applied not only to molded clay of leather-like hardness but also to coated surfaces. The reason this type of printing is referred to as high-relief printing is that the design intended to be created forms on the raised portions of the mold's surface (Figure 7) [31].

Figure 7. Cylinder seal from the BC era and applications from 13th century medieval floor tile applications [32,33].

5.3 Relief printing

The seal printing examples mentioned in the previous section can be given as examples of this type of printing. In artistic productions, patterns can be transferred onto many materials such as clay, wood, and sponge. The main purpose of printing methods applied to Tokat manuscripts using wooden molds, metal molds that give patterns to leather used for book covers, and other materials obtained from nature or prepared later is to transfer the desired pattern to the opposite surface. Relief printing on clay has been a production method used throughout the history of ceramics, particularly because it is easy to apply.

When examining works from the 13th and 14th centuries, it is possible to find many examples of patterns produced by printing on ceramic surfaces, especially in palaces and religious architecture. Products made from red clay and ceramic tile works using different colors, as seen in Figure 8, have a significant place in the works of this period. Relief is the creation of patterns and figures on a flat surface or in its depths. These are named according to their depth from the ground. They can be classified as high relief, medium relief, cut relief, and low relief. While some examples are close to the surface, others protrude from it, and those carved into the base are among the techniques applied in many artistic works (Figure 8).

Figure 8. Relief examples from different cultures obtained from archaeological excavations [34–36].

In this method, the clay is first prepared for shaping and obtained in plastic or liquid form. It is shaped with mold and then tiles and other pieces with this molded surface are dried. After drying, the patterned part is cut without distorting the shape or it is removed by carving. The patterns obtained using red or white killers are clearly visible (Figure 9). In this method, it is possible to obtain prints by pressing the mold by hand or by casting. Today, methods such as CNC machines or digital processing are used in the mold preparation stage.

Figure 9. 3D printer ceramic and outdoor ceramic relief applications [37,38].

Nowadays, after transferring the design onto a plaster mold using CNC technology and stamping it onto semi-dry clay, it is possible to produce ceramic works in the desired shape, type, and size. With the advancement of technology, the visual impact of designs is enhanced not only on the glazed surface but also at every stage of production (Figure 10).

Figure 10. CNC seal design, plate shaping seal decoration, and seal printing on wet clay, in that order [39].

5.4 Intaglio printing technique

In this technique, thin metal sheets are used as the main mold for the desired pattern. Zinc, copper, steel, and brass sheets are preferred for the mold, which is engraved (by carving and/or acid treatment) using appropriate processing techniques. Metal sheets are used as molds, and paint is placed inside them to transfer the shape of the recessed areas to the ceramic surface. The engraving technique on copper sheets is referred to as engraving print in the field of painting, and artists such as REMBRANDT and DÜRER have produced works using this technique. The reason this method is not widely used in ceramics and painting is that the engraving process is long and laborious. Additionally, the need for extra materials also contributes to the limited use of this technique. However, it was widely used in the 18th century and remains one of the techniques still in use today [40],[41].

5.5 Lithographic printing

This printing technique is called stone printing because the material used is stone, and it is preferred because the desired pattern can be easily engraved on limestone. Limestone is also preferred not only for its ease of engraving but also for its ability to absorb the applied material. The selected stone surface is first roughened with sand or sandpaper to achieve a smooth finish, preparing it for engraving. After the design is drawn on this surface, the pattern is allowed to penetrate the mold. Subsequently, the desired image is obtained within the pattern that has penetrated the stone using various chemicals (such as acids and adhesives). Paint is applied to the stone mold ready for printing, and the pattern is transferred to the desired surface from there. However, in ceramics, this process is carried out by applying the design to the ceramic surface after the printing paper (decal), and the contents and quantities of the paints used vary. After this process, the material proceeds to the sintering stage. When examined commercially, while its reusability is an advantage, the color restrictions make it less preferred in industrial applications (Figure 10) [42].

Figure 11. *Lithography-Stone printing and application stages [43].*

5.6 Stone printing

Another type of printing is lithography. It is a printing technique in which designs made on limestone using oil-based ink are transferred onto paper. This technique is based on the repulsive force between oil and water. In lithography, which is used in many different industrial fields, it is not necessary to use only limestone as the plate material. The essential requirement for this printing method is that the material used as the plate has absorbent properties. With the advancement of this technique, materials such as plaster and paper have also been used as plates. After the plates used in this technique undergo several processes, designs are applied to them using oil-based ink, and then the plates are moistened with wet sponges. The reason oil-based ink is used in this technique is that the paint applied to the plates only adheres to the areas where oil-based ink has been applied.

5.7 Sublimation printing

With this method, the design is transferred to the ceramic surface using a special transfer paper, but the paints applied to this paper are sprayed using special printers. Sublimation is known as the process of a material changing directly from a solid to a gas state when heated, and the design is transferred to the ceramic surface by heating the paint transferred to the paper so that it turns into a gas. This method offers advantages in terms of cost and production speed, and it involves a system consisting of a printer, scanner, press, and a computer that controls the system, facilitates the transfer of the design. This method, which involves applying temperatures of up to approximately 200°C, is widely used in the mass production of cups and similar products. In this method, printing can also be done on the special lacquer applied to the ceramic [3,44].

5.8 Decal printing

In the decal printing method, printing is applied to the ceramic surface through a decal. Decals are prepared using the paints used in screen printing, and this method, which is applied to the glazed ceramic surface, involves sublimation to apply the design to the ceramic surface. This method is used very frequently and offers the option of using many colors. However, a film must be used for each color when using these colors. These films are subjected to a process called exposure, and after this stage, each color is printed separately. The lacquered paper is soaked in warm water until the design separates from the lacquered paper. The partially separated design is applied to the glazed ceramic surface by creating a slippery intermediate layer with water, and then the surface to which the design is applied is rinsed with water. The surface is then subjected to another sintering process to ensure that the design diffuses into the glazed surface. Figure 12 shows the application stages of the print in sequence [44,45].

Figure 12. Decal printing application stages and resulting ceramic works [46].

These techniques can be used to transfer an image onto a ceramic surface, but they can also be used to transfer digitally created drawings, enabling artists to communicate their designs to a wider audience with higher quality and greater stability through mass production (Figure 13).

Figure 13. Examples of printed cups made from the thesis work designs [44].

5.9 Screen printing

When examined in general terms, printing refers to methods that involve a mold and the process of transferring a pattern to another place using this mold. This method is suitable for mass production and artistic It is one of the preferred ceramic decoration techniques due to its ability to support the narrative. In this method, the surface to be applied is decisive. In addition to application on glazed surfaces, it can be examined in three different categories: underglaze, overglaze, and in-glaze. This method has a history dating back to the first seal prints, and it is a method that can be used to print on uncooked clay, cooked ceramic surfaces, plaster plates, and decal paper. In addition to being applied directly to the ceramic surface, it is also possible to transfer it through an intermediate element with the output. This method, also referred to as screen printing, stencil printing, or silk screen printing, uses screens made of silk. The process involves transferring the design onto the screen through exposure, leaving the areas where the design is to be applied open, and then transferring the ink onto the substrate material scraped off from these areas using a squeegee.

However, with today's advancing technology, a method called roto color is used to transfer patterns to surfaces in a serial manner, particularly in tile production, using a rotary printing mold, ensuring continuity in both labor and product quality. This method, which is also used in fields such as graphics, painting, and textiles, differs from those used in ceramics in terms of the content of paint. Applied by mixing with a medium on glaze or primer and subjected to sintering, this method differs from other fields in terms of its thermal properties (Figure 14) [45,47].

Figure 14. Examples of industrial and artistic designs printed using screen printing [48].

Printing methods, in their most general definition, are methods that enable the transfer of an image from one place to another using a prepared mold. These methods enable the transfer of patterns, sketches, and photographs, allowing plastic elements such as color, stain, symbol, and texture to be used in artistic expression, enabling its use in a supportive manner. The printing methods used in the decoration of ceramic surfaces are divided into two main groups in terms of application: direct and indirect.

According to the definition in the sources, serigraphy, also known as "screen printing," is one of the leading "direct" printing techniques in which the design is transferred to the surface in a single step without being copied from another surface. This technique, which is widely used in graphic arts and printmaking, is also commonly seen in the textile and ceramic industries. The screen-printing technique, which is used to transfer the desired design onto a surface, is also frequently used in the field of art.

The screen-printing technique begins with mounting and stretching silk (or specially woven nylon/synthetic material) with pores of different sizes onto a metal or wooden frame. The color paste is transferred to the surface through the screen using a knife. Rotary screen printing (Rotocolor) uses a rotating screen to apply this principle automatically.

With the screen-printing method, it is possible to make single-color and multi-color prints on ceramic products directly or indirectly (transfer/decal). While the direct screen-printing technique is mostly used for single-color work, the indirect method can easily be used for complex multi-color prints.

It is the most popular technique used in the ceramic industry. It is also known in literature as screen printing, stencil printing, or silk-screen printing. The history of the screen-printing method dates back to the first use of stencils. The screen-printing method, which is the most preferred by ceramic artists, is also used in graphic processes and printed images in the textile field. With this method, printing can be applied to uncooked clay, cooked ceramic surfaces, plaster plates, and decal paper. When printing on ceramic surfaces, as in the lithography printing method, certain changes are made to the chemistry of the paints used. The paints used in this method are different from those used in other methods. This is because the preferred paints in this method are glaze-top paints mixed with medium.

When referring to ceramic decoration, the preferred terms generally relate to where the desired design will be applied, and this applies to transfer techniques as well. Water-based transfer techniques used to transfer images onto ceramic surfaces are performed in three different ways: glaze-in, glaze-under, and glaze-top.

5.9.1 Removing the cover

In ceramics, and especially in porcelain, it is difficult to maintain vivid colors at high temperatures, so applications are made at lower temperatures on glazed products. The most important and suitable method for mass production is the decal method. Patterns designed using specific software programs are printed onto water-soluble transfer paper and then transferred onto the glazed ceramic and/or porcelain surface. In this transfer process, the firing temperature of the transfer or paint must be lower than the firing temperature of the glaze on the product. This temperature value, typically around 800-900°C, is achieved through the third firing, which fixes the paint to the glazed surface through diffusion (Figure 15). Compared to standard underglaze products, the main disadvantage of these products is that the designs may fade over time due to washing and use [49].

Figure 15. *Overglaze decal applications [50].*

Ceramics undergo a production stage involving heat, and applications continue the final product during the decoration stage, requiring the firing process to be carried out with great precision. These productions are generally carried out in chamber kilns and must be in an oxygen atmosphere. During the application process, the area around the design may vaporize and evaporate due to heat, causing gas to accumulate in the environment, which is then vented through a chimney. If this process is not carried out correctly, it may result in a rough, incomplete, or visually flawed surface, which may prevent customer requests from being fulfilled. Such productions are concentrated during special times such as New Year's, Valentine's Day, and graduations, but they are also preferred for employee gift requests, promotions for room members, or events such as symposiums and conferences, and they are used as a production method that can respond to local designs and requests.

5.9.2 Removal of the core

This method is achieved by applying pre-firing removal to the glazed surface during vitrification and sintering, while high-grade glaze paints are applied to the glazed surface and fired into the glaze, resulting in a more durable method. In such productions, since only a single firing is performed, the raw body is dried, retouched, and glazed, and this application is also applied to the glaze, enabling the process to be carried out without altering the standard production format. Considering that energy is one of the largest expenses in the ceramics industry, applying decals such as logos and quality codes using this method and ensuring they are suitable for the desired temperature provides an advantage. This process is generally outsourced to companies specializing in decal work, and the desired parameters are provided to ensure that the decals and designs are produced in accordance with the company's production conditions. In our country, such work requires infrastructure, and since companies do not use decals continuously, they prefer to outsource such tasks to specialized firms in the field.

5.9.3 Under-glaze removal

When examining this method, it is known that it involves applying designs to the baked body using suitable paints or heat-sensitive fixing papers, followed by sintering with a transparent glaze [51]. Underglaze applications are particularly preferred in ceramic products, and the application of certain custom-made designs using decals provides convenience compared to hand decoration, thereby enabling the visual preferences of customers to be met in handmade products. In ceramic production known as "samur," the process involves engraving and painting on a coated surface. Although it is a valuable production method, it is very labor-intensive and time-consuming, making it unsuitable for mass production. This has led to the search for new methods, and underglaze decals have emerged as a guiding solution in this regard (Figure 16).

Figure 16. *Examples of ceramic underglaze printing applications [52,53].*

Today, due to the widespread use of digital designs, increasing demand, and the ease of application, digital printing techniques are used in all areas. It is widely used. The ability to transfer designs created in the program directly to the product also offers different possibilities. This method increases the applicability of vivid colors and designs and enhances the appeal of the ceramic surface. The application of the evolving and transforming understanding of art with digital developments to ceramic surfaces is also possible with these printing techniques.

Especially in the tile industry, feeding multi-colored designs to continuously operate kilns and meeting market demands can only be achieved through these methods. Market demand emerging in many regions such as the US, Europe, and Africa has also supported the design of devices capable of working with multiple colors, enabling the transfer of desired patterns or images onto ceramic surfaces with a color scale of up to 12 colors, while also bringing improvements in resolution and different ink productions [8]. All these developments are also opening up new areas for artistic work in industrial ceramics.

Laser printing: Although transfer paper is also used in laser printing, the iron oxide content of the cartridges is important, and laser printer outputs ranging from brown to black are also used (Figure 17).

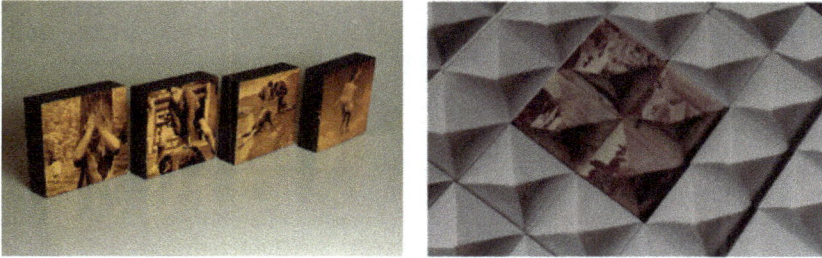

Figure 17. Examples of laser printing applications [54],[55].

In printing processes, two types of transfer paper are used, known as decals. In overglaze applications, lacquered paper is used, while in laser printing, transfer media with lacquer applied to it is used. In these decal applications, if there is no intermediate layer at the point of contact with water, color distribution may occur, so the design is dried after it is transferred to the paper. Application on ceramic surfaces is performed after the printing on this paper has dried, and after heating, the paint pigments become permanently fixed on the ceramic surface. It occurs as a result of the arrival of heat and the vaporization of the intermediate layer. The applied temperature levels vary. The most important reasons for this are the iron oxide ratio and the type of glaze. During processes carried out at temperatures between 800-1000°C, the iron oxide is absorbed by the glazed surface due to the increased heat, causing the pigment to settle into the amorphous glassy surface, thereby ensuring the permanence of the print. The correct adjustment of this temperature is essential. Firing at temperatures higher or lower than the recommended firing temperature can cause issues such as lightness or darkness in the print color or the print not diffusing properly into the surface [8,54,56].

Rotocolor printing: As mentioned in previous sections, printing techniques have come a long way since they were first applied using traditional methods, and advancing technology has had a particularly significant impact on these methods. Initially, flat printing was done using silk screens, followed by drum printing, and subsequently rotocolor printing, which has created a significant industrial-scale application area. In this printing method, patterns are printed onto silicone-based screens using laser technology, and different screen applications are created depending on the desired application. If we break down this process into stages:

➤ Creating gaps in the silicone surface and filling these gaps with paint.

➤ After applying paint to this screen, feed it through the openings to transfer the pattern onto the tile.

➤ In the final stage, the paint fed into the screen is brought into contact with the roll on the screen to complete the printing process.

It consists of the following processes.

Inkjet printing: Patented in the 1950s, this printing technique is used in many areas. In addition to enabling patterns prepared on a computer to be applied to ceramic surfaces, its most important advantage over other methods is that it is applied by spraying ink without meeting the biscuit body. This reduces the error rate and minimizes factors such as pattern loss and product loss in mass production. Additionally, unlike rotocolor printing, which requires equipment such as screens, stencils, and films, inkjet printing allows for easy changes to the design without the need for such setup, making it suitable for any type of design or artistic application on tiles. It is an important and preferred method due to its ease of transfer, low error rate, and high resolution values [1,57].

In ceramics, the decal technique is preferred in businesses where digital printing is not feasible in terms of investment costs and application areas, but many materials are used in the application of the methods. Among these materials, the two most important categories are metal oxide-based paint used inside, on top of, and beneath the glaze, and intermediate decal papers that facilitate the transfer of the design [58]. The paints used in decal papers are composed of water-insoluble ceramic paints. Whether digital or screen printing is preferred, it is possible to obtain decal paper during the decal process. In this section, in addition to the surface application material we call lacquer, kerosene, medium, thinner, elephant ear, and solvent are used. These materials are commercially available, but care must be taken when using them. The continuity of the design in application is possible through the use of these products and the controlled determination of production parameters.

In ceramics, patterns are transferred using two types of application: direct (indirect) and indirect (indirect). Direct (indirect) printing is preferred for designs with fewer colors and simpler patterns, while indirect (indirect) printing is used for more difficult and complex designs. The screen-printing method is particularly preferred for indirect printing. The reason for this is that this method yields better visual results [54,55,57,59].

In the direct (indirect) printing method, the ceramic surface is applied through screens, and a separate screen is used for each color. The main difficulties of this method are the preparation of separate screens, high costs, and the difficulty of stacking the screens. The large amount of waste generated at all stages is also one of the biggest disadvantages of this method. The non-transparency of the paints and inks used, the necessity of cleaning the ceramic surface before application, and the need to establish a mechanical system in screen printing according to the dimensions of the forms and shapes to be applied are also among the challenges of this method [4]–[59]. In this method, which can be done manually or with a device, the printing process can vary between 400 and 2500 depending on the device's specifications [55,60,61].

When transfer printing is desired on ceramic surfaces using toner-filled devices, decal paper is used, and this technique exhibits similar characteristics to the transfer methods used with laser printers. This method is preferred in designs where printing with similar colors is difficult, and its development accelerated between the 1990s and 2000s, taking on its current form. Today, two different ceramic toner sets—the Magenta Color Set and the Red Color Set—are commercially available, enabling three-dimensional designs prepared via software on digital platforms to be produced using rapid prototyping devices [3],[62].

The indirect printing method is used to apply complex patterns with different dimensions and minimal sizes that cannot be achieved with direct printing. The global paint industry has developed and continues to develop special paints for screen printing, and decals are produced using the screen-printing method. These decals are then transferred to the ceramic, porcelain, or glass surface to be printed. In this method, where printing in different colors is possible, the design feasibility study is carried out to determine how many different colors will be printed, after which the screen density is adjusted. Subsequently, the screen weaving process and exposure processes are carried out according to the design. After exposure, the printing of all colors onto the paper is done separately and in sequence. The number and quality of colors affect the cost [63,64]. If we list these processes, the first stage consists of preparing the decal, preparing the template and pattern, preparing the film according to the design and color separation, adjusting the screen according to the colors, and exposing. Afterwards, the second stage consists of paint selection, printing, drying, transferring the prepared decal onto the desired product, and baking [3,63]. After these processes are completed, the printed area is coated with a lacquer. The purpose of this process is to transfer the print without smudging the paint. The advantage of this method is that it is easy to apply concave or flat surfaces. However, the cleanliness and smoothness of the surface to be applied is important for print quality. The transfer process must be done by sliding in water so that the lacquer is on top and the paint is on the bottom. This process is carried out in the following stages:

➢ Cleaning the ceramic surface to prepare it for printing

➢ Place the printed paper in a container filled with water and leave it for 1-2 minutes to ensure that the lacquered pattern separates from the paper.

➢ Transferring the design onto the ceramic surface by sliding the design paper [65].

➢ Points to note in practice

➢ Ensure that there is no dust or air between the area to be printed and the print.

➢ After applying the decal, remove any air and water from the ceramic decal surface (using a toothbrush, sponge, cloth, etc.).

If these steps are not performed correctly, the ceramic will not adhere to the print. The design will not produce the desired color and texture on the surface [66]. The firing stage is also very important after this stage. Fingerprints left on the product during application and subsequent placement in the kiln, or firing in a reductive environment, can prevent the pattern from forming on the ceramic surface as designed. Firing must take place in an environment with plenty of oxygen to allow the burning parts to vaporize. The formation of each color is achieved through the addition of a different condoxide, and the reaction of the elements with oxygen results in significant effects on the colors. For example, while it is known that iron, which has different valences, produces colors ranging from red to brown, the oxygen ratio can cause the formation of Fe_2O_3 or FeO, demonstrating that the desired color cannot be obtained without oxygen. In ceramic production, temperature and content affect surface properties and color at all stages, from raw materials to colored sintered decorative products [67–69].

6. Conclusion

Throughout human history, change has affected the products that are produced, and this study examines the possibilities that have arisen in the transfer of both industrial and artistic designs into ceramic objects. Developments in printing methods have kept pace with the unstoppable advances in the digital field and the transformation of design, and each new method has made it easier for ceramic artists and designers to work with ceramics and continues to do so.

Every kind of development in society influences artists, and artists influenced by many phenomena, from Greek philosophy to everyday life, from esoteric approaches to movements, express these approaches in their own unique language through different forms of expression, whether abstract or concrete. With the beginning of the departure from traditional artistic understanding in the 17th and 18th centuries, the art movements that began with the impressionist artists were very quickly influenced by the changes brought about by the French Revolution and the Industrial Revolution. The invention of steam engines and many technological developments that followed resulted in the invention of printing machines, and art has been heading in a different direction ever since. The invention of the camera followed these developments and has now reached a point where it affects not only printing but also digital technologies. Ceramics, being a product that undergoes many stages and thermal processes, has been one of the most challenging application areas in terms of color and design. Therefore, at the current stage, the transfer of the desired design to the ceramic surface without alteration is of great importance, and in mass production, it provides artists with the opportunity to create richer productions by overcoming the limitations of the material. Digital-based productions are not only artistic but are also used intensively on an industrial scale, and equipment such as three-dimensional printers that produce ceramics using deposition techniques create a much broader field of application for imagination, visual impact, and the journey of transforming the imagined into a product. However, will all these technologies affect the artist's productive power? This is where the real question arises. How will the emotional bond that ceramic artists have established with their works, from hand-kneading clay throughout history to today's digital productions, be preserved? Repairing this bond, which was severed with industrialization, is also achieved by adapting to these developments. The progress of ceramic artists is solved not by looking at these technologies with prejudice, but by finding the best way to express themselves in these new conditions. Unlike all other art forms, ceramics involves more technical aspects than other art forms in terms of thermal processes and raw material recipes, and its industrial aspect can adapt more easily to new technologies. Transforming the universe created in the artist's mind into art using digital technologies yields a more vibrant result. When we look at the past, present, and future of ceramic art and the ceramic industry today, we see that the interest in and progress of this field are due to the primitive methods used in the past, which have evolved over time. This has been achieved through advances in technology and its acceptance of revision. Just as the transition from open-fire cooking to natural gas ovens and from manual grinding to continuous grinding mills has been accepted and has facilitated production, digital developments in printing methods have also provided many conveniences and created opportunities for progress. The industrial production of ceramic art is technologically important today. The direction of the sector shows that technology will continue to develop in the future, bringing new revisions to ceramic works. When the current stage is examined, it is thought that not only production, but also digital printing will influence designs, and that this development will be achieved by following digital technologies and embracing this transformation.

References

[1]   Sibel SEVİM, S. *et al.* Günümüz Serami Endüstrisinde Ve Artistik Seramik Yüzeylerde

Kullanilan BaskTekniklerinden Örnekler.

[2] Gör, Ö. and Öztürk, B. (2010). Seramik Sanatında İmaj Transfer Teknikleri. *Seramik Sanatında İmaj Transfer Teknikleri.*

[3] Ye, Y. *et al.* (2021). 3D Printing of Integrated Ceramic Membranes by the DLP Method. *Industrial and Engineering Chemistry Research.* https://doi.org/10.1021/ACS.IECR.1C02224/SUPPL_FILE/IE1C02224_SI_001.PDF

[4] Lee, J.Y. *et al.* (2017). Fundamentals and applications of 3D printing for novel materials. *Applied Materials Today.* https://doi.org/10.1016/J.APMT.2017.02.004

[5] Huang, K. *et al.* (2020). Complex SiOC ceramics from 2D structures by 3D printing and origami. *Additive Manufacturing.* https://doi.org/10.1016/J.ADDMA.2020.101144

[6] Layani, M. *et al.* (2018). Novel Materials for 3D Printing by Photopolymerization. *Advanced Materials.* https://doi.org/10.1002/ADMA.201706344

[7] Petrie, K. (2011). *Ceramic Transfer Printing.*

[8] ÖZGÜVEN, S. (2019). Seramik Yüzeylerde Dijital Baskı Uygulamaları. *Sanat ve Tasarım Dergisi.* https://doi.org/10.20488/sanattasarim.600038

[9] Hilditch, J. (2016). Ceramic analysis in Greece. *Archaeological Reports.* https://doi.org/10.1017/S0570608416000089

[10] Sibel SEVİM, S. *et al.* (2013). Günümüz Seramik Endüstrisinde ve Artistik Seramik Yüzeylerde Kullanılan Baskı Tekniklerinden Örnekler. *Sanat ve Tasarım Dergisi.*

[11] MERCİN, L. *et al.* (2021). Sofra Seramiklerinde Dekorlama Teknikleri, Ink-Jet Dekorlama Tekniği İle Diğer Tekniklerin Birlikte Kullanımı. *İnönü Üniversitesi Sanat ve Tasarım Dergisi.* https://doi.org/10.16950/iujad.937255

[12] Hartmann, P. *et al.* (2010). Optical glass and glass ceramic historical aspects and recent developments: a Schott view. *Applied Optics, Vol. 49, Issue 16, pp. D157-D176.* https://doi.org/10.1364/AO.49.00D157

[13] García, I.S. de S. et al. (2021). Archaeometry and analysis of ceramic materials from Ávila (Spain): Late-vetton evidence. Sustainability (Switzerland). https://doi.org/10.3390/su13115910

[14] Göksu, H. *et al.* (2020). Oxidation of Benzyl Alcohol Compounds in the Presence of Carbon Hybrid Supported Platinum Nanoparticles (Pt@CHs) in Oxygen Atmosphere. *Scientific Reports.* https://doi.org/10.1038/S41598-020-62400-5;TECHMETA=131,140,145,146;SUBJMETA=45,638,639,77,884;KWRD=BIOCHEMISTRY,CATALYST+SYNTHESIS

[15] Nagraik, R. *et al.* (2021). Amalgamation of biosensors and nanotechnology in disease diagnosis: Mini-review. *Sensors International.* https://doi.org/10.1016/J.SINTL.2021.100089

[16] Altuner, E.E. *et al.* (2021). Dendrimer-based nanocomposites for alcohol fuel cells, in *Nanomaterials for Direct Alcohol Fuel Cells*, Elsevier, pp. 337–352.

[17] Altuner, E.E. *et al.* (2021). Commercial aspects of direct alcohol fuel cells. *Nanomaterials for Direct Alcohol Fuel Cells: Characterization, Design, and Electrocatalysis.* https://doi.org/10.1016/B978-0-12-821713-9.00012-3

[18] Altuner, E.E. *et al.* (2022). Development of electrochemical aptasensors detecting

phosphate ions on TMB substrate with epoxy-based mesoporous silica nanoparticles. *Chemosphere*. https://doi.org/10.1016/J.CHEMOSPHERE.2022.134077

[19] Şen, B. *et al.* (2018). High-performance graphite-supported ruthenium nanocatalyst for hydrogen evolution reaction. *Journal of Molecular Liquids*. https://doi.org/10.1016/J.MOLLIQ.2018.07.117

[20] Jiang, Y. *et al.* (2020). Ultrahigh Breakdown Strength and Improved Energy Density of Polymer Nanocomposites with Gradient Distribution of Ceramic Nanoparticles. *Advanced Functional Materials*. https://doi.org/10.1002/ADFM.201906112;PAGEGROUP:STRING:PUBLICATION

[21] Halabian, R. *et al.* (2019). Composite Nanoscaffolds Modified with Bio-ceramic Nanoparticles (Zn2SiO4) Prompted Osteogenic Differentiation of Human Induced Pluripotent Stem Cells. *International Journal of Molecular and Cellular Medicine*. https://doi.org/10.22088/IJMCM.BUMS.8.1.24

[22] Gol, F. *et al.* (2022). Coloring effect of iron oxide content on ceramic glazes and their comparison with the similar waste containing materials. *Ceramics International*. https://doi.org/10.1016/J.CERAMINT.2021.10.001

[23] Sharath, B.N. *et al.* (2021). Tribological Suitability of aluminium hybrid composite above atmospheric temperature. *IOP Conference Series: Materials Science and Engineering*. https://doi.org/10.1088/1757-899X/1189/1/012018

[24] Kuncewicz, K. (2016). Figural Anatolian Stamp Seals From Three Assyrian Colony Period Sites : Karahöyük - Konya , Acemhöyük and Kültepe.

[25] Hussain, M.I. *et al.* (2024). Digital light processing 3D printing of ceramic materials: a review on basic concept, challenges, and applications. *The International Journal of Advanced Manufacturing Technology*. https://doi.org/10.1007/s00170-023-12847-3

[26] Arndt, D. *et al.* (2021). The deal.II finite element library: Design, features, and insights. *Computers & Mathematics with Applications*. https://doi.org/10.1016/J.CAMWA.2020.02.022

[27] Rashed, M.G. *et al.* (2016). Metallic microlattice materials: A current state of the art on manufacturing, mechanical properties and applications. *Materials & Design*. https://doi.org/10.1016/J.MATDES.2016.01.146

[28] YÜKSEL, İ. (2018). SERAMİK YÜZEYLERDE GEOMETRİK UNSURLARIN KULLANIMI. *Journal of International Social Research*. https://doi.org/10.17719/JISR.2018.2664

[29] Abdullah, M. *et al.* (2016). Monoprint Technique: Medium and Expression. *Proceedings of the 2nd International Colloquium of Art and Design Education Research (i-CADER 2015)*. https://doi.org/10.1007/978-981-10-0237-3_41

[30] Parraman, C. (2017). Colour printing techniques and new developments in colour printing. *Colour Design: Theories and Applications: Second Edition*. https://doi.org/10.1016/B978-0-08-101270-3.00025-4

[31] Song, X. *et al.* (2018). Influence of song porcelain aesthetics on modern product design. *Lecture Notes in Computer Science (including subseries Lecture Notes in Artificial Intelligence and Lecture Notes in Bioinformatics)*. https://doi.org/10.1007/978-3-319-92141-9_13/FIGURES/20

[32] Yücel, Ç. and Parlıtı, U. (2023). Ancient Cylinder Seals from Upper Mesopotamia. *Anatolian Research*. https://doi.org/10.26650/ANAR.2023.29.1334041

[33] Whatley, L.J. (2019). Introduction: Approaches to Medieval Seals and Sealing Practices, in *A Companion to Seals in the Middle Ages*, BRILL, pp. 1–16.

[34] Hesse, R. (2010). LiDAR-derived Local Relief Models – a new tool for archaeological prospection. *Archaeological Prospection*. https://doi.org/10.1002/ARP.374

[35] Lossada, E. (2018). Revista de la Universidad del Zulia Ciencias Sociales y Artes cambios en los ú ltimos tiempos . Un an á lisis.

[36] Al Sayegh, S. (2023). Study Of Seljuq Textile Patterns. https://doi.org/10.26756/TH.2022.531

[37] Randhawa, K.S. (2024). A state-of-the-art review on advanced ceramic materials: fabrication, characteristics, applications, and wettability. *Pigment & Resin Technology*. https://doi.org/10.1108/PRT-12-2022-0144

[38] SEVİM, S.S. and YILDIRIM, Ö. (2016). Çağdaş Türk Seramik Sanatında Resimsel Anlatım. *Anadolu Üniversitesi Sanat & Tasarım Dergisi*. https://doi.org/10.20488/austd.06496

[39] Hu, C. *et al.* (2020). Performance of Waterborne Epoxy Emulsion Sand Fog Seal as a Preventive Pavement Maintenance Method: From Laboratory to Field. *Advances in Materials Science and Engineering*. https://doi.org/10.1155/2020/6425817

[40] Karakuyu, M. (2016). The Effect of Western Understanding of Fine Arts Education on the Printmaking Artists of Turkey. *The International Journal of Interdisciplinary Cultural Studies*. https://doi.org/10.18848/2327-008X/CGP/v11i03/34-43

[41] Shah, M.A. *et al.* (2021). Classifications and Applications of Inkjet Printing Technology: A Review. *IEEE Access*. https://doi.org/10.1109/ACCESS.2021.3119219.

[42] Naile, D. and Özet, Ç. (2018). Disiplinler Arası Etkileşimler Kapsamında Alterntif Malzemeler Ve Seramik-Baskı Resim Yakınlaşmaları Üzerine Bireysel Uygulamalar.

[43] azaz, hend (2022). Lithography as a printing technique and its impact on form and content in the field of contemporary ceramics. *International Journal of Design and Fashion Studies*. https://doi.org/10.21608/ijdfs.2023.212979.1013

[44] Enst, M. *et al.* (2021). GRAFİK TASARIMDA KÜLTÜREL ESİNLENME VE SERAMİK BASKI YÖNTEMİ : ADANA BÖLGESİ HEYBE MOTİFLERİ UYGULAMASI.

[45] Gökçe, C. and Feyzoğlu, E. (2024). A GENERAL OVERVIEW OF PORTRAIT WORKS ON CERAMIC SURFACES. *Turkish Online Journal of Design Art and Communication*. https://doi.org/10.7456/tojdac.1464181

[46] Dermeik, B. and Travitzky, N. (2020). Laminated Object Manufacturing of Ceramic-Based Materials. *Advanced Engineering Materials*. https://doi.org/10.1002/adem.202000256

[47] Sesli, Y. (2024). Serigrafi Baskı Temel Terimlerinin Tanımlanması. *Avrasya Terim Dergisi*. https://doi.org/10.31451/ejatd.1417615

[48] Özkartal, M. and Acar, A. (2015). SERİGRAFİ BaskiYöntemi KullanilmişSeramik Yüzeylerde SagPişirim Uygulamalari.

[49] Zhang, Z. and Lu, H. (2021). Influence of ceramic molding technology on its decorative techniques by Analysis of Computer Software. *Journal of Physics: Conference Series.* https://doi.org/10.1088/1742-6596/1915/3/032082

[50] Yu, B.F. (2012). The Research on New and High-Grade Screen-Printing Ceramic Decal Paper under Glaze Color of once Firing. *Applied Mechanics and Materials.* https://doi.org/10.4028/www.scientific.net/AMM.192.275

[51] Moreno, A. *et al.* (2010). Innovations and New Trends in Ceramic Tile Decoration. https://doi.org/10.4028/www.scientific.net/AST.68.165

[52] Doç, Y. and Şan, P. (2016). Seramik Yüzeylerde Monobaskı Uygulamaları Öz.

[53] Ayşe Pamuk, N.R.O. (2016). Tür Çini Sanatinda Kullanilan Hayvansa Figürlerin Seramik Yüzeyler Üzerinde Üç Boyutlu Uygulanmasi.

[54] Kalay Leman (2009). Seramik yüzeylerde kullanılan baskı teknikleri ve uygulamaları.

[55] Hwa, L.C. *et al.* (2017). Recent advances in 3D printing of porous ceramics: A review. *Current Opinion in Solid State and Materials Science.* https://doi.org/10.1016/j.cossms.2017.08.002

[56] Enstit, S. *et al.* (2013). Seramik sanatinda minimalist düzenlemeler.

[57] Chen, X. *et al.* (2023). 3D printing for precision construction of ceramic membranes: Current status, challenges, and prospects. *Advanced Membranes.* https://doi.org/10.1016/j.advmem.2023.100068

[58] Khand, N.H. *et al.* (2021). A new electrochemical method for the detection of quercetin in onion, honey and green tea using Co3O4 modified GCE. *Journal of Food Measurement and Characterization.* https://doi.org/10.1007/s11694-021-00956-0

[59] Sauer, M. *et al.* (2004). Screen Printing. *Sol-Gel Technologies for Glass Producers and Users.* https://doi.org/10.1007/978-0-387-88953-5_14

[60] Pinto, T. V. *et al.* (2016). Screen-Printed Photochromic Textiles through New Inks Based on SiO2@naphthopyran Nanoparticles. *ACS Applied Materials and Interfaces.* https://doi.org/10.1021/acsami.6b06686

[61] Sel, E. (2006). Dünya Ve Türkiye Ölçeğinde Doğal Taş Ve Seramik Kaplama Malzemelerinin Sektörel Analizi.

[62] Cavalcante, P.M.T. *et al.* (2009). Colour performance of ceramic nano-pigments. *Dyes and Pigments.* https://doi.org/10.1016/j.dyepig.2008.07.004

[63] Chen, H. *et al.* (2018). 3D printing of SiC ceramic: Direct ink writing with a solution of preceramic polymers. *Journal of the European Ceramic Society.* https://doi.org/10.1016/J.JEURCERAMSOC.2018.08.009

[64] Pinargote, N.W.S. *et al.* (2020). Direct Ink Writing Technology (3D Printing) of Graphene-Based Ceramic Nanocomposites: A Review. *Nanomaterials 2020, Vol. 10, Page 1300.* https://doi.org/10.3390/NANO10071300

[65] Wang, G. *et al.* (2023). Ceramic 3D Printing via Dye-Sensitized Photopolymerization Under Green LED. *3D Printing and Additive Manufacturing.* https://doi.org/10.1089/3dp.2021.0204

[66] HAKAN VERDU MARTİNEZ, E. and CAN, E. (2016). Bilgisayar Destekli Seramik Üretim Yöntemi Olarak Üç Boyutlu Yazıcılar ve Günümüz Koşullarında Uygulama

Örneği. *Anadolu Üniversitesi Sanat & Tasarım Dergisi.* https://doi.org/10.20488/www-std-anadolu-edu-tr.290760

[67] Acharjee, N. *et al.* (2023). A review of various ceramic pigment preparation and characterization methodologies for applications. *Journal of the Australian Ceramic Society.* https://doi.org/10.1007/s41779-023-00853-3

[68] (2017). Different methods of printing transfers for ceramic bases and their ecological aspect. *European Scientific e-Journal.* https://doi.org/10.47451/her2020-11-001

[69] TITE, M.S. *et al.* (1998). LEAD GLAZES IN ANTIQUITY—METHODS OF PRODUCTION AND REASONS FOR USE*. *Archaeometry.* https://doi.org/10.1111/j.1475-4754.1998.tb00836.x

A Journey from Raw Materials to Ceramics

Materials Research Foundations 184 (2025)

Materials Research Forum LLC

https://doi.org/21741/9781644903834

Chapter 13

Contemporary Digital Printing Methods in Industrial Ceramics

Nihal Derin Coskun[1*], Ebru Halvaci[2], Irem Turk[2], Selcuk Erdogan[2], Cumhur Eren Isık[3], Fatih Sen[2*]

[1]Ceramic and Glass Department, Ordu University, 52200 Ordu, Türkiye

[2]Sen Research Group, Department of Biochemistry, Kutahya Dumlupinar University, Kutahya 43000, Türkiye

[3]Department of Handicrafts, Kutahya Fine Arts Vocational School, Kutahya Dumlupınar University, Evliya Çelebi Campus, 43000 Kutahya, Türkiye

nihalderincoskun@odu.edu.tr, fatihsen1980@gmail.com

Abstract

In this chapter, current data and studies on the processing of ceramic materials in digital printing processes on industrial platforms are discussed and presented. In particular, technologies based on slurry material, which is one of the main raw materials of ceramics, are discussed. Stereolithography (SL), digital light process (DLP), two-photon polymerisation (TPP), inkjet printing (IJP), and direct in writing (DIW) are among the slurry-based printing technologies. Selective laser sintering and selective laser melting are powder-based techniques and the LOM technique is a bulk solid technique. Each method is detailed in this chapter. As a result, it is determined that these methods are particularly suitable for use in high-tech products and allow the production of materials with superior properties such as strength, porosity, biocompatibility, etc. by overcoming the difficulties of production of materials with desired superior properties, allowing the production of building materials at lower costs and providing variable properties according to the place of use.

Keywords

Digital Technique, Printing Technique, 3D Printing, Ceramics

1. Introduction

Ceramic materials are advantageous materials due to their high temperature and wear resistance, high hardness, inertness, optical and magnetic properties. As a result of these advantages, they are used in a wide range of applications such as dental, biomaterials, defence industry, porcelain, tile, chemical industry, machinery, electronics, aerospace and biomedical engineering [1]. Ceramic materials are shaped by various technological methods during production. These methods include injection [2], mould pressing [3], tape casting [2], gel casting [4], etc. Ceramic slip consisting of stoichiometrically prepared mixtures with or without binders and various auxiliary additives at the application stage is prepared by traditional methods depending on the requested product [1,5]. However, increasing technological developments every day lead to the elimination of expensive and difficult production methods. Modern machines increase the number and types of products obtained with environmentally friendly and more economical alternative energy. In this way, developing production changes customer demands and makes design an integral part of production. With the developing technology at the design stage; the patterns made on the surfaces of traditional ceramics have changed and developed over the years with flat screen printing, rotary screen printing and rotary gravure printing methods on the raw product. These developments have reduced the losses caused by production errors in manual applications. Additive manufacturing (AM) and three-dimensional (3D) printing technologies developed in this direction are gaining importance day by day. In addition to visual effects, the production of three-dimensional products has increased the design perspective and the use of computers. The necessity of ceramics to be hard and durable leads to the development of this field.

Three-dimensional printing is a series of advanced manufacturing technologies that add a digital layer to two-dimensional sections. Three-dimensional printing is a technique that has been implemented using traditional production methods such as casting and machining [1,6]. The 3D printing technique, which is the continuation of the AM technique that first appeared in 1892, has developed to the present day. Although 3D printing is used in many fields, it is an important method used in the production of complex products, especially in the science and engineering communities. Additive manufacturing (AM) 3D printing techniques are also a preferred system for the production of ceramic components with a large industrial market [1]. Ceramic materials produced with the use of 3D printing technique were first produced by Marcus et al. in the nineties and took their place in the literature [1,7].

In the preliminary preparations made before printing with 3D printing method, classification is made according to the starting recipe. Liquid mud-based, powder-based and plastic mud-based printing methods are preferred [1]. A summary of these methods is shown in Table 1.

Tablo 1. 3D printing techniques of ceramic materials [1].

Feedstock form	3D ceramic technology	Abbreviation
Slurry based	Stereolithography	SL
	Digital light processing	DLP
	Two-photon polymerisation	TPP
	Inject printing	IJP
	Direct in writing	DIW
Powder based	Three-dimensional printing	3DP
	Selective laser sintering	SLS
	Selective laser melting	SLM
Bulk solid-based	Dominated object manufacturing	LOM
	Fused deposition modelling	FDM

231

2. Slurry Based Technologies

Liquid slurry-based ceramic materials in 3D printing technology systems depend on the solid-liquid ratio, viscosity, thixotropy, ink or paste shape of the mixture and include collodial liquid systems in which ceramic particles are dispersed as feedstock. The content of the liquid slurry is obtained by photopolymerisation, extrusion or printing by inkjet methods [1,8]. However, in these systems, the rheology should be controlled very well and the optimum amounts of the chemicals to be used should be adjusted very well.

2.1 Stereolithography (SL)

Stereolithography (SL) technique is one of the most popular slurry based techniques in the world [1,9]. The SL technique was first popularised by 3D systems Inc. in 1986, supported by the body [1,10]. In this method, the amount of solid used in the system and the electrolyte balance are important, and the dry and wet strength value requires calculation according to the size of the sample produced. This technique is based on the light-activated polymerisation process. The polymerisation proceeds from layer to layer and from layer to layer with the help of light. The polymerisation technique is initiated by the progression from layer to layer. In the polymerisation process, the photopolymerised monomers are dyed with a special dye to distinguish them. These dyes usually consist of specially selected plant essences. The SL technique should be capable of producing samples down to micrometres [1,11]. As it can be understood from this point, the SL technique applied in ceramic materials covers the studies in the micro or nanometre size of ceramic samples. This helps to obtain technical ceramic products with high added value from bioceramics, scaffolds, automotive parts, electronic ceramics, filters and energy materials. SL technique applied to ceramic samples is in aqueous or non-aqueous environment [1,12,13]. To accelerate the realisation of this process, a ceramic suspension is obtained using the necessary surfactants and additives. Ceramic particles are inert to light emission. Polymerisation under light irradiation occurs by stimulation of active monomers. SL resin moulds for the casting of complex ceramic samples can also be prepared and used in other applications [1,14]. Figure 1 shows the process of the SL technique.

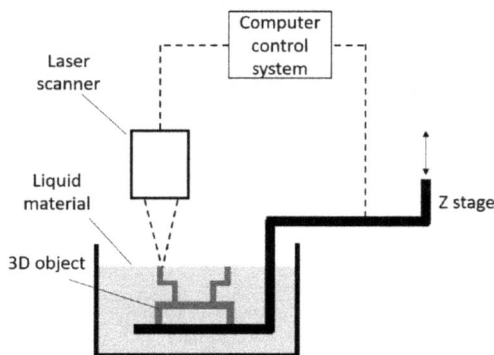

Figure 1. Schematic process of the SL technique [15], reprinted with permission from MDPI.

According to the reports in the literature, a study on the SL technique from ceramic suspensions (v/V 65%) consisting of silica, alumina and silicon contents has been investigated since 1994 [9,16–18]. Long-term stabilisation and appropriate rheological behaviour as well as suitable viscosity factors are important in this procedure. The ceramic particles are dispersed homogeneously and efficiently by photopolymerisation. The particles then undergo segregation. Segregated particles are unstable particles with very high energy. Therefore, a non-homogeneity can be seen in the process. For this reason, the system should be kept under control and a good flow and viscosity should be maintained [1]. The main problem to be considered in the application of ceramic samples in the SL technique is that the scattering of ceramic particles against irradiation affects the suspension and is undesirable [19]. The hardening depth depends on the volume fraction of the ceramic particles, particle size, light exposure and the reflectivity index of the materials. The refractive indices of photopolymerised ceramic particles and particles of ceramic samples are different. Therefore, the wavelength of photopolymerisation can be difficult [1,20]. The bonding between atoms in ceramic products affects the crystal structure. Shaping, which cannot be done in a controlled manner while in solution, affects the subsequent production stages. In photopolymerisation, the wavelength also varies according to the distance between atoms. This control is possible with kinetic polymerisation. When the studies carried out for this purpose are examined, it is seen that different particles are used as auxiliary agents for kinetic polymerisation of ceramic suspensions according to *Badev et al.* These auxiliary agents are particles such as SiO_2, Al_2O_3, ZrO_2 and SiC [21]. The refractive index ratio in ceramic particles and organic phase is the main criterion influencing the viscosity reaction. Thus, polymerisation affects the transformation [1,21]. The ceramic SL technique is used and applied in wide areas. These applications include cored casting moulds [22,23], various microelectronic components (e.g. sensors) [24,25], photonic crystals [26,27] and various dental components [28], such areas are applications. Figure 2 shows examples of these applications. According to reports in the literature, it is stated that the residual samples remaining in the SL technique affect the segregation of the particles and the light-curing effect and the segregation effect [18,29,30]. In this method, the quality of the light, i.e. the wavelength, is important rather than the amount of light. UV (200-400 nm) and visible light (400-700 nm) ranges are used in photopolymerisation.

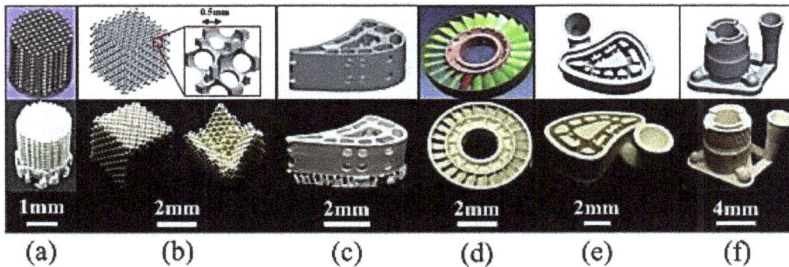

Figure 2. *Some examples produced by SL technique, **(a)** porous bioceramic scaffold, **(b)** photonic crystal, **(c)** hollow turbine blade, **(d)** impeller **(e-f)** investment casting moulds [31], reprinted with permission from MDPI.*

In the production of structures where pore size is important, the production started with suitable precursors and under photopolymerisation conditions allows the production of materials with the desired properties. In SL technique, aqueous suspensions have a negative effect on the strength. For this reason, suspensions in SL technique are used as anhydrous, resin-based or acrylamide-based [1]. In the studies in the literature, it is seen that these problems are tried to be overcome by preparing suspension by using silica instead of deionised water [13]. In materials produced by SL technique, it is also difficult to clean the raw body. Removal of the chemicals used here is important in terms of final product quality [32]. In other studies, instead of homogeneous suspensions, other forms of photopolymerisation are increasingly used in the SL technique. For example, silicon oxycarbide is an example of these forms [33]. Comparison of some silicon carbide parts prepared by SL technique with printed and sintered samples and the image of SiCw/Al₂O₃ ceramic matrix composite materials produced by SL technique.

2.2 Digital light processing technique (DLP)

In the products obtained by this method, samples are sintered at temperatures determined by thermal analyses after shaping and high strength and high temperature resistant products are obtained in accordance with the intended use. Compared to traditional methods, additive manufacturing methods allow the application of designs with high production numbers, complex shapes and large dimensions. In this respect, it provides advantages in terms of both financial gain and time saving [34].

Digital light processing is based on the technique of initiating polymerisation by photons at specific time intervals. Photomerisation is initiated by using a projector on a certain part of the entire segments of the resin. Patterns can be revealed with this technique, resulting in a more complex, independent and serial result [35,36]. In the DLP method, ceramic samples are filled with resin to approximately 40-60% for the printing process. The filled ceramic samples are then exposed to light and hardened to initiate photopolymerisation. After the exposure to light and hardening process is completed, the sintering process is started. The photo polymerised green coloured samples are sintered last [35–38]. Since the rheology of SLA and DLP techniques depends on chemicals, their stability and the production process are affected by all kinds of external factors. One of the main challenges is the inability to maintain viscosity when printing ceramic samples and the need for dispersion through special dispensers [36], [37]. These techniques are among the methods preferred by many branches of the industry due to their advantages. In particular, high turnover material markets such as medicine, military applications, dentistry, automotive sector, which require high-tech materials, have been entered thanks to these techniques [40].

The DLP technique was first proposed by Nakamoto et al. [41] in 1996. For this, the authors used physical masks [1]. In 1997, Bertsch et al. [42] developed and elaborated the DLP technique using liquid crystal display (LCD). The replacement of LCD devices with micro mirror devices (DMD) by Texas Instruments since 2001 has increased competitiveness. DLPs are known to have higher contrast and resolution than LCDs [43,44]. DMD is a chip containing one hundred thousand or more microscopic mirrors corresponding to the pixels in the cross-section of the sample to be imaged. The micro mirrors are moved by forces based on electrostatic attraction. The micro mirrors are rectangular and can be individually rotated between 10 and 12 degrees. Thus, the mirrors can be turned on or off. Micro-structured mirrors function as ultra-fast light switches. They reflect the incoming rays with precise resolution on the projection and make it appear bright or dark [1]. Figure 3 shows the DLP diagram. As can be seen in Figure 3, the exposure of the single layers is mediated by the transparent material underneath. Printing with DLP technique is very fast and convenient [45–49].

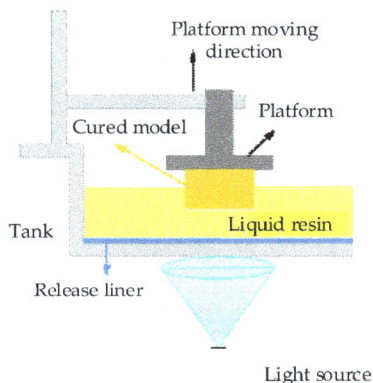

Figure 3. Schematic diagram of DLP 3D printing mechanism [50], reprinted with permission from MDPI.

The difference between SLA and DLP techniques is the use of the laser beam. While no support material is required in DLP, a point source laser beam is required in SLA [51]. Various traditional investigations on the DLP technique as a three-dimensional printing tool have taken place in the literature. For example, parts containing zirconia and alumina with Vickers hardness between 97-99% have been compared with conventional methods [51,52]. The DLP printing technique is highly advantageous and versatile for the production of products with complex, fine and sensitive properties such as alumina and bioactive glass. In addition, the DLP printing technique shows a high strength of over 90% [53–56]. Studies with this method continue and better results are obtained day by day with developmental studies such as UV curing and the effects of different dispersants on improving the final product properties. From zirconia to alumina, Si_3N_4 to SiCN materials, important commercial products such as $BaTiO_3$ piezoelectric are produced by this method and support the market position of advanced technology ceramics [51].

2.2.1 Continuous digital light processing (CDLP)

Continuous digital light processing technique (CDLP) is known as continuous liquid interface manufacturing. According to the CDLP method, light projection is a technique used to produce high-resolution parts with high mechanical properties using oxygen transparent lenses and liquid resins [57]. The CDLP technique is similar to the DLP technique, but since the platform on the z-axis moves continuously, the process is very fast. In this way, the desired parts can be completed in a short time. According to reports in the literature, in 2015, monolithic polymeric sections with a resolution of 10 μm and below were consistently produced using CLIP technology [47], [56]. The fact that the layer-by-layer production applied in DLP is continuous in CDLP has provided an advantage in the production of scaffold interface biomaterials and has provided the production of equivalent materials to replace damaged bones in the medical field.

Materials Research Forum LLC
https://doi.org/21741/9781644903834

2.3 Two photon polymerisation (TPP)

Today, there is a significant demand for nanotechnological studies. Nano means dwarf in Latin and covers very small sections equivalent to one billionth of a metre [60]. Numerous reports on nanotechnological studies have appeared in the literatüre [58]-[74]. TPP techniques are in great demand in nanotechnology. Therefore, they are used in areas such as nano electronics, nano mechanics and nano biomedicine. TPP technique has become widespread in materials science and laser applications. It is an activated process for polymerisation by absorption of two photons from a 780 nm or 515 nm laser [42]. Polymerisation process starts with the excitation of the particles by the laser beam. This method occurs in several steps. These steps are;

- Two photon absorption
- Radical formation
- Initiation of cross linking
- Polymerization chain growth
- Inhibition

It occurs under five headings, including [78].

The advantages of the TPP technique, which acts as an intermediary for nanostructured materials, are its ability to initiate polymerisation of foci in polymer liquids via TPA. Polymerisation takes place only on the surface of the liquid product. Therefore, the process does not work like single photon techniques such as SL [79]. TPP resolution has a value of 200 nm or less. TPA speed is density related [80]. Its use in the production of TPP 3D was reported in the literature in 1990 and 1992. The first production samples consisted of 7 μm and 3D microstructures [81]. The production of TPP technique with the help of urethane acrylate resins was carried out in the studies [79]-[84]. At the same time, extensive research has been conducted on this subject worldwide [87,88]. Very important studies have been carried out by obtaining printed silica glasses with surface roughness values below 200 nm. Materials produced by this production method show optically superior properties.

2.4 Inkjet printing process (IJP)

Inkjet printing process (IJP) is a widely preferred method in two-dimensional digital printing. Although this technique is highly preferred today, it is a highly preferred method for spray processing paper, plastic and various materials between cheap printers, various compacts and printer types on the market [89]. While the rate of development of IJP technology increased in the 1950s, this technique was further advanced in the 1970s with brands such as Epson and Canon [90]. Production with ink materials has a very important place in this technique. Especially polymers or metals for electronic modelling and two-dimensional cells for microelectronic soldering have been developed [91]. In addition, small droplets can be printed by means of thermal stimulation or piezoelectric effect in three-dimensional printing by means of the inkjet printing process. The ink is limited to the printing of miniaturised cross-sections due to its very small volume [90,92]. The ceramic particles are deposited directly onto the substrate with a print head or by the deposition method in a liquid solvent to ensure a good dispersion of the ceramic inks. Especially in piezoelectric materials and 3D micro moulds, positive results are obtained with this method [93,94].

2.5 Direct ink writing (DIW)

Direct inject writing (DIW) is known as robocasting and was first patented in 1997 by Cesarano et al. [95]. DIW, which is an environmentally friendly method among sustainable production methods, is especially preferred in energy storage applications [96–98]. With its controlled and adjustable parameters such as temperature, viscosity and crosslinking time, it is a method that allows diversity in obtaining different and difficult to produce complex structures [96]. This method is also advantageous in the production of ceramic, polymer, metal and composite materials as it can be adapted to robotic casting and production by extrusion from a pressurised nozzle. The materials obtained by DIW method are also preferred because they provide functions such as biocompatibility, optical properties, conductivity, colour change etc. [97,98]. In this technique, the sludge is enriched with simple and basic organic compounds for the concentration of the samples [99]. The viscosity of the materials used for deposition is high. The high viscosity of the materials is reflected in the extrusion of the materials. Samples are formed by moving from layer to layer in direct typing until the part is completed. The DIW diagram is obtained by sintering and decomposing the parts (Figure 4).

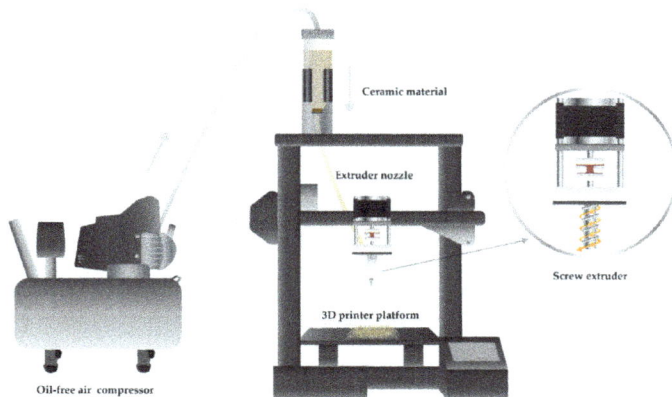

Figure 4. *Diagram of the ceramic feed DIW process [100], reprinted with permission from MDPI.*

2.6 Multiple spraying technique (Polyjet-PJ)

In this method, parameters such as the material to be applied on, the angle between the spray system and the surface etc. are important. Resin is used as photopolymer. Multiple spraying technique works similarly to the Inkjet technology system. Thanks to these properties, the hardness or softness of the samples can be adjusted. The photopolymers are cured while the resins are sprayed onto the printing template. The applied voltage and nozzle diameter shape the formation of the sample during this curing phase [101].

3. Powder Based Technologies

3.1 Selective laser melting (SLM)

Selective laser melting (SLM) system, which is one of the powder bed fusion techniques, is an additive manufacturing technology system. Sintering is performed in SLM. Popular in the aerospace and medical sectors, this technology is not ideal for the home due to its high cost. Good temperature control of the process is essential. It is mostly preferred in part manufacturing [102].

3.1.1 Electron beam melting (EBM)

The difference of EBM from SLM is that it works with electron beams and parts can be manufactured with this method. The EBM system is based on the principle of complete melting of metal powders, alloys and metals under a high pressure atmosphere by means of synergistic energy of a beam from an electron providing high energy and heat. Layers are obtained from geometrically shaped samples. The pressure provided by the vacuum prevents the collision of gas molecules and electrons. Significant energy consumption is also avoided. In addition, strength and accurate production are among its advantages [102,103].

3.1.2 Binder jetting (Binder JettingBJ)

In this method, a liquid binding material is used to bond the materials together. Special adhesives are added to the powder materials and bonding is achieved. A solid layer is obtained [104]. Then special sands, metals, ceramics are sprinkled on the layers and the process is completed. The metals produced by the BJ method give low values compared to the mechanical properties of the materials obtained by fusion and direct energy deposition methods. In this method, sintering is required to remove the adhesive from the structure and to obtain a denser structure [105]. In addition, all parameters from the material used in production, to the application, to the device that makes the printing applied to the product obtained affect the final product. For this reason, correct planning of the process and application experience are important. All these methods are preferred according to the material content to be used and the desired properties in the final product. It provides a great advantage especially in complex parts that are difficult to manufacture and in areas where the production quantity increases.

3.1.3 Electron beam additive manufacturing (EBAM)

Electron beam additive manufacturing (EBAM) is an additive manufacturing process obtained by welding metal powders or cable materials of metals through the beam of electrons. It provides more intense energy compared to laser methods, especially in materials obtained with components of Fe, Ni, Al, Cu alloys. EBAM is among the preferred methods Decently compared to other methods (SLS) due to its low cost, easy production and advantages of complex products. Powdered materials are sprayed through a nozzle, and metal cables are used instead of powdered metals [106].

4. Bulk solid based technologies

4.1 Laminated object manufacturing (LOM)

The laminated printing technique (LOM) system is a system based on the addition and removal of apparatus. Paper, plastic film, metallic sheet and ceramic tape are the most common materials used in LOM technique. In this method, production is carried out using melting filament materials. It allows the fabrication of materials that require high hardness and strength such as SiC [107]. CO_2 based lasers are used for cutting according to geometrical shapes. After cutting, the thickness of the sheet with the materials is between 0.05 mm and 5 mm. The layers are glued together by parameters such as pressure and heating. Disadvantages include low resolution of the cut materials, complex internal cavities and channels [108].

5. Conclusion

In particular, three-dimensional printing techniques are highly preferred methods in digital printing. Mud-based technologies are at the heart of digital printing processes. Various types of printing are categorized by techniques such as SL, DLP, CDLP, IJP, DIW, LOM, EBAM and EBM. The most basic of the techniques is the SL process in general. The SL technique is based on photopolymerization. Through photopolymerization, monomers, which are primarily the smallest building blocks of polymers, are stimulated and radically activated via photons. Thus, the activated monomer initiates the reaction chain. With the initiated reaction, photo polymerization begins. DLP technique is very similar to SL technique, but the most basic difference is that photopolymerization starts at certain and December time intervals. CDLP, on the other hand, is known as a liquid interface and works with transparent, colorless samples. In the DIW technique, printing is achieved by condensing the mud with the simplest organic compounds. The IJP technique is one of the most widely used methods in digital printers. In the same way, the multiple spraying technique is similar to IJP [107]. In the LOM technique, there is a state of addition and subtraction. The production is carried out by using layer by layer and sintering. In SLM, EBM and EBAM, the melting and excitation process of laser and electron is observed, so printing is provided. Bonding process is performed by including binders with special materials such as metals and alloys into the sample. As a result, it can be said that digital printing techniques used in industrial production, which have an important place in many sectors in meeting the increasing world population and changing product demands, are important in many branches of the industry and are becoming increasingly important in the future.

All these methods are used in many types of materials and engineering fields, but since interdisciplinary cooperation has not increased Decently, it seems that some productions that remain on a local basis will have more efficient use by working together of teams. These methods, which provide advantages from an industrial point of view, are used in the defense industry, biomaterials, renewable energy, etc. although its use in promising areas is not sufficient, the studies carried out in this area remain in the company's inventories and prevent teams that do not have sufficient infrastructure from catching up in these areas. For this reason, it is necessary to train more people by going to the training provided in these areas and to cooperate with the manufacturers of the sector. Although the primary investment inputs are not clear in the literature, it is necessary to create an information infrastructure and calculations related to depreciation expenses and the use of these methods.

References

[1]   Chen, Z. *et al.* (2019). 3D printing of ceramics: A review. *Journal of the European Ceramic Society*. https://doi.org/10.1016/j.jeurceramsoc.2018.11.013

[2]   Zhang, Y. *et al.* (2010). Preparation and properties of bimodal porous apatite ceramics through slip casting using different hydroxyapatite powders. *Ceramics International*.

https://doi.org/10.1016/J.CERAMINT.2009.07.008

[3]   Duan, X. *et al.* (2016). Influence of hot-press sintering parameters on microstructures and mechanical properties of h-BN ceramics. *Journal of Alloys and Compounds.* https://doi.org/10.1016/J.JALLCOM.2016.05.153

[4]   Tulliani, J.M. *et al.* (2013). Development and mechanical characterization of novel ceramic foams fabricated by gel-casting. *Journal of the European Ceramic Society.* https://doi.org/10.1016/J.JEURCERAMSOC.2013.01.038

[5]   Thümmler, F. (1990). Engineering ceramics. *Journal of the European Ceramic Society.* https://doi.org/10.1016/0955-2219(90)90011-4

[6]   Edgar, J. and Tint, S. (2015). “Additive Manufacturing Technologies: 3D Printing, Rapid Prototyping, and Direct Digital Manufacturing”, 2nd Edition. *Johnson Matthey Technology Review.* https://doi.org/10.1595/205651315X688406

[7]   Marcus, H.L. *et al.* (1990). Solid freeform fabrication. Powder processing. *American Ceramic Society Bulletin.*

[8]   Tang, H.H. and Yen, H.C. (2015). Slurry-based additive manufacturing of ceramic parts by selective laser burn-out. *Journal of the European Ceramic Society.* https://doi.org/10.1016/J.JEURCERAMSOC.2014.10.019

[9]   Brady, G.A. and Halloran, J.W. (1997). Stereolithography of ceramic suspensions. *Rapid Prototyping Journal.* https://doi.org/10.1108/13552549710176680/FULL/XML

[10]   Kruth, J.P. *et al.* (1998). Progress in additive manufacturing and rapid prototyping. *CIRP Annals - Manufacturing Technology.* https://doi.org/10.1016/S0007-8506(07)63240-5

[11]   Jacobs, P.F. (1992). Fundamentals of Stereolithography. https://doi.org/10.15781/T24M91T5H

[12]   Griffith, M.L. and Halloran, J.W. (1996). Freeform Fabrication of Ceramics via Stereolithography. *Journal of the American Ceramic Society.* https://doi.org/10.1111/J.1151-2916.1996.TB09022.X

[13]   Chen, Z. *et al.* (2010). Curing characteristics of ceramic stereolithography for an aqueous-based silica suspension. *Proceedings of the Institution of Mechanical Engineers, Part B: Journal of Engineering Manufacture.* https://doi.org/10.1243/09544054JEM1751

[14]   Lombardo, S.J. (2015). Minimum Time Heating Cycles for Diffusion-Controlled Binder Removal from Ceramic Green Bodies. *Journal of the American Ceramic Society.* https://doi.org/10.1111/JACE.13284

[15]   Huang, J. *et al.* (2020). A Review of Stereolithography: Processes and Systems. *Processes.* https://doi.org/10.3390/pr8091138

[16]   Griffith, M.L. and Halloran, J. (1994). Ultraviolet curable ceramic suspensions for stereolithography of ceramics.

[17]   Halloran, J.W. *et al.* (2011). Photopolymerization of powder suspensions for shaping ceramics. *Journal of the European Ceramic Society.* https://doi.org/10.1016/j.jeurceramsoc.2010.12.003

[18]   Bae, C.J. *et al.* (2018). Quantifying particle segregation in sequential layers fabricated by additive manufacturing. *Journal of the European Ceramic Society.* https://doi.org/10.1016/J.JEURCERAMSOC.2018.02.008

[19]   Gentry, S.P. and Halloran, J.W. (2013). Depth and width of cured lines in photopolymerizable ceramic suspensions. *Journal of the European Ceramic Society*. https://doi.org/10.1016/J.JEURCERAMSOC.2013.02.033

[20]   de Hazan, Y. and Penner, D. (2017). SiC and SiOC ceramic articles produced by stereolithography of acrylate modified polycarbosilane systems. *Journal of the European Ceramic Society*. https://doi.org/10.1016/J.JEURCERAMSOC.2017.03.021

[21]   Badev, A. *et al.* (2011). Photopolymerization kinetics of a polyether acrylate in the presence of ceramic fillers used in stereolithography. *Journal of Photochemistry and Photobiology A: Chemistry*. https://doi.org/10.1016/J.JPHOTOCHEM.2011.05.010

[22]   Zhou, W.Z. *et al.* (2009). Direct fabrication of an integral ceramic mould by stereolithography: *http://dx.doi.org/10.1243/09544054JEM1628*. https://doi.org/10.1243/09544054JEM1628

[23]   Chen, Z. *et al.* (2012). Process parameters appraisal of fabricating ceramic parts based on stereolithography using the Taguchi method: *http://dx.doi.org/10.1177/0954405412442607*. https://doi.org/10.1177/0954405412442607

[24]   Nguyen, N.T. *et al.* (2010). Design and characterization of 60-GHz integrated lens antennas fabricated through ceramic stereolithography. *IEEE Transactions on Antennas and Propagation*. https://doi.org/10.1109/TAP.2010.2050447

[25]   Leigh, S.J. *et al.* (2011). A miniature flow sensor fabricated by micro-stereolithography employing a magnetite/acrylic nanocomposite resin. *Sensors and Actuators A: Physical*. https://doi.org/10.1016/J.SNA.2011.03.058

[26]   Chen, W. *et al.* (2007). Fabrication and Measurement of Micro Three-Dimensional Photonic Crystals of SiO2 Ceramic for Terahertz Wave Applications. *Journal of the American Ceramic Society*. https://doi.org/10.1111/J.1551-2916.2007.01676.X

[27]   Kirihara, S. and Niki, T. (2015). Three-Dimensional Stereolithography of Alumina Photonic Crystals for Terahertz Wave Localization. *International Journal of Applied Ceramic Technology*. https://doi.org/10.1111/IJAC.12320

[28]   Lian, Q. *et al.* (2018). Additive manufacturing of ZrO2 ceramic dental bridges by stereolithography. *Rapid Prototyping Journal*. https://doi.org/10.1108/RPJ-09-2016-0144/FULL/XML

[29]   Bae, C.J. and Halloran, J.W. (2011). Influence of Residual Monomer on Cracking in Ceramics Fabricated by Stereolithography. *International Journal of Applied Ceramic Technology*. https://doi.org/10.1111/J.1744-7402.2010.02578.X

[30]   Mitteramskogler, G. *et al.* (2014). Light curing strategies for lithography-based additive manufacturing of customized ceramics. *Additive Manufacturing*. https://doi.org/10.1016/J.ADDMA.2014.08.003

[31]   Gao, B. *et al.* (2022). A Review of Research Progress in Selective Laser Melting (SLM). *Micromachines*. https://doi.org/10.3390/mi14010057

[32]   Xing, Z. *et al.* (2022). Efficient cleaning of ceramic green bodies with complex architectures fabricated by stereolithography-based additive manufacturing via high viscoelastic paste. *Additive Manufacturing*. https://doi.org/10.1016/j.addma.2022.102809

[33]   Colombo, P. *et al.* (2010). Polymer-Derived Ceramics: 40 Years of Research and

Innovation in Advanced Ceramics. *Journal of the American Ceramic Society.* https://doi.org/10.1111/j.1551-2916.2010.03876.x

[34] Chaudhary, R. *et al.* (2023). Additive manufacturing by digital light processing: a review. *Progress in Additive Manufacturing.* https://doi.org/10.1007/s40964-022-00336-0

[35] Schmidt, J. and Colombo, P. (2018). Digital light processing of ceramic components from polysiloxanes. *Journal of the European Ceramic Society.* https://doi.org/10.1016/J.JEURCERAMSOC.2017.07.033

[36] Zocca, A. *et al.* (2015). Additive Manufacturing of Ceramics: Issues, Potentialities, and Opportunities. *Journal of the American Ceramic Society.* https://doi.org/10.1111/JACE.13700

[37] Dadkhah, M. *et al.* (2023). Additive manufacturing of ceramics: Advances, challenges, and outlook. *Journal of the European Ceramic Society.* https://doi.org/10.1016/j.jeurceramsoc.2023.07.033

[38] Jang, J.H. *et al.* (2000). Preparation and Characterization of Barium Titanate Suspensions for Stereolithography. *Journal of the American Ceramic Society.* https://doi.org/10.1111/J.1151-2916.2000.TB01467.X

[39] Travitzky, N. *et al.* (2014). Additive Manufacturing of Ceramic-Based Materials. *Advanced Engineering Materials.* https://doi.org/10.1002/ADEM.201400097

[40] Shahzadi, L. *et al.* (2022). Functional Materials for DLP-SLA 3D Printing Using Thiol–Acrylate Chemistry: Resin Design and Postprint Applications. *ACS Applied Polymer Materials.* https://doi.org/10.1021/acsapm.2c00358

[41] Nakamoto, T. *et al.* Consideration on the producing of high aspect ratio micro parts using UV sensitive photopolymer. *MHS'96 Proceedings of the Seventh International Symposium on Micro Machine and Human Science.* https://doi.org/10.1109/MHS.1996.563401

[42] Bertsch, A. *et al.* (1997). Microstereophotolithography using a liquid crystal display as dynamic mask-generator. *Microsystem Technologies 1997 3:2.* https://doi.org/10.1007/S005420050053

[43] Kaneko and Y. (2001). UV Exposure System for Photolithography and Rapid Prototyping Using DMD Projector. *Asia Display/IDW'01.*

[44] Hadipoespito, G.W. *et al.* (2003). Digital Micromirror Device Based Microstereolithography for Micro Structures of Transparent Photopolymer and Nanocomposites. https://doi.org/10.26153/TSW/5409

[45] Zhou, C. and Chen, Y. (2009). Calibrating Large-area Mask Projection Stereolithography for Its Accuracy and Resolution Improvements. *University of Texas at Austin.* https://doi.org/10.26153/tsw/15091

[46] Choi, J.W. *et al.* (2009). Fabrication of 3D biocompatible/biodegradable micro-scaffolds using dynamic mask projection microstereolithography. *Journal of Materials Processing Technology.* https://doi.org/10.1016/J.JMATPROTEC.2009.05.004

[47] Pan, Y. *et al.* (2012). A fast mask projection stereolithography process for fabricating digital models in minutes. *Journal of Manufacturing Science and Engineering, Transactions of the ASME.* https://doi.org/10.1115/1.4007465/474915

[48] Zhang, A.P. *et al.* (2012). Rapid Fabrication of Complex 3D Extracellular Microenvironments by Dynamic Optical Projection Stereolithography. *Advanced Materials.* https://doi.org/10.1002/ADMA.201202024

[49] Lee, M.P. *et al.* (2015). Development of a 3D printer using scanning projection stereolithography. *Scientific Reports 2015 5:1.* https://doi.org/10.1038/srep09875

[50] Jiang, T. *et al.* (2022). Enhanced Adhesion Efficient Demolding Integration DLP 3D Printing Device. *Applied Sciences.* https://doi.org/10.3390/app12157373

[51] He, R. *et al.* (2018). Fabrication of complex-shaped zirconia ceramic parts via a DLP-stereolithography-based 3D printing method. *Ceramics International.* https://doi.org/10.1016/J.CERAMINT.2017.11.135

[52] Zhou, M. *et al.* (2016). Preparation of a defect-free alumina cutting tool via additive manufacturing based on stereolithography – Optimization of the drying and debinding processes. *Ceramics International.* https://doi.org/10.1016/J.CERAMINT.2016.04.050

[53] Felzmann, R. *et al.* (2012). Lithography-Based Additive Manufacturing of Cellular Ceramic Structures. *Advanced Engineering Materials.* https://doi.org/10.1002/ADEM.201200010

[54] Tesavibul, P. *et al.* (2012). Processing of 45S5 Bioglass® by lithography-based additive manufacturing. *Materials Letters.* https://doi.org/10.1016/J.MATLET.2012.01.019

[55] Hatzenbichler, M. *et al.* (2012). DLP-based light engines for additive manufacturing of ceramic parts. *https://doi.org/10.1117/12.907113.* https://doi.org/10.1117/12.907113

[56] Gmeiner, R. *et al.* (2015). Stereolithographic Ceramic Manufacturing of High Strength Bioactive Glass. *International Journal of Applied Ceramic Technology.* https://doi.org/10.1111/IJAC.12325

[57] Surmen, H.K. (2019). Eklemeli İmalat (3 B baskı) : teknolojiler ve uygulamalar. *Uludağ Üniversitesi Mühendislik Fakültesi Dergisi.* https://doi.org/10.17482/uumfd.519147

[58] Park, J. *et al.* (2023). A Review on Recent Advances in Piezoelectric Ceramic 3D Printing. *Actuators.* https://doi.org/10.3390/act12040177

[59] Enbergs, S. *et al.* (2023). 3D Printing of Bone Substitutes Based on Vat Photopolymerization Processes: A Systematic Review. *Journal of Tissue Engineering and Regenerative Medicine.* https://doi.org/10.1155/2023/3901448

[60] Altuner, E.E. *et al.* (2021). Dendrimer-based nanocomposites for alcohol fuel cells, in *Nanomaterials for Direct Alcohol Fuel Cells*, Elsevier, pp. 337–352.

[61] Altuner, E.E. *et al.* (2021). Ternary/quaternary nanomaterials for direct alcohol fuel cells, in *Nanomaterials for Direct Alcohol Fuel Cells*, Elsevier, pp. 157–172.

[62] Aygun, A. *et al.* (2023). Highly active PdPt bimetallic nanoparticles synthesized by one-step bioreduction method: Characterizations, anticancer, antibacterial activities and evaluation of their catalytic effect for hydrogen generation. *International Journal of Hydrogen Energy.* https://doi.org/10.1016/j.ijhydene.2021.12.144

[63] Altuner, E.E. *et al.* (2022). High-efficiency application of CTS-Co NPs mimicking peroxidase enzyme on TMB(ox). *Chemosphere.* https://doi.org/10.1016/j.chemosphere.2021.133429

[64] Goksu, H. *et al.* (2021). The synthesis and characterization of size-controlled bimetallic

nanoparticles, in *Nanomaterials for Direct Alcohol Fuel Cells*, Elsevier, pp. 433–447.

[65]   Akin, M. *et al.* (2021). Fundamentals of alcohol fuel cells, in *Nanomaterials for Direct Alcohol Fuel Cells*, Elsevier, pp. 75–94.

[66]   Bekmezci, M. *et al.* (2021). Fundamentals of electrochemistry. *Nanomaterials for Direct Alcohol Fuel Cells*. https://doi.org/10.1016/B978-0-12-821713-9.00023-8

[67]   Nas, M.S. *et al.* (2019). Magnetic nanocomposites decorated on multiwalled carbon nanotube for removal of Maxilon Blue 5G using the sono-Fenton method. *Scientific Reports*. https://doi.org/10.1038/s41598-019-47393-0

[68]   Unal, F.A. *et al.* (2020). Synthesis, characterization, and application of transition metals (Ni, Zr, and Fe) doped TiO2 photoelectrodes for dye-sensitized solar cells. *Journal of Molecular Liquids*. https://doi.org/10.1016/j.molliq.2019.112177

[69]   Gulbagca, F. *et al.* (2021). The synthesis and characterization of Pt-based catalysts for hydrogen storage applications. *Nanomaterials for Hydrogen Storage Applications*. https://doi.org/10.1016/B978-0-12-819476-8.00005-0

[70]   Acidereli, H. *et al.* (2020). Palladium/ruthenium supported on graphene oxide (PdRu@GO) as an efficient, stable and rapid catalyst for hydrogen production from DMAB under room conditions. *Renewable Energy*. https://doi.org/10.1016/j.renene.2020.07.105

[71]   Arikan, K. *et al.* (2022). Glucose nano biosensor with non-enzymatic excellent sensitivity prepared with nickel–cobalt nanocomposites on f-MWCNT. *Chemosphere*.

[72]   Sen, B. *et al.* (2019). Monodisperse ruthenium–copper alloy nanoparticles decorated on reduced graphene oxide for dehydrogenation of DMAB. *International Journal of Hydrogen Energy*. https://doi.org/10.1016/j.ijhydene.2019.02.176

[73]   Alptekin, O. *et al.* (2020). Use of silica-based homogeneously distributed gold nickel nanohybrid as a stable nanocatalyst for the hydrogen production from the dimethylamine borane. *Scientific Reports 2020 10:1*. https://doi.org/10.1038/s41598-020-64221-y

[74]   Aygün, A. *et al.* (2020). Biological synthesis of silver nanoparticles using Rheum ribes and evaluation of their anticarcinogenic and antimicrobial potential: A novel approach in phytonanotechnology. *Journal of Pharmaceutical and Biomedical Analysis*. https://doi.org/10.1016/J.JPBA.2019.113012

[75]   Bilgicli, H.G. *et al.* (2020). Composites of palladium nanoparticles and graphene oxide as a highly active and reusable catalyst for the hydrogenation of nitroarenes. *Microporous and Mesoporous Materials*. https://doi.org/10.1016/J.MICROMESO.2020.110014

[76]   Korkmaz, N. *et al.* (2020). Biogenic nano silver: Synthesis, characterization, antibacterial, antibiofilms, and enzymatic activity. *Advanced Powder Technology*. https://doi.org/10.1016/J.APT.2020.05.020

[77]   Hernandez, F.J. and Ozalp, V.C. (2012). Graphene and Other Nanomaterial-Based Electrochemical Aptasensors. *Biosensors 2012, Vol. 2, Pages 1-14*. https://doi.org/10.3390/BIOS2010001

[78]   Jaiswal, A. *et al.* (2023). Two decades of two-photon lithography: Materials science perspective for additive manufacturing of 2D/3D nano-microstructures. *iScience*. https://doi.org/10.1016/j.isci.2023.106374

[79] Sun, H.B. and Kawata, S. (2004). Two-Photon Photopolymerization and 3D Lithographic Microfabrication. *Advances in Polymer Science*. https://doi.org/10.1007/B94405

[80] Lee, K.S. *et al.* (2008). Advances in 3D nano/microfabrication using two-photon initiated polymerization. *Progress in Polymer Science (Oxford)*. https://doi.org/10.1016/j.progpolymsci.2008.01.001

[81] Wu, E.-S. *et al.* (1992). Two-photon lithography for microelectronic application. *https://doi.org/10.1117/12.130367*. https://doi.org/10.1117/12.130367

[82] Sänger, J.C. et al. (2020). First time additively manufactured advanced ceramics by using two-photon polymerization for powder processing. Open Ceramics. https://doi.org/10.1016/j.oceram.2020.100040

[83] Kawata, S. *et al.* (2001). Finer features for functional microdevices. *Nature 2001 412:6848*. https://doi.org/10.1038/35089130

[84] Sun, H.B. *et al.* (1999). Three-dimensional photonic crystal structures achieved with two-photon-absorption photopolymerization of resin. *Applied Physics Letters*. https://doi.org/10.1063/1.123367

[85] Cumpston, B.H. *et al.* (1999). Two-photon polymerization initiators for three-dimensional optical data storage and microfabrication. *Nature 1999 398:6722*. https://doi.org/10.1038/17989

[86] Coenjarts, C.A. and Ober, C.K. (2004). Two-Photon Three-Dimensional Microfabrication of Poly(Dimethylsiloxane) Elastomers. *Chemistry of Materials*. https://doi.org/10.1021/CM048717Z

[87] Takada, K. *et al.* (2005). Improved spatial resolution and surface roughness in photopolymerization-based laser nanowriting. *Applied Physics Letters*. https://doi.org/10.1063/1.1864249

[88] Schizas, C. et al. (2010). On the design and fabrication by two-photon polymerization of a readily assembled micro-valve. The International Journal of Advanced Manufacturing Technology. https://doi.org/10.1007/s00170-009-2320-4

[89] Cao, T. *et al.* (2024). Inkjet printing quality improvement research progress: A review. *Heliyon*. https://doi.org/10.1016/j.heliyon.2024.e30163

[90] Singh, M. *et al.* (2010). Inkjet Printing—Process and Its Applications. *Advanced Materials*. https://doi.org/10.1002/ADMA.200901141

[91] Dong, H. *et al.* (2006). An experimental study of drop-on-demand drop formation. *Physics of Fluids*. https://doi.org/10.1063/1.2217929

[92] Nakamura, M. *et al.* (2006). Biocompatible Inkjet Printing Technique for Designed Seeding of Individual Living Cells. *https://home.liebertpub.com/ten*. https://doi.org/10.1089/TEN.2005.11.1658

[93] Das, B. and Mohanty, S. (2025). Synthesis Of Ferroelectric Bi0.5Na0.5TiO3 Ink Towards Fabrication of a Flexible Piezoelectric Nanogenerator for Bio-Mechanical Energy Harvesting. https://doi.org/10.2139/ssrn.5208533

[94] Park, J. *et al.* (2025). Additive manufacturing of water-soluble 3D micro molds for complex-shaped lipid microparticles. *Nature Communications*. https://doi.org/10.1038/s41467-025-56984-7

[95]   Cesarano, J. and Calvert, P.D. (2000). Freeforming objects with low-binder slurry. issued Oct. 2000.

[96]   Tagliaferri, S. *et al.* (2021). Direct ink writing of energy materials. *Materials Advances.* https://doi.org/10.1039/D0MA00753F

[97]   Wan, X. *et al.* (2020). Direct Ink Writing Based 4D Printing of Materials and Their Applications. *Advanced Science.* https://doi.org/10.1002/advs.202001000

[98]   Paul D L, B. *et al.* (2023). Rheological Behavior and Printability Study of Tri-Calcium Phosphate Ceramic Inks for Direct Ink Writing Method. *Polymers.* https://doi.org/10.3390/polym15061433

[99]   Lewis, J.A. *et al.* (2006). Direct Ink Writing of Three-Dimensional Ceramic Structures. *Journal of the American Ceramic Society.* https://doi.org/10.1111/J.1551-2916.2006.01382.X

[100]   Lin, T. *et al.* (2023). Three-Dimensional Printing of Large Ceramic Products and Process Simulation. *Materials.* https://doi.org/10.3390/ma16103815

[101]   Wong, K. V and Hernandez, A. (2012). A Review of Additive Manufacturing. *International Scholarly Research Network ISRN Mechanical Engineering.* https://doi.org/10.5402/2012/208760

[102]   ÖZMEN, E. and ERTEK, C. (2022). Eklemeli İmalat Teknolojilerinde Kullanılan Biyomalzemeler ve Biyomedikal Uygulamaları. *Gazi Üniversitesi Fen Bilimleri Dergisi Part C: Tasarım ve Teknoloji.* https://doi.org/10.29109/gujsc.1150799

[103]   Chua, C.K. and Leong, K.F. (2014). 3D Printing and additive manufacturing: Principles and applications (with companion media pack) - fourth edition of rapid prototyping. *3D Printing and Additive Manufacturing: Principles and Applications (With Companion Media Pack) - Fourth Edition of Rapid Prototyping.* https://doi.org/10.1142/9008

[104]   Surmen, H.K. *et al.* (2020). Fundamentals of 3D Printing and Its Applications in Biomedical Engineering. https://doi.org/10.1007/978-981-15-5424-7_2

[105]   BAŞ, H. *et al.* (2022). Yapışkan Püskürtme Yöntemimde Üretim Parametrelerinin Önemi Ve Örnek Uygulama. *Mühendislik Bilimleri ve Tasarım Dergisi.* https://doi.org/10.21923/jesd.1103808

[106]   Zhang, X. *et al.* (2023). Processing, microstructure, and mechanical behavior of AZ31 magnesium alloy fabricated by electron beam additive manufacturing. *Journal of Alloys and Compounds.* https://doi.org/10.1016/j.jallcom.2022.168567

[107]   Kachaev, A.A. *et al.* (2023). Use of Additive Technologies for Making Silicon Carbide Ceramic Materials: A Review. *Glass and Ceramics.* https://doi.org/10.1007/s10717-023-00533-5

[108]   Shahrubudin, N. et al. (2019). An overview on 3D printing technology: Technological, materials, and applications. Procedia Manufacturing. https://doi.org/10.1016/j.promfg.2019.06.089

Materials Research Forum LLC
https://doi.org/21741/9781644903834

Chapter 14

Laboratory Tests for Industrial Ceramics

Farah Mutlag[1,2*], Hussein Elaibi[1], Mustafa Ucar[3], Abdullah Seyrankaya[4], Ebru Halvaci[1], Selcuk Erdogan[1], Fatih Sen[1*]

[1]Sen Research Group, Department of Biochemistry, Kutahya Dumlupinar University, Kutahya 43000, Türkiye

[2]Ministry of Education, Karbala Education Directorate, Karbala, 56001, Iraq

[3]Chemistry Department, Faculty of Arts and Science, Afyon Kocatepe University, Ahmet Necdet Sezer Campus, 03200, Afyon, Türkiye

[4]Faculty of Engineering, Department of Mining Engineering, Dokuz Eylul University, Izmir, Türkiye

farahfakhir28@gmail.com, fatihsen1980@gmail.com

Abstract

Industrial ceramics are essential components in challenging applications owing to their excellent mechanical, thermal and chemical properties. To control performance and reliability, a variety of laboratory tests are adopted in the process of development and production. In this review, we present for testing the physical, mechanical, thermal, chemical and microstructural properties of hydrogels. Every single test is an important piece of information for quality control, an optimized design and choosing the best material. In addition, the results are important for predicting life in service and for the development of new advanced ceramic technology. Future perspectives of ceramic testing, including nondestructive testing, automated data processing, and artificial intelligence are presented and their contribution to advance quality management of materials engineering is discussed.

Keywords

Industrial Ceramics, Laboratory Testing, Material Performance

1. Introduction

Laboratory testing of industrial ceramics is difficult, despite advances in their production. One of the biggest issues is handling fragile materials during sample preparation; analyzing mechanical and thermal properties requires complex equipment, and certain measuring methods can't detect small imperfections. Manufacturing methods, firing temperatures, and raw material variances can potentially affect test results, requiring precise protocols to ensure data quality and consistency [1–4].

Laboratory testing is essential for assessing industrial ceramics' quality and performance under different service situations. Heat, pressure, corrosion, and chemical resistance are crucial in these severe settings. Even slight structural faults can cause application failures. Thus, laboratory testing is essential for quality, industry compliance, and service maintenance [1,5,6].

The objectives and benefits of laboratory testing are varied. They ensure that ceramic materials meet technical specifications, help engineers select the right material for each application, and guide the development of new products based on a clear understanding of material behavior. They also reduce costs through early detection of defects, enhance design reliability, and support long-term performance in critical industrial environments.

2. Types of Industrial Ceramics

Industrial ceramics are broadly classified into two main categories based on their composition, processing methods, and end-use applications: traditional ceramics and advanced ceramics. Understanding the differences between these types is essential for selecting the appropriate material for specific industrial requirements.

2.1 Traditional ceramics

Traditional ceramics are primarily made from natural raw materials such as clay, silica, and feldspar. They are usually produced through traditional forming techniques such as pressing or extrusion, followed by firing at high temperatures. Common ceramic materials include bricks, tiles, porcelain, and sanitary ware. These materials are often used in construction, household applications, and low-tech industrial processes. Although they offer good strength and durability, they are generally limited in their performance under severe mechanical or thermal stress [1,4,7–10].

Table 1. *Comparison of laboratory testing for traditional vs. advanced ceramics.*

	Traditional Ceramics	Advanced Ceramics
Basic Test Types	Simple physical tests such as density, water absorption, thermal shrinkage	Comprehensive tests including mechanical, thermal, chemical, and microstructural analysis
Complexity Level	Low Can be performed using standard laboratory equipment	High Requires advanced, high-precision instruments and protocols
Testing Equipment	Basic tools: balance, furnace, water bath	Specialized tools: SEM, XRD, mechanical testers, thermal analyzers, spectrometers
Purpose of Testing	General quality control, verifying standard specifications	Performance assessment under extreme conditions, failure analysis, material optimization
Applicable Standards	Basic ASTM standards (e.g., ASTM C373 – Water Absorption)	Advanced ASTM/ISO standards (e.g., ASTM C1161 – Flexural Strength, ASTM C1421 – Fracture Toughness)
Microscopic Analysis	Rarely required, mostly for quality assurance	Essential – used for grain size analysis, defect detection, phase identification
Time and Cost	Fast and cost-effective	More time-consuming and expensive due to specialized procedures and tools

2.2 Advanced Ceramics

Technical or engineering ceramics are made from high-purity, synthetic materials. Examples include alumina (Al_2O_3), zirconia (ZrO_2), silicon carbide (SiC), and silicon nitride (Si_3N_4). High mechanical strength, heat resistance, corrosion resistance, and chemical stability are designed into this ceramic. They are employed in high-performance applications such as aerospace

A Journey from Raw Materials to Ceramics
Materials Research Foundations 184 (2025)

Materials Research Forum LLC
https://doi.org/21741/9781644903834

components, medical implants, electronics, and cutting tools. Advanced ceramics demand careful production and quality testing [8,11–13].

The distinction between traditional and advanced ceramics highlights the wide diversity in this field and the need for specialized laboratory testing to ensure each type performs efficiently for its intended use. Table 1 summarizes the studies in Section 2, which systematically illustrate their properties and areas of application.

3. Laboratory Tests for Industrial Ceramics

Laboratory testing of industrial ceramics is a critical step in evaluating their performance, reliability, and suitability for demanding applications. These tests are typically classified into five main categories: physical, mechanical, thermal, chemical, and microstructural. Each category targets specific material properties that are essential for industrial use. The following is a detailed overview of the most important types of tests. Figure 1 illustrates laboratory test for industrial ceramics.

Figure 1. Laboratory test for industrial ceramics.

3.1 Physical tests uncovering the core properties of industrial ceramics

Physical tests are fundamental to understanding how ceramic materials behave under real-world conditions. These tests reveal fundamental properties such as density, porosity, and interaction with water, which directly influence the mechanical and thermal performance of ceramics.

Archimedes' principle, which calculates bulk density and apparent porosity by comparing the sample's weight in air and water, is a popular approach for assessing these properties. This simple yet effective method shows how well-packed ceramic particles are and how much open pore space the structure has. High-bulk density ceramics are ideal for load-bearing or high-temperature applications due to their strength and thermal conductivity. Higher porosity may be better for weight reduction or thermal insulation, but it reduces mechanical strength. Finding the correct balance between these two factors is vital and typically depends on the ceramic component's intended function [14–18].

Beyond density and porosity, another critical aspect of ceramic performance is its interaction with moisture. The water absorption test plays a vital role in this context, determining how much water a ceramic material can take in when exposed to humid or wet conditions. This is particularly important for ceramics used in insulation, structural tiling, or filtration systems. Materials that absorb excessive moisture may suffer from long-term degradation, reduced dielectric strength, or even cracking under freezing thaw cycles. High water absorption typically indicates a porous or underfired ceramic, suggesting that either the raw material blend or the sintering process may require adjustment to improve durability [19–22].

Equally important is the material's dimensional stability, which is assessed through the linear shrinkage test. During the high-temperature firing process, ceramic particles undergo sintering a process where individual grains bond and the structure densifies. This causes the material to contract, sometimes significantly. Measuring the amount of shrinkage between the shaped (green) and fired body gives engineers a sense of how predictable and stable the material is. Uniform shrinkage across batches reflects a well-controlled production process, while uneven or excessive shrinkage can signal inconsistencies in composition or processing issues that might lead to warping, cracking, or failure in the final application [23–26].

3.2 Mechanical tests evaluating strength and structural integrity

While physical tests reveal how a ceramic is built, mechanical tests show how well it can stand up to stress. In industrial applications ranging from aerospace components to cutting tools ceramic materials must maintain their integrity under high loads, impact, or constant vibration. Mechanical testing provides a direct window into the material's reliability in such conditions, making it a vital part of the quality control and development process [21,27,28].

Among the most important metrics is flexural strength, often assessed using three- or four-point bending tests. This test measures the ceramic's resistance to breaking when force is applied perpendicular to its surface. Since ceramics are brittle by nature, their flexural strength is a key indicator of how much mechanical stress they can handle before catastrophic failure. Even small internal flaws like microcracks or uneven porosity can significantly reduce strength, making consistent quality essential [29–31].

Hardness is another fundamental property. Typically measured using the Vickers or Knoop methods, hardness reflects a material's resistance to localized plastic deformation or scratching. In intensive wear environments such as machining, electronics, or protective armor ceramics with high hardness values perform better and last longer. However, hardness must be balanced with toughness; extremely hard ceramics may be prone to cracking under sudden impact if their fracture toughness is too low [32–34].

Fracture toughness quantifies a ceramic's ability to resist crack propagation once a flaw is present. Unlike metals, ceramics cannot plastically deform to absorb energy, so once a crack starts, it may grow rapidly. High fracture toughness means the ceramic can better contain or slow crack growth, which is essential in applications where safety and durability are critical [35–38].

3.3 Thermal tests understanding behavior under heat

Industrial ceramics are frequently selected for their ability to withstand high temperatures, making thermal testing a central pillar in evaluating their suitability for heat-intensive environments. From furnace linings to engine components and electronic substrates, ceramics must maintain stability, resist thermal shock, and function reliably despite rapid or sustained exposure to heat [19,39–41].

A core property examined in thermal testing is thermal conductivity, a measure of how efficiently heat passes through the ceramic. In some applications, such as heat sinks or substrates in electronics, high thermal conductivity is desirable to disperse heat quickly and avoid localized overheating. In others, such as thermal insulation or protective coatings, low conductivity is essential to contain or block heat. Engineers select ceramics based on this performance balance, optimizing energy efficiency and system safety [1,42,43].

Closely related is thermal expansion, which describes how much ceramic expands when heated. This is quantified by the coefficient of thermal expansion (CTE). Materials with low, stable CTE are ideal for environments with fluctuating temperatures, as they're less likely to crack or warp due to repeated expansion and contraction. Matching the coefficient of thermal expansion CTE of ceramic components with adjoining materials like metals or composites is critical in multi-material systems to prevent delamination or mechanical failure [44–46].

Another evaluation is thermal shock resistance, which tests a ceramic's ability to endure sudden changes in temperature without fracturing. This is particularly important in applications like kilns, exhaust systems, or aerospace shielding, where rapid heating or cooling is common. A ceramic that scores well in thermal shock testing typically combines low thermal expansion with good fracture toughness [1,47–49].

3.4 Chemical tests assessing durability in reactive environments

Over time, these substances can degrade even the most mechanically sound materials. In many Ceramics industrial settings, ceramics are exposed not only to heat and pressure but also to aggressive chemical environments acids, alkalis, salts, and reactive gases. That's chemical testing plays a crucial role in determining whether a ceramic will maintain its integrity and functionality over the course of its service life [1,50–52].

The most fundamental evaluation is chemical resistance, which measures how well ceramic withstands corrosion or reaction when exposed to specific chemicals. Depending on the application, this might include immersion tests in acidic or basic solutions, gas exposure at elevated temperatures, or salt spray tests. Ceramics used in chemical reactors, biomedical implants, or environmental systems must be able to retain their structure and properties despite prolonged contact with corrosive agents [6,53–55].

3.5 Microstructural tests revealing the internal architecture

While physical, mechanical, thermal, and chemical tests evaluate the external behavior of ceramic materials, microstructural analysis dives beneath the surface literally to examine the internal makeup of the ceramic at the microscopic and even nanoscopic levels. These tests are essential for understanding why a material behaves the way it does, and they often uncover flaws or patterns that are invisible to conventional testing [21,56–58].

A primary tool in this category is scanning electron microscopy (SEM), which provides high-resolution imaging of a ceramic's grain structure, pore distribution, and surface features. Through SEM, engineers can assess grain size uniformity, the presence of microcracks, or abnormal voids that may weaken the material. Finer, well-bonded grains typically lead to higher strength and better thermal stability, while irregularities can indicate processing defects or raw material issues [6,59–61].

In addition to imaging, X-ray diffraction (XRD) is used to identify and quantify the crystalline phases present within the ceramic. Different phases can have vastly different properties; some may enhance mechanical strength, while others may compromise thermal or chemical performance. Understanding the phase composition allows materials scientists to optimize the firing process or adjust the raw formulation to achieve the desired balance of properties [62–64].

Energy-dispersive X-ray spectroscopy (EDS), often integrated with SEM, provides elemental mapping and composition analysis. This helps detect contaminants, inhomogeneities, or unreacted components that could affect the material's performance or consistency across batches [6,63,65,66].

Table 2 provides a detailed overview of the main laboratory tests used in evaluating industrial ceramics, summarizing their objectives, tools, and measurable outcomes across different test categories discussed in Section 3.

Table 2. *Laboratory tests for industrial ceramics.*

Test Category	Objective	Instrumentation / Techniques	Measured Parameters & Output
Physical Tests	Characterize bulk material properties, structure uniformity, and dimensional integrity	- Archimedes setup- Digital caliper- Precision oven	- Bulk and apparent density (g/cm^3)- Water absorption (%)- Linear shrinkage (%)
Mechanical Tests	Assess strength, toughness, wear resistance under load or impact	- Vickers/Mohs hardness tester- Universal Testing Machine (UTM)- Pin-on-disk rig	- Vickers hardness (HV)- Flexural strength (MPa)- Fracture toughness (K_IC, $MPa \cdot m^{1/2}$)- Wear rate ($mm^3/N \cdot m$)
Thermal Tests	Evaluate response to heat, thermal cycling, and expansion	- Thermal shock chamber- Laser flash apparatus- Dilatometer	- Thermal shock endurance (ΔT cycles)- Thermal conductivity ($W/m \cdot K$)- Coefficient of thermal expansion (CTE, $\mu m/m \cdot K$)
Chemical Tests	Determine resistance to corrosion, and elemental purity/composition	- X-ray Fluorescence (XRF)- Inductively Coupled Plasma (ICP)- Acid/alkali immersion setup	- Elemental analysis (wt%)- Corrosion rate (mass loss % or g/cm^2)- Residue profile after chemical attack
Microstructural / Phase Tests	Examine internal structure, grain morphology, and crystalline phase distribution	- Scanning Electron Microscope (SEM)- X-ray Diffraction (XRD)- Optical microscope	- Grain size (μm)- Porosity mapping- Crystal phase identification- Defect visualization and fracture surface analysis

4. Standards and Testing Protocols Ensuring Consistency and Reliability

To translate laboratory results into real-world performance, standardized testing protocols are essential. They ensure that ceramic materials are evaluated under consistent conditions, allowing for reliable comparisons between different products, batches, and manufacturers. These standards not only promote scientific rigor but also help industries meet regulatory requirements and international quality benchmarks. Figure 2 illustrates the Standards. and testing protocols ensuring consistency and reliability.

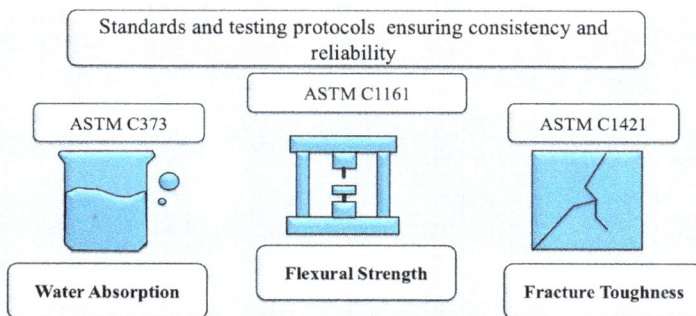

Figure 2. Standards and testing protocols ensuring consistency and reliability.

4.1 ASTM Standards

Two of the most widely recognized organizations in this space are the American Society for Testing and Materials (ASTM) and the International Organization for Standardization. Each provides detailed procedures for testing various physical, mechanical, thermal, and microstructural properties of ceramic materials [32,54,67,68].

Among the most widely adopted frameworks for ceramic testing are the ASTM International standards, which provide clear, detailed protocols to ensure precision and consistency across laboratories and industries. One example is ASTM C373, which outlines the standard method for evaluating water absorption, bulk density, apparent porosity, and specific gravity in fired ceramic products. This test is especially valuable for ceramics used in construction and insulation, where moisture interaction and structural porosity directly impact performance and durability [17,69–71].

For mechanical assessment, ASTM C1161 offers a standardized procedure to measure the flexural strength of advanced ceramics at an ambitious temperature. This test is critical in determining how much mechanical stress a material can endure before failure, and is widely used across industrial sectors where mechanical reliability is paramount [72–74].

Complementing this, ASTM C1421 focuses on the fracture toughness of ceramics specifically, their ability to resist crack propagation. This standard is essential in evaluating materials intended for high-stress or impact-prone environments, ensuring that ceramics not only perform under load but maintain long-term durability even in the presence of inherent flaws [32,74].

4.2 ISO standards

In addition to ASTM International processes, ISO standards help ensure that advanced ceramics testing is standardized and relevant worldwide in industry and research settings. The ISO 14704 standard measures bulk density and open porosity in fine or advanced ceramics. Physical attributes are crucial to understanding the material's structure, mechanical performance, and thermal insulation. ISO 14704 helps manufacturers and laboratories assess quality consistently, especially when components are part of global supply chains that require cross-border comparability [6,75–77].

Table 3 summarizes ASTM International, and International Organization for Standardization (ISO) standards used in the testing of advanced ceramics. It outlines each organization's test methods, their technical focus, and the practical relevance of these evaluations in real-world applications.

Table 3. Standards and testing protocols ASTM & ISO.

Organization	Standard	Test Focus	Application
ASTM	ASTM C373	Water Absorption, Bulk Density, Apparent Porosity	Key for evaluating moisture interaction and porosity in construction and insulation ceramics.
ASTM	ASTM C1161	Flexural Strength	Determines how much stress ceramics can endure before breaking; crucial for structural applications.
ASTM	ASTM C1421	Fracture Toughness	Measures crack resistance; essential for high-stress and impact-prone environments.
ISO	ISO 14704	Bulk Density and Open Porosity	Ensures consistency in assessing structure and porosity across international supply chains.
ISO	ISO 17561	Thermal Diffusivity (Laser Flash Method)	Evaluates heat response; vital for aerospace, thermal barriers, and electronic systems.

5. Applications for Test Results

Testing in a lab is an important part of making ceramics for industry. It is used to test the physical, mechanical, thermal, and chemical properties of ceramic materials. These tests do more than only check the quality of a product; they also have many other uses in the industry. Figure 3 shows this job with a clock under a shield, which stands for how important it is to estimate durability and analyze behavior over time.

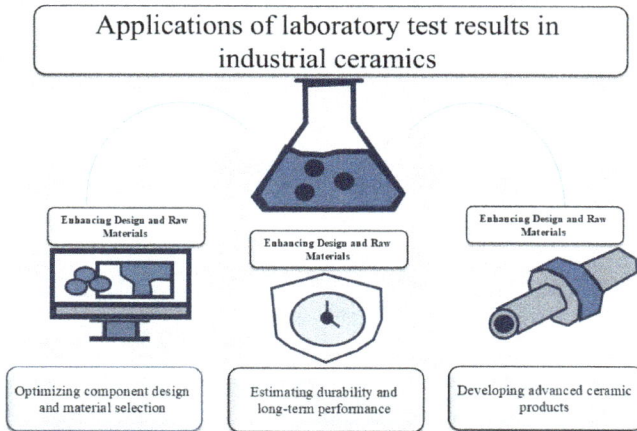

Figure 3. Applications for laboratory test results in industrial ceramics.

5.1 Enhancing design and raw material selection

The results obtained from laboratory tests are essential in optimizing the design of ceramic components. By analyzing parameters such as flexural strength, porosity, and thermal expansion, engineers can make informed decisions about geometry, wall thickness, and tolerances to prevent structural failure. Furthermore, test data helps in selecting or modifying raw materials to achieve better sintering behavior, lower defect rates, and improved consistency. This contributes to cost efficiency and greater reliability in mass production [4,78–80].

5.2 Predicting product lifespan and performance

Long-term durability and performance are critical in industrial applications where ceramics are exposed to harsh environments. Laboratory tests, especially those focused on fracture toughness, thermal shock resistance, and chemical stability, enable accurate prediction of how a ceramic component will behave over time. These insights allow manufacturers to estimate service life, identify failure thresholds, and establish preventive maintenance schedules, which are vital for sectors like aerospace, electronics, and energy [81–83].

5.3 Advancing the development of high-performance ceramics

The continuous development of advanced ceramics relies heavily on detailed laboratory evaluation. Through the integration of microstructural analysis, phase identification, and thermal/mechanical testing, researchers can tailor materials for specific high-tech applications. For example, test-driven development has enabled the creation of bio-inert ceramics for implants, ultra-hard materials for cutting tools, and thermally stable components for jet engines. Testing not only validates innovation but also accelerates material evolution toward next-generation ceramic technologies [8,12,84].

Table 4 is a summary of Section 5, illustrating the main applications of laboratory test results in the industrial ceramics sector and how they contribute to design improvement, performance prediction, and advanced material development.

Table 4. *Applications for laboratory test results in industrial ceramics.*

Application Area	Purpose of Test Results	Industrial Benefit	Typical Industry
Enhancing Design and Raw Material Selection	Optimize geometry, wall thickness, and select proper ceramic composition	Reduced failure rate, improved processing, better product consistency	Structural ceramics, refractories
Predicting Product Lifespan and Performance	Estimate long-term durability under mechanical, thermal, or chemical stress	Lower maintenance cost, improved reliability, extended service life	Aerospace, electronics, medical devices
Advancing High-Performance Ceramic Development	Tailor materials for extreme environments or cutting-edge applications	Enables innovation in next-gen products, functional material design	Energy, automotive, biomedical, defense

6. Conclusion

Testing of ceramics in the laboratory is an indispensable element in the ceramics industry and can provide crucial information concerning physical, mechanical, thermal, chemical, and microstructural properties of ceramic materials. Not only quality control and product certification is essential, but these tests are the basis for success for material selection, parameter optimization and design validation. With a consistent and repeatable testing process, manufacturers can have confidence that ceramic components will hold up to the most severe conditions in a wide range of industrial applications today, from construction and electronics to biomedical and aerospace equipment.

In the future, the development of ceramic testing will likely to be strongly affected by the development of smart manufacturing and advanced material science. Current developments are the combination of non-destructive testing (NDT) methods, in-line quality control during production, AI-based data processing and automated microstructure imaging. It enables the users to predict the system's behavior with high accuracy coupled with less prototyping requirements using digital twins and simulation-based performance modeling. These developments may let ceramic testing become more efficient and accurate and increasingly fully integrated with the full life cycle of material development and application.

References

[1] M.F. Zawrah, M.A. Taha, R.A. Youness, Advanced ceramics: stages of development, in: Adv. Ceram., Springer, 2023: pp. 1–46.

[2] Z. Bačkalić, M. V. Vasić, B.B. Holló, D. Jovanović, S. Sremac, P. Awoyera, From lab to industry: Analysing the performance of ceramic batch modified with sodium silicate and phosphoric acid-based additive in ceramic floor tile production, Next Mater. 8 (2025) 100891. https://doi.org/10.1016/j.nxmate.2025.100891

[3] L. Balli, A. Atifi, M. Hlimi, B. Hamri, Y. Achenani, Experimental Study and Energy Optimization of a Prototype Furnace for the Pottery-Ceramic Industry, J. Adv. Res. Fluid Mech. Therm. Sci. 98 (2022) 44–57. https://doi.org/10.37934/arfmts.98.2.4457

[4]   J. García-Ten, M. Dondi, J.V.M.B. Vieira Lisboa, M. Vicent Cabedo, L. Pérez-Villarejo,
        E. Rambaldi, C. Zanelli, Critical raw materials in the global high-throughput ceramic
        industry, Sustain. Mater. Technol. 39 (2024).
        https://doi.org/10.1016/j.susmat.2024.e00832

[5]   A. Maged, S.A. Abu El-Magd, A.E. Radwan, S. Kharbish, S. Zamzam, Evaluation insight
        into Abu Zenima clay deposits as a prospective raw material source for ceramics industry:
        Remote Sensing and Characterization, Sci. Rep. 13 (2023) 58.
        https://doi.org/10.1038/s41598-022-26484-5

[6]   N. Acharjee, S. Ganguly, P. Biswas, B. Sarangi, Uses of black pigmented CAS glass in
        conventional ceramic industry: production, impediments to metallization, and startling
        approaches for color evaluation in ceramic tech-artistry, Pigment Resin Technol. 54
        (2024) 689–701. https://doi.org/10.1108/PRT-02-2024-0018

[7]   I. Djobov, E. Karamanova, G. Avdeev, A. Karamanov, Characterization of Clays From
        "Mines Maritsa Iztok" As Raw Materials for Ceramic Industry, J. Chem. Technol. Metall.
        59 (2024) 1331–1340. https://doi.org/10.59957/jctm.v59.i6.2024.7

[8]   G. Kiradoo, The transition of traditional pottery-making into advanced ceramics in context
        to the indian ceramic industry, Turkish J. Physiother. Rehabil. 32 (2021) 546–556.

[9]   M.J. Ribeiro, D. Tulyaganov, Traditional Ceramics Manufacturing, in: PoliTO Springer
        Ser., Springer, 2021: pp. 75–118. https://doi.org/10.1007/978-3-030-85776-9_3

[10]   M. Vidak Vasić, P. Muñoz Velasco, S. Bueno-Rodríguez, I. Netinger Grubeša, M. Dondi,
        L. Pérez Villarejo, D. Eliche-Quesada, C. Zanelli, State and perspectives of sustainable
        production of traditional silicate ceramics, Open Ceram. 17 (2024) 100537.
        https://doi.org/10.1016/j.oceram.2024.100537

[11]   Japan, Recent Trends of Advanced Ceramics Industry and Fine Ceramics Roadmap 2050,
        CFI Ceram. Forum Int. 100 (2023) E 40-E 43.

[12]   J. Liu, Y. Liu, Y. Zheng, X. Li, Progress in Green and Low-carbon Technologies
        development of Building Ceramics Industry, in: J. Phys. Conf. Ser., IOP Publishing, 2023:
        p. 12008. https://doi.org/10.1088/1742-6596/2639/1/012008

[13]   J. Zhou, W. Cao, Y. Zhang, M. Omran, G. Chen, L. Gao, F. Zhang, J. Tang, Effect of
        microwave drying technology on drying kinetics of Al2O3–ZrO2 composite ceramic
        powder, J. Mater. Res. Technol. 24 (2023) 6762–6778.
        https://doi.org/10.1016/j.jmrt.2023.04.229

[14]   C.B. Emrullahoğlu Abi, H.Ş. Çoban Tetik, E. Abi, Utilization of Slip Casting Process for
        Recycling CAD/CAM Dental Zirconia Wastes, Black Sea J. Eng. Sci. 7 (2024) 401–408.
        https://doi.org/10.34248/bsengineering.1435621

[15]   M.H. Ibrahim, M.I. Mustaffar, A.N. Ismail, S.A. Ismail, N. Othman, Study of sintering
        behaviour and mechanical-physical properties of porous glass ceramic, in: J. Phys. Conf.
        Ser., IOP Publishing, 2025: p. 12007. https://doi.org/10.1088/1742-6596/3003/1/012007

[16]   H.M.H. Zakaly, H.A. Saudi, H.O. Tekin, M. Rashad, S.A.M. Issa, Y.S. Rammah, A.I.
        Elazaka, M.M. Hessien, A. Ene, Glass fabrication using ceramic and porcelain recycled
        waste and lithium niobate: physical, structural, optical and nuclear radiation attenuation

properties, J. Mater. Res. Technol. 15 (2021) 4074–4085.
https://doi.org/10.1016/j.jmrt.2021.09.138

[17] S.A. Ismail, M.I. Mustaffar, M.H. Ibrahim, N. Othman, Sea sand as a fine aggregate for lightweight glass ceramic tiles, in: J. Phys. Conf. Ser., IOP Publishing, 2025: p. 12005. https://doi.org/10.1088/1742-6596/3003/1/012005

[18] R. Elias, A. Zaccaron, O.R.K. Montedo, F. Raupp-Pereira, A.M. Bernardin, S. Arcaro, J.B. Rodrigues Neto, Pyroplastic deformation analysis of Brazilian porcelain tile formulations using mixture design, Int. J. Appl. Ceram. Technol. 21 (2024) 4241–4256. https://doi.org/10.1111/ijac.14813

[19] S. Carvalheira, M. Oliveira, M. Robaina, J.C.O. Matias, Energy Efficiency Improvements in a Portuguese Ceramic Industry: Case Study, Appl. Sci. 13 (2023) 5028. https://doi.org/10.3390/app13085028

[20] Giacomo Boschi, A. Tugnoli, M. Chiara Bignozzi, Sustainability and environmental impact of the Italian ceramic tile industry, (2023).

[21] L. Li, Z. Liu, W. Liu, Z. Wang, C. Zhou, Q. Zeng, Microstructure and transport properties of cement mortar made with recycled fine ceramic aggregates, Dev. Built Environ. 22 (2025) 100643. https://doi.org/10.1016/j.dibe.2025.100643

[22] G.U. Kiran, G. Nakkeeran, D. Roy, G.U. Alaneme, Optimization and prediction of paver block properties with ceramic waste as fine aggregate using response surface methodology, Sci. Rep. 14 (2024) 23416. https://doi.org/10.1038/s41598-024-74797-4

[23] S. Bose, C. Das, Introduction to Ceramics: Fabrication, Characterizations, and Applications, CRC Press, 2024.

[24] L. Wang, H. Tian, W. Lei, N. Dai, H. Wang, Development of high-strength ceramsite via sintering of iron ore tailings: Process optimization and properties, Constr. Build. Mater. 457 (2024) 139440. https://doi.org/10.1016/j.conbuildmat.2024.139440

[25] C.A. de Oliveira Guimarães, G.C.G. Delaqua, A.R.G. de Azevedo, S.N. Monteiro, L.F. Amaral, C.L.M. Souza, A.G.P. da Silva, J.N.F. de Holanda, C.M.F. Vieira, Heating rate effect during sintering on the technological properties of Brazilian red ceramics, Int. J. Adv. Manuf. Technol. 119 (2022) 8125–8135. https://doi.org/10.1007/s00170-022-08793-1

[26] W. Klunghirun, K. Serivalsatit, Development of cordierite-mullite refractory castables bonded with magnesium silicate hydrate cement, J. Thai Ceram. Soc. 1 (2025) 18–29.

[27] M. Indumathi, G. Nakkeeran, G.U. kiran, D. Roy, A. Al-Fakih, Mechanical and sustainability performance of concrete incorporated limestone powder, recycled ceramic aggregates, and coconut fibers, Innov. Infrastruct. Solut. 10 (2025) 197. https://doi.org/10.1007/s41062-025-02018-6

[28] S. Rajesh, K.S. Anitha, C. Sivanarayana, Impact of Ceramic Insulator Waste as Partial Replacement of Coarse Aggregate in Concrete: An Experimental Study and Evaluation of Mechanical Properties, Int. J. Emerg. Res. Eng. Sci. Manag. 3 (2024). https://doi.org/10.58482/ijeresm.v3i4.2

[29] R. Kashyap, M. Saxena, A. Gautam, A. Kushwaha, K. Priyanka, A. Patel, R.K. Maurya,

Exploring sustainable construction through experimental analysis and AI predictive modelling of ceramic waste powder concrete, Asian J. Civ. Eng. 25 (2024) 4789–4801. https://doi.org/10.1007/s42107-024-01080-2

[30] S.E. Avram, L. Barbu Tudoran, S. Cuc, G. Borodi, B.V. Birle, I. Petean, Conditioning Influence of Kaolinite Matrices on Flexural Strength of Raw Pressed Slurry Collected from Ceramic Tile Production Wastewater, J. Compos. Sci. 8 (2024) 219. https://doi.org/10.3390/jcs8060219

[31] X. Yang, Y. Xing, Z. Wang, S. Duan, G. Zhao, J. Song, Z. Xiao, Research on the Influence Law and Mechanism of Regenerated Ceramic Tile Form and Replacement Rate on the Mechanical Properties of Ultra-High-Performance Concrete, Materials (Basel). 18 (2025) 3028. https://doi.org/10.3390/ma18133028

[32] T.G.T. Nindhia, D.N.K.P. Negara, I.W.B. Adnyana, Mechanical Properties Of Advance Ceramic, Mega Press Nusantara, 2023.

[33] W. Jia, C. Su, H. Zhang, X. Zou, G. Ren, Q. Wei, M. Zhao, C. Ma, Quantitative relationship between microstructure and mechanical properties in Nd: YAG transparent ceramics, Ceram. Int. 47 (2021) 12144–12152. https://doi.org/10.1016/j.ceramint.2021.01.061

[34] L. Wahl, Robocasting of Reaction-Bonded Carbides, Friedrich-Alexander-Universitaet Erlangen-Nuernberg (Germany), 2023.

[35] A.N. Wijayanto, R.B. Rashonda, Praptana, K. Wahyudi, B. Pribadi, Application of non-destructive test method with ultrasonic for testing of flexural strength (Modulus of Rupture) on ceramic tile, in: J. Phys. Conf. Ser., IOP Publishing, 2025: p. 12022. https://doi.org/10.1088/1742-6596/2945/1/012022

[36] J. Yang, X. Gao, J. Xu, H. Zhu, M.M. Hasan, J. Shao, S.I. Haruna, A multi-scale investigation on recycled ceramic and rubber composite cement-based materials: Acoustic emission, NMR, molecular dynamics simulation, Constr. Build. Mater. 412 (2024) 134881. https://doi.org/10.1016/j.conbuildmat.2024.134881

[37] F.A. Costa Oliveira, M. Sardinha, J.M.J. Netto, M. Farinha, M. Leite, M.A. Barreiros, S. Abanades, J.C. Fernandes, Mechanical Performance of Ceria-Coated 3D-Printed Black Zirconia Cellular Structures After Solar Thermochemical CO/H2 Fuel Production Cycles, Crystals 15 (2025) 629.

[38] A.J. Ruys, Silicon Carbide Ceramics: Structure, Properties and Manufacturing, Elsevier, 2023. https://doi.org/10.1016/C2020-0-03298-6

[39] D.R. Macedo Ramos De Afonso Matias, Optimization of energy consumption in the ceramics industry: a techno-economic analysis, (2023).

[40] A.Ç. Arianpour, F. Arianpour, Characterization, technological properties, and ceramic applications of Kastamonu alluvial clays (Northern Turkey) in building materials, Constr. Build. Mater. 356 (2022) 129304. https://doi.org/10.1016/j.conbuildmat.2022.129304

[41] J. Gilabert, E. Zumaquero, E.M. Díaz-Canales, M.J. Ventura, M.P. Gómez-Tena, A Tailored Approach in the Bentonite-like Raw Material Characterization Using Thermal and Water Sorption Techniques, Minerals 13 (2023) 1313.

https://doi.org/10.3390/min13101313

[42] K. Singh, M. Kaur, A. Kumar, Progress in Advanced Ceramics: Energy and Environmental Perspective, in: Adv. Ceram. Energy Environ. Appl., CRC Press, 2021: pp. 1–12. https://doi.org/10.1201/9781003005155-1

[43] S. Valayampillil Narayanan, M. Kumar, A. Solaiappan, NIR reflective lanthanum zirconate (La2Zr2O7) ceramic colorant and its solar heat reflective glaze coatings, Opt. Mater. (Amst). 142 (2023) 114000. https://doi.org/10.1016/j.optmat.2023.114000

[44] A. Turkmani, Lower Cretaceous Clay in Anti-Lebanon Mountains, Syria And Their Importance in Ceramic Manufacturing, Earth Environ. Sci. Res. Rev. 5 (2022). https://doi.org/10.33140/eesrr.05.01.01

[45] Y. Sun, T. Wu, Y. Bao, Y. Li, D. Wan, K. Li, L. He, Preparation and strengthening mechanism of prestressed ceramic tile components, Int. J. Appl. Ceram. Technol. 19 (2022) 604–611. https://doi.org/10.1111/ijac.13757

[46] W.S. Mohamed, E.M.A. Hamzawy, A.F. Osman, H.A. Eliwa, K.G. El-Gameel, A. Firstauthor, T.B. Secondauthor, J.Q. Thirdauthor, Utilization of rhyolite and red syenite rocks as an alternative of potash feldspars in manufacturing the sanitary ware bodies, Egypt. J. Chem. 66 (2023) 425–433. https://doi.org/10.21608/EJCHEM.2022.165676.7037

[47] W. Yin, K. Wang, Analysis of energy saving in domestic ceramic industry kilns, in: E3S Web Conf., EDP Sciences, 2021: p. 1011. https://doi.org/10.1051/e3sconf/202126701011

[48] L. Ruivo, M. Russo, R. Lourenço, D. Pio, Energy management in the Portuguese ceramic industry: Analysis of real-world factories, Energy 237 (2021) 121628. https://doi.org/10.1016/j.energy.2021.121628

[49] R.B. Heimann, On the nature of ceramics technology: from Empedocles to Dawkins, Archaeometry 67 (2025) 55–71. https://doi.org/10.1111/arcm.12997

[50] D.R. Darul, M.J.H. Dowlath, S.K. Karuppannan, K.D. Arunachalam, Industrial applications of nanoceramics: from lab to real-time utilization in the biomedical industry, in: Ind. Appl. Nanoceramics, Elsevier, 2024: pp. 143–158. https://doi.org/10.1016/B978-0-323-88654-3.00010-X

[51] M.O. Ajadi, Pozzolanic Reactivity of Pulverized Ceramic Waste as Partial Substitute for Cement and Supplementary Cementitious Materials in Concrete, J. Appl. Sci. Environ. Manag. 28 (2024) 4425–4435.

[52] S. Nandipati, N.K. Degloorkar, R. Gobinath, A.N. Sai, Performance Evaluation of Alkali-Activated Masonry Blocks With Binary Blended Composition of Flyash and Ceramic Tile Dust, Adv. Civ. Eng. 2025 (2025) 8652832. https://doi.org/10.1155/adce/8652832

[53] A.A. Hashim, R. Anaee, M.S. Nasr, Improving the Mechanical, Corrosion Resistance, Microstructural and Environmental Performance of Recycled Aggregate Concrete Using Ceramic Waste Powder as an Alternative to Cement, Ceramics 8 (2025) 11. https://doi.org/10.3390/ceramics8010011

[54] M.S. Joudi, J.N. Jaafar, H.Y. Mahdi, Evaluation of the Claystones for Ceramic Industry From Bussiya Area, Muthanna Governorate, Iraqi Bull. Geol. Min. 21 (2025) 147–166.

https://doi.org/10.59150/ibgm2101a10

[55] M.H. Al-Salmany, M.A. Al-Nuaimy, Chemical Treatment of Efflorescence in Clay Bricks During the Manufacturing Process, J. Univ. Anbar Pure Sci. 18 (2024) 271–282. https://doi.org/10.37652/juaps.2024.148903.1234

[56] G.F. Huseien, Z.H. Joudah, M.H. Baghban, N.H. A. Khalid, I. Faridmehr, K. Dong, Y. Li, X. Gu, Sustainability of Recycling Waste Ceramic Tiles in the Green Concrete Industry: A Comprehensive Review, Buildings 15 (2025) 2406.

[57] W.-T. Lin, Y.-W. Lin, B.-X. Zhang, K.-L. Lin, Utilizing billet sintering for producing environmentally friendly porous tiles from brick waste and catalyst waste: Properties and moisture control applications, Case Stud. Constr. Mater. 22 (2025) e04617.

[58] R. Carrillo Beltran, E. Picazo Camilo, G. Perea Toledo, F.A. Corpas Iglesias, Study of Novel Geopolymer Concrete Prepared with Slate Stone Cutting Sludge, Chamotte, Steel Slag and Activated with Olive Stone Bottom Ash, Materials (Basel). 18 (2025) 1974.

[59] S.J.N. Jaramillo, J.A.C. López, M.P.F. Álvarez, S.N.T. Santacruz, Chemical Characterization OF Clays Located in Palora, Morona Santiago Province of Ecuador, Int. J. Environ. Sci. 11 (2025) 1054–1064.

[60] M. Rodrigo, M.F. Gazulla, A. Moreno, J. Gilabert, P. Gómez, Blue Inkjet Inks From E-Waste: Toward A Greener Ceramic Industry, J. Eur. Ceram. Soc. (2025) 117696.

[61] B. Fatima, F. Batool, A.J. Sangi, Experimental investigation on the use of ceramic tiles as aggregates in concrete, Innov. Infrastruct. Solut. 10 (2025) 76.

[62] B.S. da Silva, N.A. Cerqueira, J. Alexandre, J. de Andrade Paula Madalena, I. do Espirito Santo Simões, L.A.F. Barreto, Physico-chemical and Mineralogical Characterization of Soils for the Production of Extruded, Pressed, and Fired Red Ceramic Blocks, in: TMS Annu. Meet. Exhib., Springer, 2025: pp. 425–433.

[63] C. Candeias, A. Gomes, F. Rocha, Assessment of Feldspars from Central Portugal Pegmatites for Sustainable Ceramic Applications, Minerals 15 (2025) 527.

[64] B. Yassen, Suitability Assessment of Fatha Formation Clayey Raw Material for Ceramic Production, in Salah Al-Din Governorate, Iraq, Iraqi Geol. J. (2025) 278–290.

[65] M. Dondi, S. Conte, C. Molinari, C. Zanelli, Mineral Resources for the Ceramic Industry: Survey of Feldspathic Raw Materials in Italy, Minerals 15 (2025) 87.

[66] N.M. Marian, M. Perotti, C. Indelicato, C. Magrini, G. Giorgetti, G. Capitani, C. Viti, From high-volume industrial waste to new ceramic material: The case of red gypsum muds in the TiO2 industry, Ceram. Int. 49 (2023) 15034–15043.

[67] O.W. OBOT, U.D. PAUL, Physicochemical And Mineralogical Characterization Of Use Abat, Ibiono Ibom Clay For Industrial Application, (n.d.).

[68] S.H. Hassan, N.K. Attia, G.I. El Diwani, R.S. Ettouney, M.A.H. El-Rifai, S.K. Amin, Preparation of inorganic catalysts from ceramic industry wastes for use in the production of biofuel, J. Eng. Appl. Sci. 70 (2023) 68.

[69] M. Awais, A. Mansoor, I. Shah, M. Hussain, M. Asif, Characterization and Development of Ceramics with Inorganic Additives, Inventions 9 (2024) 121.

[70] O.M. Nweke, A.I. Omeokachie, Ceramics properties of indurated-shale quarry wastes from Abakaliki, Southeastern Nigeria: application as raw materials in roofing-tile production, Clays Clay Miner. 71 (2023) 143–165.

[71] R. Meena, M. Mallik, R. Yadav, A. Meena, A.W. Hashmi, Investigation of physical properties of bamboo dust with marble powder filled ceramic composites: Sustainable approach for ceramic tiles manufacturing, Proc. Inst. Mech. Eng. Part L J. Mater. Des. Appl. 238 (2024) 1425–1437.

[72] L. Resullar, R.R. Aseniero, M.C.J. Galinato, Mapping and Profiling of Clay Resources found in Brgy. Bugas-Bugas, Placer and Brgy. Cabugo, Claver in Surigao Del Norte, (2025).

[73] D. Kırsever, H.Ö. Toplan, A.Ş. Demirkıran, Effect of porcelain polishing waste additive on properties and corrosion resistance of ceramic foams produced from zeolite, J. Aust. Ceram. Soc. 59 (2023) 671–683.

[74] J. Bhattarai, N. Duwal, Investigation on Physico-chemical and Sintering Properties of Ceramic Tiles and Bricks Available in Local Market of Kathmandu Valley, UGC Fac. Res. Gants-2072/73, Univ. Grants Comm. Sanothimi, Bhaktapur, Nepal (2024).

[75] Z.J. Ting, X. Meng, Z. Yang, S.A. Jiskani, L. Hu, W. Dong, M. Zhao, Solid Recovered Fuel (SRF): A Comprehensive Review of Its Origins, Production, and Industrial Utilization, Energy and Fuels 39 (2025) 9726–9761. https://doi.org/10.1021/acs.energyfuels.5c01674

[76] S.A. Khan, D. Kumar, S. Kumar, A. Isha, T.C. D'Silva, R. Chandra, V.K. Vijay, Biofuel Applications: Quality Control and Assurance, Techno-Economics and Environmental Sustainability, Biofuel Extr. Tech. (2023) 367–420.

[77] H. Son, H. Jang, Y.-K. Moon, J. Kim, H.-A. Cha, J.-J. Choi, B.-D. Hahn, J.-W. Lee, S.-Y. Yoon, K.-H. Cho, An easy approach to adjust microstructure and physical properties in alumina, J. Korean Ceram. Soc. 61 (2024) 298–306.

[78] H. Monteiro, P.L. Cruz, B. Moura, Integrated environmental and economic life cycle assessment of improvement strategies for a ceramic industry, J. Clean. Prod. 345 (2022) 131173.

[79] M. Dondi, J. García-Ten, E. Rambaldi, C. Zanelli, M. Vicent-Cabedo, Resource efficiency versus market trends in the ceramic tile industry: Effect on the supply chain in Italy and Spain, Resour. Conserv. Recycl. 168 (2021) 105271.

[80] M. Coelho de Azevedo, M. Teixeira Marvila, G. Carla Girondi Delaqua, L. Fonseca Amaral, H. Colorado, C. Maurício Fontes Vieira, Economy analysis of the implementation of extruded tiles fabrication in a ceramic industry containing ornamental rock waste, Int. J. Appl. Ceram. Technol. 18 (2021) 1876–1890.

[81] M.S. Medina-Salgado, F.E. García-Muiña, M. Cucchi, D. Settembre-Blundo, Adaptive life cycle costing (LCC) modeling and applying to Italy ceramic tile manufacturing sector: Its implication of open innovation, J. Open Innov. Technol. Mark. Complex. 7 (2021) 101.

[82] S. Ozcan, U. Sengul, The use of supervised artificial intelligence methods in quality determination in continuous production lines: a case study of ceramic industry, Int. J.

Interact. Des. Manuf. (2025) 1–20. https://doi.org/10.1007/s12008-025-02310-w

[83] A.M.E. Khalil, A.P.M. Velenturf, M. Ahmadinia, S. Zhang, Context Analysis for
      Transformative Change in the Ceramic Industry, Sustain. 15 (2023) 12230.
      https://doi.org/10.3390/su151612230

[84] M.B. Islam, H.J. Sumona, M.J. Haque, A. Al Mahmood, A progressive overview of the
      mainstream additive manufacturing of ceramic components for industrial advancement,
      Mater. Open Res. 3 (2024) 8. https://doi.org/10.12688/materialsopenres.17669.1

A Journey from Raw Materials to Ceramics | Materials Research Forum LLC
Materials Research Foundations 184 (2025) | https://doi.org/21741/9781644903834

Chapter 15

R&D in Industrial Ceramics Applications

Eray Casin[1], Fatma Nur Maran[2], Ebru Halvaci[2], Iskender Isık[3*], Fatih Sen[2*]

[1]Genesis Technological Products Industry and Trade Limited Company, Corum, Türkiye

[2]Sen Research Group, Department of Biochemistry, Kutahya Dumlupinar University, Kutahya 43000, Türkiye

[3]Department of Materials Science & Engineering, Faculty of Engineering, Kutahya Dumlupinar University, Evliya Çelebi Campus, 43100 Kutahya, Türkiye

iskender.isik@dpu.edu.tr, fatihsen1980@gmail.com

Abstract

Industrial ceramics is a term used for ceramic products manufactured in series production. R&D applications in industrial ceramics involve experimental studies based on scientific research. Conducting research and development activities in line with companies' desire to produce customer-oriented products enables both production stability and the manufacture of products that meet customer demands. The units managing research and development activities should be organized to make improvements in all processes involved in production, starting with the incorporation of ceramic raw materials into production. Since research and development activities are based on data collection and processing, the integration of experimental studies conducted in process control into production should generally be aimed at data stability and improvement. Furthermore, conducting development work in line with planning techniques that vary according to production efficiency and customer demands increases the company's ability to maintain quality products in the market. This section will describe how R&D applications are carried out in industrial enterprises producing ceramic materials, which methods are used, and how development activities are conducted in light of these methods.

Keywords

Industrial Ceramics, Mechanical Tests, Microscopic Analysis, Thermal

1. Introduction

Research and development encompasses all activities undertaken publicly or privately to create new or improved products and processes. It is a comprehensive process that spans from fundamental scientific research, through the transformation of products or processes derived from scientific research into added value, to the testing of all these activities. Expenditure on R&D activities reveals indicators of innovation performance [1]. R&D activities involve discovering new information, investigating the applications of research findings or other information, and developing alternative conceptual designs for potential products or processes. This encompasses the use of new technologies in product or process development stages, the adaptation of industrial developments to processes, and ensuring that all engineering activities are economically viable, functional, and suitable for human use [2].

Advanced countries that design R&D strategies focus their research on technology-intensive sectors while maintaining their primary focus on policies aimed at greater technological advancement. Developing countries, on the other hand, conduct studies to identify technology-intensive sectors. The path followed by developing countries in these studies is determined by developing policies to develop sectors where the country's resources can be utilized, taking into account the resources the country possesses. In addition, market strategies and the export-oriented work of private enterprises in the country require that R&D activities be developed in these areas and that products offered to the market be those that can provide added value [3].

In general, ceramic materials can be examined under two headings: traditional ceramics and technical ceramics [4]. Today, ceramic materials have a wide range of potential applications, from household items to high-performance ceramic materials for industrial use [5]. Researchers are conducting their research topics with a growing focus on traditional and technical ceramics [4]. Traditional ceramic material production in the sense of sustainable production is increasingly being carried out worldwide. Research and development activities in companies producing traditional ceramics are conducted both within the company and in collaboration with public industry partnerships.

Three key factors stand out in understanding the relationship between innovation and development. The first is the exponential increase in R&D expenditures alongside the level of development measured by per capita gross domestic product (GDP), along with the significant increase in R&D activities in developing countries and the implementation of developments by following the literature. The third factor stems from the quality of academic institutions in countries with low GDP affecting the quality of R&D activities [6]. Furthermore, the employment of individuals with academic backgrounds in the private sector also enhances R&D activities. This facilitates the widespread use of technological analysis devices and improves inputs that impact production. Over the past 10 years, the establishment of state-supported R&D Centers and the execution of original research and development activities from production units have contributed to increased production efficiency and new product manufacturing processes.

2. Industrial Ceramic Raw Materials R&D Applications

Industrial ceramics are defined as alumina silicate-based ceramics, including cement, ceramics and refractory materials produced from clay and clay, ceramic sanitary ware, ceramic wall and floor covering materials, clay-based roofing materials, and bricks [7]. Two main inputs are used in the production of industrial ceramic materials. One is clay, and the other is glaze, which gives the product its aesthetic appearance. Developments in the field of clay and glaze are made not only to increase the efficiency of production processes but also to reduce costs and to enable the use of alternative raw materials in production processes. In order to carry out research and development activities in these areas, it is necessary to scientifically investigate and apply the properties of the raw materials used in the ceramic production process, their behavior under normal conditions, and their interactions under heat.

In industrial terms, the most important factor in traditional ceramic material production is the raw material. Ceramic raw materials are divided into two categories: plastic and non-plastic raw materials. Porous raw materials are those that can be kneaded and shaped when they react with water and retain their shape when dry. Non-porous raw materials, on the other hand, cannot be shaped with water and lose their shape and disintegrate when shaped, even when shaped, in the direction of the force applied from outside [8]. The main raw materials used in industrial ceramics are processed clays, kaolin, feldspar-quartz, and silica sand. These raw materials have different effects throughout the production process. The presence of low colorants, glass formers, and earth alkali oxides is a common feature in each raw material group. The presence of kaolinite promotes mullite formation, while the formation of illite and smectite affects glassy phase formation and the proper densification of the ceramic body during firing [9].

The term "clay" has three distinct meanings: a geological term describing a natural, earthy, fine-grained material that develops plasticity when mixed with a limited amount of water; a particle size term for the smallest particles (less than 2 μm); and the name of a group of sheet-like silicate minerals [10]. The three main mineralogical components in ceramic killers are kaolinite, illite, and quartz structures [9]. Illite/smectite structures are also sometimes observed in plastic clays. The physical properties of ceramic materials vary depending on the content of clay minerals in plastic clays, as well as the crystal structure of kaolinite and illite. These properties affect the particle size distribution and specific surface area (BET) of clays [11]. Their high solid content and high dry strength ensure that the shaped semi-finished product remains stable and provides plasticity during shaping, making clays critically important in ceramic systems [12]. Additionally, in order to conduct technical studies on the investigation of the properties of clays and other raw materials and their use in the development of industrial ceramic clays, it is essential to perform chemical and mineralogical analysis (XRF and XRD), thermogravimetric analysis (TG-DTA) to determine thermal properties, and thermal expansion analysis to determine expansion values during sintering processes.

One of the points to consider when conducting raw material research in the ceramics industry is to investigate the raw materials available in the region where the industrial production facility is located and ensure they are incorporated into the system. In line with this type of work, the cost of the recipe is reduced by lowering the costs arising from raw material freight charges. Academic studies also address the detailed specifications of the raw materials required for ceramic production.

Findings obtained in the laboratories of industrial production facilities and in academic studies, although they provide numerical data and accuracy of use, the raw material trials to be conducted in production facilities and the efficiency values obtained from these trials play an important role in determining the raw materials suitable for production facilities. Even if appropriate results are obtained in academic studies or laboratory tests, not every good result will provide the same efficiency in every production facility. The most important factors here are systemic factors, and another and one of the most important parameters is environmental factors. Furthermore, in sectors where manual labor is intensively used, such as ceramic sanitary ware, worker performance and manual dexterity are also among the factors affecting raw material efficiency. Generally, academic studies do not take these types of factors into account. However, production trials conducted in the later stages of studies carried out in industrial production facility R&D laboratories are important for understanding whether raw materials can be used in production and for choosing the raw material.

When conducting raw material research, one of the most important parameters for industrial production facilities is the proximity of the raw material to the production facility. Although this parameter is not always the first consideration, it is important in efforts to reduce costs or increase the use of alternative raw materials. Looking at academic studies on raw materials, the availability of local resources in the region or country where the study is conducted for use in production is investigated [13]. Research and development activities at industrial ceramic production facilities are primarily focused on and controlled in the area of raw materials.

One of the most fundamental raw materials in ceramic production, clays, are expected to exhibit different properties depending on the specific ceramic product being manufactured. The properties of clay generally reflect the properties of the different clay minerals present in the raw material. In the production of roof tiles, which are used as roofing materials, it is necessary to know the technical requirements and acceptance criteria of the final product according to the standards and to ensure that the product formed as a result of the production processes of the selected raw material or raw materials meets these criteria. In their 2019 study, Pavlova et al. investigated the suitability of the properties specified in the standards as a result of the use of montmorillonitic clays found in the Ural region in tile production. The proximity of the raw material to the production facility is particularly important in the production of tiles and ceramic coating materials due to the high production quantities.

In their 2018 study, Uz et al. examined the usage characteristics of five different processed clay raw materials used in Turkey in the ceramic sanitary ware industry. The study determined the casting properties, chemical, mineralogical, and physical-mechanical properties of the raw materials.

In another study conducted by Laursen et al. in 2019, while characterizing clay minerals found in northeastern Brazil, it was explained that these plastic clays could be used as alternative clays. Applying such studies to ceramic materials can contribute to the industrial development of ceramic materials.

To understand whether a raw material is technically suitable for production, its composition must first be known. Various characterization analyses are used to understand the composition of raw materials. The most commonly used of these analysis methods is chemical analysis (XRF). XRF is an analysis method that enables the quantitative determination of the elemental components present in almost any material [14]. When the atoms that make up the material are excited by X-rays, they emit different energies. The excited photons enable the qualitative and quantitative analysis of many elements in a given sample.

3. Experimental Methods

In a basic industrial ceramics laboratory, experiments conducted on raw materials are divided into two main methods. For clay and kaolin group raw materials, since they are cohesive raw materials that can be kneaded with water, an aqueous suspension is prepared to control their rheological properties. When performing these analyses, it is necessary to understand the concept of rheology and the reaction of clay particles during the experimental work. Briefly, rheology is a branch of science that studies the deformation and fluidity of a substance in response to external forces [15],[16]. The concept of rheology is based on the principles of elasticity developed by Robert Hooke in 1678 for solid materials and viscosity proposed by Isaac Newton in 1687 [15]. According to Newton, viscosity arises when one layer of fluid is moved relative to another layer. Therefore, the greater the frictional force, the greater the force per unit area required to advance this movement [17].

Ceramic suspensions are used as starting materials in colloidal processing techniques such as slip casting, tape casting, and spray coating. Controlling the rheological behavior of ceramic suspensions and obtaining the desired microstructural properties of the ceramic structure is very important. In order to obtain a flawless green body in ceramic materials shaped using colloidal processing methods, the highest possible ceramic particle loading at the lowest possible suspension viscosity is desired. Ceramic suspensions are suspensions with high solid concentrations, typically characterized by shear-dependent viscosities exhibiting non-Newtonian flow properties. Furthermore, high-solid-concentration ceramic suspensions typically exhibit a time-dependent decrease in apparent viscosity at increasing shear rates, but the viscosity recovers at decreasing shear rates. This reversible flow property is called thixotropy. A decrease in the particle size of the solid particles within the ceramic suspension leads to an increase in the number of solid particles per unit volume, which consequently increases the viscosity of the suspension [18]. Therefore, the rheological properties of ceramic suspensions vary depending on the physical and chemical properties of the solvent used in suspension preparation, pH level, solid mass ratio, particle size and distribution, particle shape and surface properties, and suspension and ambient temperature. This means that when preparing ceramic suspensions, it is important to bear in mind the effect of different parameters on the suspension and to carry out development work on the production inputs. Before moving on to what will be done in the experimental work, the environment in which the experiment will be carried out, the experimental equipment, and the solvent liquids to be used in the experiment must be checked and verified to be within the desired criteria ranges.

Shear stress measurements are performed by adding sodium silicate or similar electrolytes to clay and kaolin suspensions. Shear stress (viscosity) and thixotropy values, typically measured using a Gallengamp viscometer, are also determined using Brookfield or Lehmann viscometers. The dispersion of clay and kaolin in water defines a quantitative relationship between ions undergoing different reactions, namely cation exchange capacity. Cation exchange capacity is defined as the amount of cations expressed in terms of dry matter at a specific pH value. Negatively charged ions in clays and kaolins bind with positively charged ions [19]. The cation exchange capacity of a clay or kaolin represents the maximum possible amount of electrolyte that can be used in industrial ceramic production. The most important parameter to be determined in production is the amount of electrolyte absorbed by the prepared ceramic suspension at a Brookfield viscometer measurement of 500 cP. This ratio represents the amount of electrolyte required for the complete dispersion of a single clay or kaolin raw material in the preparation tanks, especially when preparing ceramic slurry.

The examination of the casting properties of prepared clay or kaolin suspensions at a specific thixotropic point plays an important role in determining the thickness ratio that the raw material can achieve based on its rheological properties, as well as in determining the physical properties (dry shrinkage, firing shrinkage, dry bond strength, amount of deformation) in these rheological properties. Knowing these values is important for raw material selection in all types of industrial ceramic production. In research and development activities, the correct selection of clay and kaolin in particular has a direct positive impact on production efficiency.

Plasticity is a particularly important factor in the production of industrial ceramics, especially in clay raw materials. Defined as the ability of clay to change shape under finite forces, plasticity is a material property that allows the material to deform repeatedly without breaking when a force sufficient to cause deformation is applied, and to retain its shape after the applied force is removed [19]. Plasticity contributes significantly to the processing of molded industrial ceramic materials, especially those shaped by casting.

4. Heat Treatments

Sintering at high temperatures causes ceramic materials to undergo shape changes. These shape changes are referred to as pyroplastic deformation. Increased stress with increasing product size also increases the product's tendency to deform. Deformation behavior varies with changes in many parameters, such as chemical composition, density of the ceramic suspension, sintering process speed, particle size of the raw materials used, and the reaction of the raw materials with each other at high temperatures [20]. The tests conducted to determine the deformation rates of clay and kaolin will simulate the product's behavior under high temperatures.

Firing color determination and loss on ignition tests are performed on basic inert raw materials used in ceramics, such as feldspar and quartz. Color determination in inert raw materials directly affects the color properties of the ceramic body or glaze composition. The lower the porosity determined in ceramic bodies, the higher the color stability [21]. The stability and consistency of firing color and loss on ignition are key factors in determining the use of raw materials in ceramic suspensions. The determination of the firing loss and firing colors of raw materials selected for use in the production facility is the first parameter in the selection of alternative or equivalent raw materials. After these analyses are performed, small-scale clay or glaze samples should be prepared and the general appearance of the ceramic sample should be compared with that of the ceramic material.

Glazes not only change the appearance of ceramics but also enhance their physical properties. Glazes that make ceramic materials waterproof allow liquids or foodstuffs to be placed inside the ceramic material. Glazes have added value to ceramic objects, made them fashionable with color and pattern options, and in some cases even turned them into luxury products and thus objects of trade. Research and development work on ceramic glazes provides an understanding of their chemical composition, microstructure, and sintering technologies. Characterization analyses performed during the development of glazes help to understand glaze behavior during sintering [22].

5. The Importance of Glazing Ceramics

Ceramic glazes vary depending on the industrial ceramic material to be produced. Ceramic glazes can be divided into two groups: opaque and transparent. Differences in the proportions of raw materials used also affect the physical properties of ceramic glazes. The pigments used in the glaze cause the glaze colors to change and give the ceramic material an aesthetic appearance.

To understand the properties imparted by raw materials to glaze compositions and to enable mathematical modeling and application in production when formulating glazes, Seger formulations must be used. The Seger approach is a system developed by Hermann August Seger that allows the behavior of oxides within a glaze to be calculated mathematically. When creating a ceramic glaze, the oxides present in the raw materials must be divided into three main groups. RO and R2O are defined as fluxing agents, R2O3 as amphoteric agents, and RO2 oxides as acidic oxides. According to the Seger approach, when the molar quantity of each oxide is known, it can be calculated how much of this quantity can be produced from a different alternative raw material or in different quantities [23]. This calculation method enables the pre-production calculation of ceramic glazes, allowing an understanding of how the glaze may behave at the sintering temperatures used in the facility. Additionally, knowing the molar quantity of oxides present in the glaze theoretically allows the calculation of the glaze's melting temperature, thermal expansion coefficient, and surface tension.

Figure 1. Industrial Ceramic Glazing Application

The choice of glaze for ceramic materials requiring a smooth texture depends on the impurity levels of the raw materials used in the clay body: opaque glazes are selected when impurity levels are high, and transparent glazes when the body color is white. The opacity of a glaze stems from the different refractions between the crystallization phase and the vitreous phase. This difference indicates that the opacity obtained depends on the refractive index, grain size, amount of crystals, and their distribution. The opacity of the glaze is provided by crystals with high refractive indices, such as $ZrSiO_4$, TiO_2, spinel, and feldspars. In the production of ceramic sanitary ware, opaque white glazes are mostly used to protect the products from contamination and to give them an aesthetic appearance. Generally, these ceramics are sintered at approximately 1200°C. Zircon silicate is more commonly used in this type of glaze due to its high temperature stability. It also imparts high refractive index, hardness, chemical and mechanical resistance to the glaze body, providing high whiteness to the glaze surface [24]. Fluctuations in zircon prices and their continuous impact on prescription costs in industrial applications have also led to the use and development of alternative opacifying agents based on the Na_2O-K_2O-MgO-ZnO-Al_2O_3-(TiO_2)-B_2O_3-SiO_2 system [25].

The quartz used in glaze enables the formation of a glassy structure. In addition to feldspars, which are alkalis and alkaline earths, raw materials such as calcite, wollastonite, and dolomite, which are sources of CaO, are the basic raw materials used as vitrifiers. These raw materials reduce the viscosity of the glaze and provide hardness and durability after sintering [26]. Wollastonite is a white natural calcium silicate with a theoretical composition of $CaSiO_3$. Its chemical composition is approximately 48% calcium oxide and 52% silica. Wollastonite used in the ceramics industry accounts for 30-40% of total wollastonite sales. Although it occurs naturally, synthetic wollastonite is also produced from diatomaceous earth, which is rich in diato [27]. The use of raw materials such as wollastonite in glazes increases brightness, facilitates spreading in the molten phase, and improves the glassy structure of glazes [28]. The percentage of CaO-containing raw materials plays an important role in determining the surface properties of ceramic glazes. In some studies, it has been observed that using 16-22% wollastonite in the glaze results in glossiness, while using 30% or more results in a matte finish. Adding MgO to glazes in the form of dolomite or talc causes diopside crystals to crystallize, resulting in a matte surface [29].

Zinc oxide is used as a flux because it dissolves faster than soil alkalis and has a lower melting point [30]. This enables the glaze to begin melting at a temperature lower than its sintering temperature, in conjunction with earth alkalis. Furthermore, the lower melting temperature facilitates and enhances the spreading of the glaze over the body with low additions. When used in high quantities within the glaze, it promotes the formation of willemite crystals, which react with other glaze components to create a matte surface [31]. Recently, reducing the particle size of zinc oxide to the nanoscale and activating the particles with silver ions has limited bacterial growth on ceramic surfaces by increasing the surface hydrophobicity of willemite crystals. This facilitates the formation of particles with antibacterial properties and enables these particles to exhibit antibacterial properties when used in ceramic glazes [31], [32].

Alumina is a synthetic raw material produced from bauxite using the Bayer process, with an Al2O3 content of approximately 99% or higher. Widely used in the production of ceramic materials, alumina acts as an opacifier in porcelain tiles and a matting agent in glaze compositions. The matting effect in the glaze depends on the particle size of the alumina used and the size of the alumina particles in the particle size distribution of the glaze suspension after the grinding process. The refractory behavior of alumina provides a glass matrix containing insoluble particles that prevent the formation of smooth surfaces. In the microstructure of the glaze, alumina particles reacting with other oxides in the glaze allow the formation of new phases such as anorthite. After this reaction, CaO and ZnO released from the glassy phase increase the viscosity of the glassy phase. Aluminas with a high specific surface area provide opacity and whiteness in the glaze, while glazes using aluminas with a low specific surface area show reduced brightness. They also increase chemical resistance and fracture toughness [33].

6. Principles of Color Grading

Pigments, which are frequently used in ceramic production and enable the coloring of ceramic surfaces, are produced with properties similar to the oxide forms contained within glazes. In recent years, pigments used in digital printing methods developed especially for ceramic coating materials have also been introduced to the ceramic industry. The coloring of industrial ceramics is a design argument that enhances the appearance and aesthetic components of ceramic products and quickly triggers the user's emotional experience [34]. Ceramic pigments with particle sizes at the nano scale have a large potential market due to their high surface area and higher reflection angle. Reducing the particle size from micron to nano scale has also enabled the use of these pigments in digital printing with different solvents. While the particle size of pigments used in ceramic glazes is expected to be between 1-10 microns, the particle size distribution of nano pigments used in digitally printed ceramic tiles or tableware is expected to be between 10-80 nm [35].

When conducting research and development studies for colored glaze applications on ceramic materials, it is essential to first understand and identify the colors that can be obtained from color mixtures. In industrial ceramic applications, patterns or colors designed according to market demands must also be understood in terms of the use of ceramic pigments to ensure consistency and color harmony in their areas of application. A color space has been defined in international color standards to represent color differences. CIELAB is one of the most commonly used color spaces. When this method is to be used, a sample is first measured with a color measurement device that measures these parameters, and the resulting values are converted into perceptual relationships such as lightness-darkness, chromaticity, and hue angle. The standard deviation of the distance between a pair of colors is calculated and determined as the color difference (ΔE). This difference is then evaluated according to a specific color region and a predetermined color tolerance that could be a product [36].

A ceramic pigment is typically a metal transition complex oxide obtained through a calcination process and exhibiting three main characteristics:

a. Ensuring thermal stability when the temperature rises;

b. Provides chemical stability when sintered with sintered or ceramic matrices;

c. Provides high colorfastness when dispersed and sintered with pigments or ceramic matrices.

As a general rule, ceramic paints and ceramic pigments look almost the same before and after sintering, but this is not the case for paints based on soluble salts such as carbonates or nitrates, in addition to raw oxides, because they decompose during sintering and dissolve within the glaze or ceramic matrix [37].

The research of ceramic pigments is a field of high importance. Red pigments produced on an industrial scale use toxic transition metal ions as chromophores, such as Al_2O_3: Cr (corundum), $ZnAl_2O_4$: Cr, $CaSnSiO_5$: Cr, $MnAl_2O_4$: Cr (spinel), and $Y_2Sn_2O_7$: Cr (pyrochlore). Chromophores such as Cr, Ni, and Co can be incorporated into these transition metals. Iron-containing, zirconium-based pigments, commonly referred to as FeZrSiO4, are typically used in red pigments. Controlling this color at high temperatures is quite difficult. In this regard, the starting materials, zirconium and silica, must be finely ground [38], [39]. The coral red pigment known as coral pink in the ceramics industry varies in particle size depending on the production process

[39]. When iron oxide-containing pigments are used in glazes, they cause color changes ranging from red to brown and from gray to black during sintering, depending on the composition of the glaze [40].

7. Conclusion

The industrial production approach benefits both workers and society. Increasing the welfare level of a country's society is directly proportional to the development of economic inputs. The development of economic and foreign trade potential must be achieved through innovative work and the transfer of technological developments to production facilities. The production stability achieved through the application of mechanization and production planning techniques in the production field continues with the digitalization of industrial facilities and the transition to green energy. The more efficient use of natural resources is achieved through the use of circular production models and supporting technologies in production. The use of industrial development activities in production facilities has also enabled the development of research and development activities in production facilities. The development of R&D activities in production facilities enables production facilities to obtain higher quality products at lower costs and, in addition, maximizes production efficiency. The state's support for the development of production facilities and R&D investments, in the sense of creating R&D centers that encourage independent research and development activities, ensures that the culture of R&D is embraced by businesses. Ensuring continuous improvement and development efforts and developing final products with innovative approaches enables new products to be introduced to the market.

Figure 2. Pressure Casting Applications in Industrial Ceramics

The principles of research and development activities in industrial ceramic applications vary depending on the company's perspective on these activities. However, the experimental procedures applied during basic process control and raw material input control operations in R&D laboratories, and the observation of the applications of these experiments in the production field, also contribute to the development of research and development activities. The analysis of the results of laboratory-scale trials enables the prior observation of positive and negative outcomes that may be encountered in production. Furthermore, the variations applied during experimental studies enable technical personnel to understand the properties of inputs (i.e., raw materials) through experimental methods.

Production planning in an industrial ceramic production facility will also account for the variability of inputs and the materials or machinery to be used. The ability to respond to production variability in laboratories that conduct research and development activities based on production also helps improve the planning stages of production. Preparing alternative clay, glaze, and colored glaze recipes based on alternative raw material sources, as well as conducting alternative research and development studies that are not based on production requirements, should be done without causing time loss in terms of meeting production demands. The continuous research of alternative inputs that can be used in production by production R&D laboratories, the conduct of experimental studies related to these inputs, and the monitoring of their trials in production are fundamental requirements of industrial ceramic companies producing high-capacity products.

In addition to controlling inputs and processes, research and development activities must also include work that is not production-oriented and that will also enable the development of processes. The personnel conducting research and development activities must be competent in both production and research, ensuring that the knowledge base that will impact the final product is used optimally and transferred to production.

References

[1]   Anon, "Research and Development.," *Automot. Eng.*, vol. 12, no. 2, pp. 40–41, 1987. https://doi.org/10.12968/npre.2007.5.5.23745

[2]   F. Moris and C. Pece, "Definitions of Research and Development: An Annotated Compilation of Official Sources | NSF - National Science Foundation," *Natl. Cent. Sci. Eng. Stat.*, 2022, [Online]. Available: https://ncses.nsf.gov/pubs/ncses22209

[3]   H. Karahan, "Determining Strategic R&D Sectors In Turkey: An Import-Based Approach," *Afro Eurasian Stud.*, vol. 1, no. 2, pp. 51–68, Dec. 2012, Accessed: Sep. 30, 2025. [Online]. Available: https://dergipark.org.tr/en/pub/afes/issue/44784/557044

[4]   L. Sánchez Muñoz *et al.*, "R&D+i for ceramic tiles in the 21st century: Competition, diversity, and functionality", Accessed: Sep. 30, 2025. [Online]. Available: https://digital.csic.es/handle/10261/41912

[5]   I. O. Odewale *et al.*, "Indrances and Prospects of Ceramic Technology Development in Nigeria," no. August 2020, pp. 13–22, 2013.

[6]   D. Lederman, W. F. Maloney, and Q. M. Jlj, "R&D and Development The World Bank LatinAnericaandCaribbeanRegicn. I Officeof theCief Eanist," 2003, Accessed: Sep. 30, 2025. [Online]. Available: http://econ.worldbank.org

[7]   "Ceramic Materials - Synthesis, Characterization, Applications and Recycling," 2019, Accessed: Sep. 30, 2025. [Online]. Available: https://books.google.com/books/about/Ceramic_Materials.html?hl=tr&id=KBT8DwAAQBAJ

[8]   A. Arcasoy, "Seramik Teknolojisi." pp. 1–277, 1983.

[9]   K. Galos, "Composition and ceramic properties of ball clays for porcelain stoneware tiles manufacture in Poland," *Appl. Clay Sci.*, vol. 51, no. 1–2, pp. 74–85, Jan. 2011. https://doi.org/10.1016/J.CLAY.2010.11.004

[10]   G. Rapp, "Pigments and Colorants," pp. 201–221, 2009. https://doi.org/10.1007/978-3-540-78594-1_9.

[11]   B. Bauluz, M. J. Mayayo, E. Laita, and A. Yuste, "Micro- and nanotexture and genesis of ball clays in the lower cretaceous (Se iberian range, ne spain)," *Minerals*, vol. 11, no. 12, p. 1339, Dec. 2021. https://doi.org/10.3390/MIN11121339/S1

[12]   P. S. A. Irabor, S. O. Jimoh, and O. J. Omowumi, "Ceramic Raw Materials Development In Nigeria," *Int. J. Sci. Technol. Res.*, vol. 3, no. 9, 2014, Accessed: Sep. 30, 2025. [Online]. Available: www.ijstr.org

[13]   S. Parveen and S. A. Jahan, "Investigation on Physico-Chemical Properties of Vitreous China Sanitary Ware from Local Clays," *Bangladesh J. Sci. Ind. Res.*, vol. 45, no. 4, pp. 387–392, Jan. 2010. https://doi.org/10.3329/BJSIR.V45I4.7386

[14]   Z. W. Chen, W. M. Gibson, and H. Huang, "High Definition X-Ray Fluorescence: Principles and Techniques," *X-Ray Opt. Instrum.*, vol. 2008, no. 1, p. 318171, Jan. 2008. https://doi.org/10.1155/2008/318171

[15]   T. Enstitüye and V. Tarih, "Seramikte Kaolen Kullanımının Polielektrolit Katkısıyla Geliştirilmesi." Fen Bilimleri Enstitüsü, Sep. 09, 2009. Accessed: Sep. 30, 2025. [Online]. Available: http://hdl.handle.net/11527/937

[16]   W. Liu, X. Liu, L. Zhang, Y. Wan, H. Li, and X. Jiao, "Rheology, mechanics, microstructure and durability of low-carbon cementitious materials based on circulating fluidized bed fly ash: A comprehensive review," *Constr. Build. Mater.*, vol. 411, Jan. 2024. https://doi.org/10.1016/j.conbuildmat.2023.134688

[17]   F. Peluso, "The Viscosity of Liquids in the Dual Model," *Thermo 2024, Vol. 4, Pages 508-539*, vol. 4, no. 4, pp. 508–539, Dec. 2024. https://doi.org/10.3390/THERMO4040028

[18]   B. A. Horri, P. Ranganathan, C. Selomulya, and H. Wang, "A new empirical viscosity model for ceramic suspensions," *Chem. Eng. Sci.*, vol. 66, no. 12, pp. 2798–2806, Jun. 2011. https://doi.org/10.1016/J.CES.2011.03.040

[19]   N. Kumari and C. Mohan, "Basics of Clay Minerals and Their Characteristic Properties," *Clay Clay Miner.*, Dec. 2021. https://doi.org/10.5772/INTECHOPEN.97672

[20]   N. Kunduracı, G. Binal, İ. Nur Gamze Şimşek, B. Ecevit Üniversitesi Metalurji ve Malzeme Mühendisliği Bölümü, and B. Üniversitesi Metalurji ve Malzeme Mühendisliği Bölümü, "Seramik Sağlık Gereçleri Fine Fire Clay Ürünlerde Şamot Alternatifi Malzemelerin Kullanımının Araştırılması ve Sentezlenmesi," *AKU J. Sci. Eng*, vol. 17, pp. 203–208, 2017. https://doi.org/10.5578/fmbd.51757

[21]   C. Ângela, M. Volpato, M. C. Fredel, M. Celso Fredel, A. G. Philippi, and C. Otávio Petter, "Ceramic Materials and Color in Dentistry Ceramic Materials and Color in Dentistry 155 x Ceramic Materials and Color in Dentistry," 2010, Accessed: Sep. 30, 2025. [Online]. Available: https://www.researchgate.net/publication/221909435

[22]   T. Pradell and J. Molera, "Ceramic technology. How to characterise ceramic glazes," *Archaeol. Anthropol. Sci.*, vol. 12, no. 8, pp. 1–28, Aug. 2020. https://doi.org/10.1007/S12520-020-01136-9/FIGURES/10

[23]   C. Romanosoglou and T. Alexandridis, "Glaze calculation software based on the Seger method with recipe mixing utilities, limit formulas and toxicity measurements," 2010. https://doi.org/10.13140/2.1.1409.8249

[24]   Y. Yu, H. Su, C. Peng, and J. Wu, "Submicro-zirconia crystal-intergrown zircon opaque glaze," *J. Eur. Ceram. Soc.*, vol. 39, no. 2–3, pp. 652–659, Feb. 2019.

https://doi.org/10.1016/J.JEURCERAMSOC.2018.09.044

[25]   J. Cai et al., "Development of spinel opaque glazes for ceramic tiles," J. Eur. Ceram. Soc., vol. 38, no. 1, pp. 297–302, Jan. 2018. https://doi.org/10.1016/J.JEURCERAMSOC.2017.07.037

[26]   L. Koroglu and C. Pekşen, "The Replacement of Seashells with Calcite in White Ceramic Glaze Preparation," Sak. Univ. J. Sci., vol. 24, no. 5, pp. 956–964, Oct. 2020. https://doi.org/10.16984/SAUFENBILDER.728340

[27]   M. Regueiro Y González-Barros, J. García-Ten, and A. Alonso-Jiménez, "Synthesis of wollastonite from diatomite-rich marls and its potential ceramic uses," Bol. la Soc. Esp. Ceram. y Vidr., vol. 61, no. 6, pp. 585–594, Nov. 2022. https://doi.org/10.1016/J.BSECV.2021.05.002

[28]   G. M. Azarov, E. V. Maiorova, M. A. Oborina, and A. V. Belyakov, "Wollastonite raw materials and their applications (a review)," Glas. Ceram., vol. 52, no. 9, pp. 237–240, 1995. https://doi.org/10.1007/BF00681090/METRICS

[29]   T. Kronberg and L. Hupa, "The impact of wollastonite and dolomite on chemical durability of matte fast-fired raw glazes," J. Eur. Ceram. Soc., vol. 40, no. 8, pp. 3327–3337, Jul. 2020. https://doi.org/10.1016/J.JEURCERAMSOC.2020.03.033

[30]   D. U. Tulyaganov, S. Agathopoulos, H. R. Fernandes, and J. M. F. Ferreira, "The influence of incorporation of ZnO-containing glazes on the properties of hard porcelains," J. Eur. Ceram. Soc., vol. 27, no. 2–3, pp. 1665–1670, Jan. 2007. https://doi.org/10.1016/J.JEURCERAMSOC.2006.05.011

[31]   J. Kim, D. Jeong, J. Choi, and U. Kim, "Changes in the glaze characteristics and moderate antibacterial activity of ceramic tile glazes with the addition of ZnO," J. Asian Ceram. Soc., vol 10, no. 1, pp. 241–252, Jan. 2022. https://doi.org/10.1080/21870764.2022.2038044;PAGE:STRING:ARTICLE/CHAPTER

[32]   R. Demirel, E. Suvacı, İ. Şahin, S. Dağ, and V. Kiliç, "Antimicrobial activity of designed undoped and doped MicNo-ZnO particles," J. Drug Deliv. Sci. Technol., vol. 47, pp. 309–321, Oct. 2018. https://doi.org/10.1016/J.JDDST.2018.07.024

[33]   E. Bou, J. Garcia-Ten, R. Perez, S. Arrufat, and G. Atichian, "Influence of alumina characteristics on glaze properties," Bol. la Soc. Esp. Ceram. y Vidr., vol. 49, 2010.

[34]   Y. Wang, Q. Zhao, J. Chen, W. Wang, S. Yu, and X. Yang, "Color Design Decisions for Ceramic Products Based on Quantification of Perceptual Characteristics," Sensors 2022, Vol. 22, Page 5415, vol. 22, no. 14, p. 5415, Jul. 2022. https://doi.org/10.3390/S22145415

[35]   P. M. T. Cavalcante, M. Dondi, G. Guarini, M. Raimondo, and G. Baldi, "Colour performance of ceramic nano-pigments," Dye. Pigment., vol. 80, no. 2, pp. 226–232, Feb. 2009. https://doi.org/10.1016/J.DYEPIG.2008.07.004

[36]   M. R. Luo and M. R. Pointer, "CIE colour appearance models: A current perspective," Light. Res. Technol., vol. 50, no. 1, pp. 129–140, Jan. 2018. https://doi.org/10.1177/1477153517722053.

[37]   G. Monros, "Pigment, Ceramic," Encycl. Color Sci. Technol., pp. 1026–1040, Jan. 2016. https://doi.org/10.1007/978-3-642-27851-8_181-3

[38]   M. Jovaní, A. Sanz, H. Beltrán-Mir, and E. Cordoncillo, "New red-shade environmental-friendly multifunctional pigment based on Tb and Fe doped Y2Zr2O7 for ceramic applications and cool roof coatings," Dye. Pigment., vol. 133, pp. 33–40, Oct. 2016. https://doi.org/10.1016/J.DYEPIG.2016.05.042

[39]   M. Jovaní, M. Fortuño-Morte, H. Beltrán-Mir, and E. Cordoncillo, "Environmental-friendly red-orange ceramic pigment based on Pr and Fe co-doped Y2Zr2O7," J. Eur. Ceram. Soc., vol. 38, no. 4, pp. 2210–2217, Apr. 2018. https://doi.org/10.1016/J.JEURCERAMSOC.2017.12.005

[40]   M. Çakı, B. Karasu, and G. Kaya, "Use of Iron and Zinc Oxide Based Pigments in Stoneware Glazes," Proc. 10th ECerS Conf, pp. 1784–1787, 2007, [Online]. Available: https://www.researchgate.net/publication/258897358

A Journey from Raw Materials to Ceramics
Materials Research Foundations 184 (2025)

Materials Research Forum LLC
https://doi.org/21741/9781644903834

Chapter 16

Quality Control and Packaging in Industrial Ceramics

Farah Mutlag[1,2*], Hussein Elaibi[1], Idris Kaynak[3], Ebru Halvaci[1], Selcuk Erdogan[1], Fatih Sen[1*]

[1]Sen Research Group, Department of Biochemistry, Kutahya Dumlupinar University, Kutahya 43000, Türkiye

[2]Ministry of Education, Karbala Education Directorate, Karbala, 56001, Iraq

[3]Machinery and Metal Technologies, Vocational School of Technical Sciences, Usak University, 1 Eylul Campus, 64200 Usak, Türkiye

farahfakhir28@gmail.com, fatihsen1980@gmail.com

Abstract

In the ceramics industry, quality control is a fundamental element to ensure the production of highly efficient products that meet technical and aesthetic standards. The quality control process begins with the inspection of raw materials, where the purity of clay and minerals is verified through chemical and physical analyses. This step is essential to prevent defects in later stages. To prevent cracks or uneven shrinkage, it is necessary to carefully monitor the size and distribution of materials during the drying and forming phases. Following the fire and glazing processes, the shrinkage is recorded, and the color, gloss, and adhesion of the glaze are assessed to identify any minute flaws. The latter steps involve a thorough visual inspection with the use of high-resolution cameras and sensors, as well as tests to determine water absorption, impact and abrasion resistance, and durability. Reducing errors is one of the objectives of quality control. Other objectives include increasing manufacturing efficiency, reducing energy and material waste, and satisfying customers with a well-made, functional, and aesthetically acceptable product. Designs can be improved, development decisions supported, and market competitiveness enhanced using data obtained during the quality control stages, which indicates the effectiveness of the manufacturing system. A quality system is a useful tool for guiding production toward higher levels of reliability and efficiency, distinguishing between inspection (detection of errors) and control (prevention of errors). Quality control in the ceramics industry is therefore more than just a technical process; it is an essential strategy for manufacturing ceramic products that are reliable in the market due to their durability, beauty, and reliability.

Keywords

Ceramics, Packaging, Control, Quality

1. Introduction

The ceramics business relies on quality as a foundation to sustain production and achieve customer satisfaction. Surface and mechanical properties like color stability, dimensional precision, and fracture resistance are essential for ceramic goods used in both industrial and residential settings. Defect rates, energy and material waste, and the production facility's overall economic efficiency are all improved by strict adherence to quality standards. When problems are caught early on in the production process through quality control, manufacturers may make quick decisions to fix them, which helps keep operational performance and the company's reputation in the market [1–3].

Packaging plays a critical role in preserving the integrity of ceramic products after manufacturing is complete. Due to their brittle nature, ceramics are prone to breakage or cracking when exposed to shock, moisture, or vibrations during transport and storage. Therefore, using appropriate packaging materials such as protective padding, durable cardboard boxes, or bubble wrap ensures the safe arrival of products to end users. Moreover, effective packaging enhances the product's marketing value by offering an attractive appearance and clear labeling, which builds consumer trust in the brand [4–6].

The relationship between quality control and packaging is shown by their complementary roles in the production and distribution cycle. It's important that a product is free from defects during production, but it's even more important that it's well protected during shipping and handling. Total quality management (TQM) systems rely on proper packaging, which affects damage rates during shipping and storage and return rates. As part of the quality control process, packaging inspection ensures that packaging is suitable, safe, and able to withstand external factors. A successful supply chain and customer satisfaction therefore, require packaging integration and quality control [3,7–9].

This study aims to highlight the integrated relationship between quality control and packaging in the industrial ceramics industry, explaining the practices and techniques used, the challenges facing this stage, and providing real-life examples of their role in increasing production efficiency and reducing losses.

2. Concept of Quality Control in Ceramics

Quality control in the ceramic industry refers to a set of technical and organizational procedures implemented to ensure that products meet defined standards in terms of shape, dimensions, physical properties, and aesthetic quality. It is an essential component of the Total Quality Management system, aiming to enhance the production process through continuous monitoring at every stage. This system contributes to reducing defects, improving operational efficiency, and increasing customer satisfaction by delivering final products that meet expectations and comply with local and international standards [10–13]. As illustrated in Figure 1, Concept of quality control in ceramics.

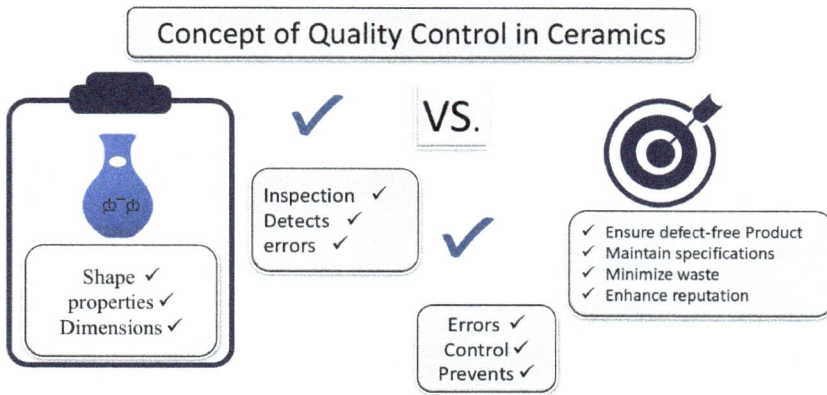

Figure 1. Concept of quality control in ceramics.

The terms "inspection" and "monitoring" are often confused, even though they perform different functions in quality systems. Inspection refers to the evaluation of a product or its components during or after production to verify that, it meets specifications. In contrast, monitoring is a proactive tool that aims to manage production conditions in real time to prevent defects from occurring in the first place. In other words, inspection detects errors after they occur, while monitoring seeks to avoid them by regulating variables during the production process [14,15].

Compliance with operational processes in ceramics manufacturing is essential. Aside from identifying damages like cracks or surface irregularities, meeting the strict procedural guidelines poses a greater challenge. Almost impossible. Even a tiny mistake changes the function and appearance of ceramics. It is the reason why precise oversight helps so much. Enhanced operational sustainability helps a company remain profitable, and the margins refine these factors; less waste, efficient material consumption, lower energy costs. A company's reputation also plays a critical role. Buyers appreciate consistent quality. Products meeting expectations reinforce trust and foster customer loyalty. Gathering data or tracking defects, and evaluating patterns is also important. It is much more than administrative work, and these figures pinpoint areas needing improvement. Small changes make processes more efficient and products more durable over time [15–17].

In Table 1, the role of quality control in the ceramics industry is depicted within the broader context of Total Quality Management (TQM). It shows the movement from inspection in a reactive way too much more proactive types of monitoring and shows how the monitoring of processes using data improves the consistency of the products, customer satisfaction, and the efficiency of operations. This systematic framework provides the foundation for sustaining technological leadership and market competitiveness in the entire production cycle.

Table 1. Conceptual framework of quality control in ceramic industry.

Category	Detail / Explanation	Strategic Impact
Definition	A structured set of preventive and corrective procedures ensuring ceramic products meet technical and aesthetic standards throughout all production stages.	Ensures consistent product quality, minimizes variation, and supports lean manufacturing.
Position in TQM	Quality control is a pillar of Total Quality Management (TQM), integrated across departments from raw materials to delivery under a feedback-driven model.	Enhances interdepartmental coordination, reduces errors at source, and boosts organizational efficiency.
Inspection vs Control	**Inspection:** Post-process checking (detects flaws) **Control:** Real-time process regulation (prevents flaws)	Moves focus from reactive quality checks to predictive and preventive strategies.
Key Objectives	- Prevent cracks, deformation, and discoloration- Ensure dimensional and structural precision- Lower production losses- Ensure compliance with standards	Improves product yield, reduces waste and reworks, and boosts environmental and financial performance.
Data Utilization	Ongoing collection and analysis of defect rates, process variables, and material behavior	Drives continuous improvement, supports R&D, and informs managerial decisions.
Customer Perspective	Delivery of durable, functional, and attractive ceramic products that meet expectations and safety regulations	Builds trust, enhances market reputation, and fosters long-term customer loyalty.

3. Quality Control Stages During Production

Quality control during various production stages is a critical step to ensure a high-quality ceramic product with stable properties. These stages begin with the inspection of raw materials and continue through to the final product, involving a series of technical checks and tests designed to identify potential defects and allow timely corrective actions.

3.1 Raw material inspection

The first step involves carefully examining the raw materials clay, and various minerals to check their quality. Impurities can cause problems later, so it's important to catch them early. We conduct chemical tests to figure out what's in these materials, looking at oxides and other elements. Then there are physical tests to see how fine the particles are, how well they hold together, and how much water they contain. All this information helps tweak the mixture so the [18–20].

3.2 During forming and drying

The current quality control process primarily involves checking the geometric dimensions and internal consistency of the formed component. We take precise measurements to verify that thickness and lengths match the specified design requirements. Simultaneously, we examine material distribution patterns to identify potential stress points within the structure. The drying parameters, time, temperature, and humidity are closely regulated. This helps prevent issues like cracking or irregular shrinkage, which could otherwise result in deformation or compromised structural integrity. The goal is to maintain consistency while minimizing defects that might affect performance [5,21,22].

3.3 After firing and glazing

Following the firing process, detailed inspections are conducted to measure shrinkage levels caused by thermal transformations and to verify that final dimensions match specifications.

Glaze quality is evaluated for color accuracy, glossiness, and uniformity. Additionally, the glaze's adhesion to the surface is tested. Laboratory and visual tools are used to detect defects such as staining, micro-cracks, or inconsistent glaze coverage that could compromise the product's functionality or appearance [15,23–25].

3.4 Final testing

The final stage of quality control includes rigorous durability testing, such as resistance to impact, abrasion, and water absorption, to ensure the product's suitability for its intended application. A final visual inspection is also carried out using advanced technologies like high-resolution digital cameras and smart sensors, enabling accurate and rapid detection of surface defects that may not be visible to the naked eye. These tests collectively ensure that only products meeting the highest standards of quality and safety reach the customer [26–28].

In Table 2, the various steps leading up to the quality control testing of ceramics are displayed, "starting from the raw material inspection and finishing with the final testing." Every stage employs various methods considering the technical aspects of purity, mechanical grading, dimensional precision, glaze adhesion, and even artificial intelligence. Inspection using these methods helps to guarantee visual precision and structural intricacies alongside the strength of the product throughout the entire production process.

Table 2. Quality control stages in ceramic production.

Stage	Inspection Focus	Tools & Technologies Used	Key Objectives	Expected Impact on Product
Raw Material Inspection	• Purity and composition of clay and minerals• Particle size, moisture, plasticity	• XRF/XRD analysis• TGA/DSC for thermal behavior• Laser particle analyzers	• Verify raw material suitability• Adjust formulation for optimal workability and firing results	• Stable body formulation• Reduction in early-stage defects
During Forming & Drying	• Dimensional accuracy• Material homogeneity• Drying rate control	• Precision calipers• Digital thickness gauges• Smart drying chambers with sensors	• Prevent internal stress and cracking• Maintain design geometry	• Lower rejection rate• Improved shape retention after drying
After Firing & Glazing	• Shrinkage control• Glaze uniformity and adhesion• Surface aesthetics	• 3D scanners• Colorimeters (CIE Lab*) • Adhesion strength testers	• Match final specs• Ensure even glazing and secure bonding	• Enhanced visual appeal• Elimination of surface defects and failures
Final Testing	• Mechanical strength• Water absorption and durability• Surface finish and precision	• ISO-standard impact/abrasion testers• Water absorption tests• AI-enabled visual systems	• Ensure functional reliability• Meet international performance standards	• High consumer satisfaction• Long-term performance and durability

4. Modern Techniques and Tools for Quality Control

With advancements in manufacturing technology, quality control in the ceramics industry has become more precise and efficient. This improvement is largely due to the integration of advanced techniques and smart tools, which have significantly reduced human error, accelerated defect detection, and enhanced the overall efficiency of production lines, as illustrated in Figure 2; Modern Techniques and Tools for Quality Control [7,29,30].

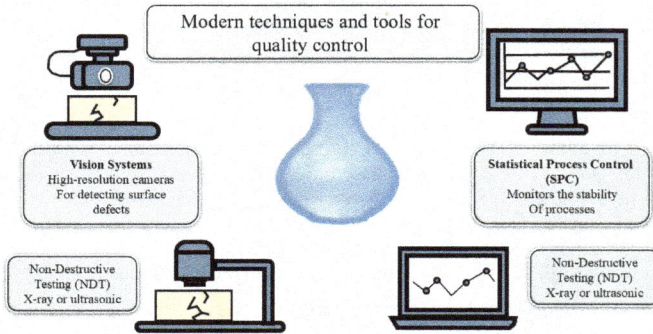

Figure 2. Modern techniques and tools for quality control.

Integrating advanced techniques and smart tools into manufacturing systems has had a positive impact on quality control practices in the ceramic industry, making it more effective and sharper. Manufacturing method improvements have progressively enhanced the speed of the production line, broaden the precision of quality checks, and heighten the efficacy of quality inspections [31–33].

In addition, Statistical Process Control SPC tools track the stability of a production process by analyzing relevant quality control data. This method identifies and mitigates unwarranted trends in quality and helps adjust before problems escalate.
[9,34,35].

Another valuable approach is the use of Non-Destructive Testing (NDT) methods, such as X-ray imaging and ultrasonic inspection, to examine the internal structure of ceramic pieces without damaging them. These techniques are especially important for high-value products or those requiring precise structural integrity [36–38].

Combined, these modern tools significantly enhance accuracy, minimize waste, and increase the reliability of ceramic products in both local and global markets.

Table 3 highlights modern techniques and tools used in quality control across the ceramic production process, emphasizing automation and precision. From smart vision systems and Non-Destructive Testing methods to thermal cameras and statistical control software, each tool plays a role in enhancing accuracy, reducing defects, and ensuring compliance with standards. These technologies are tailored to different production stages and product types, enabling efficient, data-driven ceramic manufacturing.

Table 3. Modern techniques and tools for quality control in ceramics.

Technique / Tool	Main Function	Key Advantages	Production Stage	Suitable Product Type	Automation Level
Vision Systems	Detecting surface defects using smart cameras	Fast inspection, high precision, reduces human error	Post-forming / Pre-packaging	Decorative and industrial ceramics	Fully automated
Statistical Process Control (SPC)	Monitoring and adjusting process quality in real time	Early error detection, continuous improvement, less waste	During production	All types of ceramics	Semi-automated / Software
Non-Destructive Testing (NDT)	Internal defect detection via X-ray or ultrasonic methods	Reveals hidden defects, ideal for precision ceramics	Post-firing	Technical and medical ceramics	Semi-automated
Smart Sensors & Optical Scanners	Measuring dimensions, humidity, and temperature in real-time	Instant feedback, integrates with production systems	Forming and drying stage	Industrial and engineering ceramics	Partially automated
Intelligent Analysis Software	Linking quality data with automated production decisions	Fault prediction, performance tracking, advanced quality control	All stages	Large-scale manufacturing	Automated / Cloud-based
Colorimeters	Testing color consistency and conformity to standards	Accurate results, color unification, ISO-compatible	Post-glazing	Glazed and artistic ceramics	Manual / Automatic
Precision Digital Scales	Accurate measurement of raw material ratios	Reduces mix errors, ensures consistency in properties	Pre-forming	All types of ceramics	Manual / Semi-automated
Thermal Cameras	Monitoring heat distribution inside kilns	Improves firing quality, identifies hot/cold zones	During firing	Thermal and technical ceramics	Fully automated

5. Packaging in Industrial Ceramics

Packaging plays a vital role in the industrial ceramics production chain. Its function extends beyond mere protection; it also helps preserve product quality and enhances market value. Due to the fragile nature of ceramic materials and their susceptibility to breakage, well-designed packaging is essential to ensure that products reach end users in perfect condition, free from damage or defects [5,39,40]. As illustrated in Figure 3, Packaging in industrial ceramics

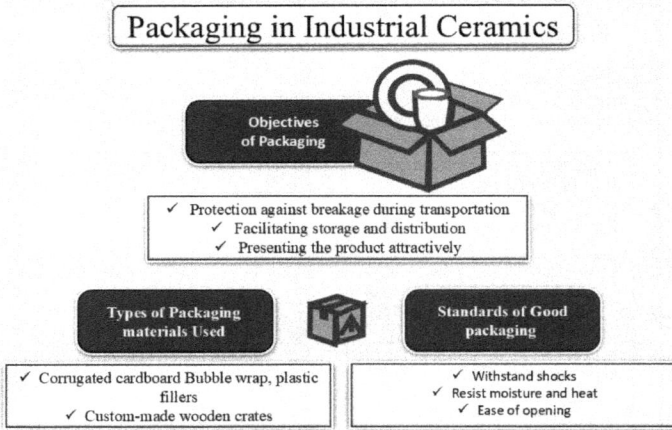

Figure 3. Packaging in industrial ceramics.

5.1 Objectives of packaging

The primary objectives of packaging in industrial ceramics revolve around three key aspects. First, protection against breakage during transportation, since ceramic items are exposed to vibrations and impacts during shipping and handling. Packaging must act as a shock-absorbing barrier. Second, facilitating storage and distribution, where efficient packaging ensures safe stacking and reduces warehouse space requirements. Third, presenting the product in a visually appealing and marketable form, as packaging serves as a branding element that enhances the product's commercial identity and creates a strong first impression for customers [6,40–42].

5.2 Types of packaging materials used

The materials used for ceramic packaging vary depending on the type, size, and value of the product. Among the most used materials are corrugated cardboard for outer boxes, and bubble wrap, which provides flexible protection against impact and vibration. Plastic fillers are also used to protect the product inside the box and prevent movement. For high-end or large items such as sinks or basins, custom-made wooden crates are employed to offer extra structural support and durability during transportation and storage [41,43–45].

5.3 Standards of good packaging

Effective packaging must meet several technical standards. These include the ability to withstand shocks and vibrations without compromising the integrity of the product. Packaging should also be resistant to moisture and heat, especially in uncontrolled storage or shipping environments. Finally, the packaging must be easy to open without damaging the product or the packaging itself, while also maintaining a clean and professional appearance [46–48].

Table 4 outlines the critical role of packaging in the industrial ceramics sector, focusing on protection, efficiency, and presentation. It details key materials like corrugated cardboard, bubble wrap, and wooden crates, highlighting their technical benefits for safe transport and storage. Adherence to packaging standards such as shock resistance and moisture protection ensures product integrity and enhances customer experience.

Table 4. Packaging in industrial ceramics.

Category	Content	Technical Details / Practical Benefits
Packaging Objectives	- Protection from breakage during transport- Ease of storage and distribution- Attractive product presentation	- Shock absorption reduces damage by up to 80%- Optimized stacking saves warehouse space- Enhances brand image and customer satisfaction
Packaging Materials Used	- Corrugated cardboard- Bubble wrap- Plastic fillers- Wooden crates	- Cardboard resists bending and moisture (up to 70%)- Bubble wrap cushions irregular shapes- Wooden crates ideal for heavy or high-value items
Good Packaging Standards	- Shock resistance- Moisture and heat protection- Easy opening	- Drop-tested from 1 meter- multi-layer packaging improves insulation- Smart design ensures customer-friendly unboxing

6. Relationship Between Quality Control and Packaging

The relationship between quality control and packaging in the ceramic industry is deeply interdependent and complementary. A product cannot be considered "high quality" simply because it has successfully passed production stages unless it is packaged in a way that preserves its integrity and characteristics until it reaches the consumer. Packaging is not merely a post-production step; it is an extension of the quality process itself and is treated as an integral part of TQM [3,7,39,49,50].

Effective packaging directly contributes to reducing loss rates due to breakage or damage during transportation and storage making it a key indicator of the efficiency of the overall quality control system. Well-designed packaging also reflects the manufacturer's commitment to industrial standards, especially in markets where distribution and safety requirements are highly regulated [51–54].

Operationally, packaging inspections are included as part of the final quality checks. These involve reviewing the quality of materials used, sealing methods, box strength, and resistance to external factors such as shock and humidity. Various technical tools are employed, including pressure and compression testers, as well as drop tests and environmental simulations [7,16,55].

Case studies from large-scale ceramic manufacturers indicate that optimizing packaging strategies can reduce return rates by 15–25%, highlighting packaging's measurable impact on overall product quality performance [39,52,56,57].

In summary, packaging is not just a protective measure; it is a strategic component of quality control, enhancing product reliability, minimizing losses, and improving the end-user experience.

Table 5 illustrates the interconnected role of quality control and packaging in ceramic manufacturing as part of an integrated Total Quality Management approach. It shows how effective packaging reduces product loss, ensures regulatory compliance, and enhances customer satisfaction. Real-world data and tools such as ISO standards, drop tests, and branding demonstrate how optimized packaging improves both operational performance and market competitiveness.

Table 5. *Quality control and packaging in ceramic manufacturing.*

	Description	Industrial Impact	Tools/Technologies Used	Real-World Example
Packaging as Part of TQM	Treated as a stage within the Total Quality Management system	Integration between production, inspection, and shipping	ISO protocols, material quality audits	Implementing ISO 9001 with packaging control included
Reducing Loss & Damage	Packaging minimizes breakage during storage and transport	Significant reduction in waste (10–20%)	Drop tests, vibration-dampening materials	18% decrease in damage-related losses annually
Compliance with Standards	Reflects professionalism and readiness for international markets	Easier export approvals and customs clearance	CE labels, safety warnings, recyclable symbols	Exporting ceramics to EU with compliant packaging
Final QC Includes Packaging	Packaging is inspected as part of final product release	Lower return rates, improved customer trust	Pressure testing, compression tests, thermal imaging	Shipment redesign after packaging failure in transit
Case Study Results	Packaging optimization led to 15–25% reduction in returns	Better reputation and lower operational costs	Comparative return data (before/after)	Local factory reduced returns from 12% to 6%
Strategic Marketing Role	Packaging reflects product value and enhances brand perception	Enhances user experience and customer loyalty	Branding design, eco-friendly materials, printed logos	Premium ceramic sets with custom packaging boosted sales by 30%

7. Conclusion

Quality control in the ceramics industry isn't just about catching defects, it's a strategic, hands-on process that touches every stage of production. From selecting raw materials to putting the final product in its packaging, quality control ensures that the shape, function, and appearance of each item meet clearly defined standards.

By closely monitoring each step, running tests, and stepping in early when problems arise, manufacturers can cut down on errors, reduce waste, and make their operations more efficient. and thanks to modern tools like digital simulations, advanced inspection systems, and data analysis, quality control has evolved. It's no longer just about fixing issues after they happen, it's now about predicting and preventing them.

This shift isn't just technical; it supports creativity, long-term sustainability, and a stronger position in a competitive market whether, in handcrafted ceramics or large-scale industrial production. At its core, quality control is what transforms ordinary production into craftsmanship. It builds trust with customers, strengthens a brand's reputation, and helps ceramic products deliver both function and beauty.

References

[1] Chernsiri, P. (2023). Business strategy analysis case study: NGK Insulators Company.

[2] Mandal, M. and Sarkar, D. (2019). Manufacturing Excellence in Ceramic Industry, in *Ceramic Processing*, CRC Press, pp. 1–36.

[3]   Ridwan, A. *et al.* (2024). Analysis of ceramic product quality: A Six Sigma approach. *Journal Industrial Servicess.*

[4]   Khalil, A.M.E. *et al.* (2023). Context analysis for transformative change in the ceramic industry. *Sustainability.*

[5]   Rotroff, S.I. (2021). 19 The Ceramic Industry. *The Cambridge Companion to Ancient Athens.*

[6]   Ozkanlısoy, O. and Akkartal, E. (2021). Evaluation Of A Ceramic Enterprise Regarding A Scope Of Strategic Management As A Part Of Supply Chain. *Journal of Management Marketing and Logistics.*

[7]   Ozcan, S. and Sengul, U. (2025). The use of supervised artificial intelligence methods in quality determination in continuous production lines: a case study of ceramic industry. *International Journal on Interactive Design and Manufacturing.* https://doi.org/10.1007/s12008-025-02310-w.

[8]   Arifianti, R. and Raharja, S.J. (2019). Analysis of Ceramic Product Attributes in a Ceramic Industrial Center in Purwakarta, Indonesia. *Review of Integrative Business and Economics Research.*

[9]   Shojaee Barjoee, S. *et al.* (2025). Noise climate assessment in ceramic industries (Iran) using acoustic indices and its control solutions. *Advances in Environmental Technology.* https://doi.org/10.22104/aet.2024.6922.1899.

[10]   Bhadu, J. *et al.* (2025). Implementation of an Integrated Framework of Lean Manufacturing and Industry 4.0 toward Sustainability: A Case Study of the Indian Ceramic Industry. *EMJ - Engineering Management Journal.* https://doi.org/10.1080/10429247.2024.2440248.

[11]   Provenzano, M. *et al.* (2025). TQM and Industry 4.0: an analysis of key enabling technologies about Critical Success Factors (CSFs). *Iaquis Conference Proceedings.*

[12]   Andalib Ardakani, D. *et al.* (2025). An analysis of the barriers to the implementation of quality 4.0 utilizing the approach fuzzy DANP. *Journal of Quality Engineering and Management.*

[13]   Jovic, F. et al. (2013). Quality control engineering in automated ceramic tile production using a signal information content approach. Advanced Engineering Informatics. https://doi.org/10.1016/j.aei.2012.11.007

[14]   Muttamara, T. *et al.* (2025). Automated Quality Inspection Control Station For Ceramic Furnace Disk Using Image Detection Techniques. *Journal of Physics: Conference Series.*

[15]   Carvalheira, S. *et al.* (2023). Energy Efficiency Improvements in a Portuguese Ceramic Industry: Case Study. *Applied Sciences.*

[16]   Bustillo Revuelta, M. (2021). Ceramic Products, in *Construction Materials: Geology, Production and Applications*, Springer, pp. 339–374.

[17]   Shojaee Barjoee, S. and Rodionov Alekseevich, V. (2024). Respirable dust in ceramic industries (Iran) and its health risk assessment using deterministic and probabilistic approaches. *Pollution.*

[18]   Shathi, A.S. *et al.* (2024). Iron removal from red clay using oxalic acid leaching for enhanced ceramic industry applications. *Heliyon.*

[19] Vasić, M.V. *et al.* (2022). Aplitic granite waste as raw material for the production of outdoor ceramic floor tiles. *Materials.*

[20] Götze, J. and Göbbels, M. (2023). Inorganic Non-metallic Raw Materials, in *Introduction to Applied Mineralogy*, Springer, pp. 23–77.

[21] Ribeiro, M.J. and Tulyaganov, D. (2021). Traditional Ceramics Manufacturing, in *PoliTO Springer Series*, Springer, pp. 75–118.

[22] Nait-Ali, B. *et al.* (2024). Perspectives in drying of ceramics. *Open Ceramics.*

[23] Izam, N.S.M.N. *et al.* (2023). The characteristics and production method between glazed and unglazed of ceramic and porcelain tiles by firing process: A review. *IOP Conference Series: Earth and Environmental Science.*

[24] Hasanuzzaman, M. *et al.* (2022). Investigation of methods to prevent pin-holing defect in tableware ceramic industry. *International Journal of Ceramic Engineering & Science.*

[25] Turkel, E. and Yayma, F.R. (2025). In-mold glaze and slip applications in ceramics. *Ceramics: Art and Perception.*

[26] Lourenco Alves, C. *et al.* (2025). Challenges and opportunities for increase sustainability and energy efficiency in ceramic tile industry. *International Journal of Applied Ceramic Technology.* https://doi.org/10.1111/ijac.15097.

[27] Sabet, N. (2021). Sustainability communications and the ceramic tile industry.

[28] Devjibhai, S.H. (2023). Performance Improvement Through Implementation Of MFCA Based System Framework In Ceramic Tiles Manufacturing.

[29] Zarafshan, E. *et al.* (2021). The analysis and interpretation of quality cost using rough set theory: a case study from the ceramic industry. *International Journal of Service and Computing Oriented Manufacturing.*

[30] Kiradoo, G. (2021). The transition of traditional pottery-making into advanced ceramics in context to the indian ceramic industry. *Turkish Journal of Physiotherapy and Rehabilitation.*

[31] Mao, W. *et al.* (2025). Artificial intelligence-enabled defect detection method and engineering application of ceramic mug. *Engineering Applications of Artificial Intelligence.*

[32] Lu, F. *et al.* (2022). HFENet: A lightweight hand-crafted feature enhanced CNN for ceramic tile surface defect detection. *International Journal of Intelligent Systems.*

[33] Schneider, M. *et al.* (2024). AI-based Quality Control for Sanding Ceramics. *Communications for Industry 4.0/5.0 (ARCI 2024).*

[34] Tu, T. *et al.* (2024). Ceramic Process Optimization and Automation Design Based on CAD and Reinforcement Learning.

[35] Pheng, T. *et al.* (2022). Prediction of process quality performance using statistical analysis and long short-term memory. *Applied Sciences.*

[36] Dong, G. *et al.* (2022). Application of machine vision-based NDT technology in ceramic surface defect detection–a review. *Materials Testing.*

[37] Wunderlich, C. *et al.* (2022). NDE in additive manufacturing of ceramic components, in *Handbook of Nondestructive Evaluation 4.0*, Springer, pp. 735–753.

[38] Wunderlich, C. *et al.* (2025). NDE in Additive Manufacturing of Ceramic Components by LSP and OCT, in *Handbook of Nondestructive Evaluation 4.0*, Springer, pp. 1057–1078.

[39] Chen, Y. (2024). Ceramic packaging design and brand development: a case study of Jingdezhen. *International Journal of Innovation and Sustainable Development.*

[40] Cheng, Y. *et al.* (2022). Investigation and Accounting Research of VOC in Daily and Specialty Ceramic Industry. *Coatings.*

[41] Patel, M.T. and Desai, D. (2023). Implementation of Six Sigma in the small-scale ceramic industry and its holistic assessment. *International Journal of Quality Engineering and Technology.*

[42] Barjoee, S.S. and Gendler, S. (2024). Sustainable illumination: Experimental and simulation analysis of illumination for workers wellbeing in the workplace. *Heliyon.*

[43] Alessandra, C. (2023). Packaging, logistics and sustainability. Exploring innovative solutions for eco-sustainable packaging. *Sinergie-SIMA 2023 Conference Proceedings– Long papers, "Rediscovering local roots and interactions in management."*

[44] Desole, M.P. *et al.* (2024). Life Cycle Assessment (LCA) of ceramic sanitaryware: focus on the production process and analysis of scenario. *International Journal of Environmental Science and Technology.*

[45] Nafe, C.M. (2024). An Electronic Waste Ecology of Knowledge: How Stakeholder Interactions Shape the Sociotechnical System of Used Electronics Management in the United States.

[46] Bose, S. and Das, C. (2024). *Introduction to Ceramics: Fabrication, Characterizations, and Applications*, CRC Press.

[47] Costa Oliveira, F.A. *et al.* (2025). Mechanical Performance of Ceria-Coated 3D-Printed Black Zirconia Cellular Structures After Solar Thermochemical CO/H2 Fuel Production Cycles. *Crystals.*

[48] Barjoee, S.S. *et al.* (2024). Noise climate assessment in ceramic industries (Iran) using acoustic indices and its control solutions.

[49] dos Santos Silva, J. and Leite, M.S.A. (2024). Analysis of a supply chain in the ceramic sector: a look at business processes. *Gestao e Producao.* https://doi.org/10.1590/1806-9649-2024v31e12021.

[50] Realyvásquez-Vargas, A. et al. (2019). Implementation of production process standardization-A case study of a publishing company from the SMEs sector. Processes. https://doi.org/10.3390/pr7100646

[51] Shojaee Barjoee, S. *et al.* (2025). Occupational injuries associated with safety climate among ceramic industry workers in Iran. *Scientific Reports.*

[52] Sawitri, Y. *et al.* (2025). Descriptive Study of Reverse Logistics (RL) Activities and Ceramic Waste Innovation at PT. Lucky Indah Keramik. *Jurnal Serambi Engineering.*

[53] Li, X. *et al.* (2024). Low-carbon and low-cost preparation of non-sintering bauxite-based solid thermal energy storage materials. *Solar Energy.*

[54] Niekurzak, M. and Lewicki, W. (2025). Optimisation of the Production Process of Ironing Refractory Products Using the OEE Indicator as Part of Innovative Solutions for Sustainable Production. *Sustainability.*

Materials Research Forum LLC
https://doi.org/21741/9781644903834

[55] Giacomo Boschi *et al.* (2023). Sustainability and environmental impact of the Italian ceramic tile industry.

[56] Julee, S.S. (2024). Strategic & cost-effective procurement process of Star Ceramics Limited Company of Bangladesh.

[57] SUMANTRI, Y. *et al.* (2022). Improving Logistics Services of LSP to Minimize Logistics Outsourcing Risks in the Ceramic Industry. *Journal of Distribution Science.*

Chapter 17

Innovations and Industrial Experience Gained in the Production of Ceramic Materials

Eray Casin[1], Cumhur Eren Isık[2], Fatma Nur Maran[3], Iskender Isık[4*], Fatih Sen[3*]

[1]Genesis Technological Products Industry and Trade Limited Company, Corum, Türkiye

[2]Department of Handicrafts, Kutahya Fine Arts Vocational School, Kutahya Dumlupınar University, Evliya Çelebi Campus, 43000 Kutahya, Türkiye

[3]Sen Research Group, Department of Biochemistry, Dumlupinar University, Kutahya 43000, Türkiye

[4]Department of Materials Science & Engineering, Faculty of Engineering, Dumlupınar University, Evliya Çelebi Campus, 43100 Kutahya, Türkiye

iskender.isik@dpu.edu.tr, fatihsen1980@gmail.com

Abstract

Industrial ceramic production systems have evolved with the advancement of technology. Ceramic material production, which has developed with Industry 4.0, has now reached the point where fully automated devices deliver products to customers without human intervention. Automation integration is being implemented across all systems from glazing robots and fully automated pressure casting machines used in ceramic healthcare product production to printer technologies used in tile production to minimize human error. The use of artificial intelligence in production technologies, due to the variability of inputs in ceramic material production, has enabled the integration of materials developed in the automation industry into production systems. This section will discuss developments in production systems and the impact of engineers' experience in production systems on production.

Keywords

İndustry 4.0, Ceramic Production Technologies, Automation And Robotic Systems, Artificial İntelligence İntegration

1. Introduction

Human development in the last decade has been accompanied by rapid changes in technology and the increasing proliferation of digitized devices and services, and it seems likely that the pace of change will accelerate as a result of "frontier technologies" such as artificial intelligence (AI), robotics, biotechnology, and nanotechnology. These technologies have already provided significant benefits, but rapid advancements can lead to serious negative consequences when they exceed society's ability to adapt. Developing countries must discuss how they can balance innovation with equity to catch the wave of cutting-edge technologies and achieve the Sustainable Development Goals [1].

Ceramic material production is a type of production that requires careful attention from the initial process to the final processing, where each stage is interconnected and any incorrect application in any stage can result in errors in other stages. Ceramic production processes are materials in which physical, chemical, and microstructural properties are effective throughout the production process. When producing a ceramic product, it is essential to understand the physical properties that change at each stage of production. If these properties, which are affected by the production processes, are not known before production begins, it becomes difficult to resolve any errors that may occur during the production process.

Ceramic material production begins with design. In a three-dimensional design context, design elements have become the main factors in creating a design that meets the criteria of beauty, aesthetics, grandeur, and functionality for a good product. The design phase of contemporary, elegant, simple, and easy-to-use products suitable for today's modern lifestyle varies according to the cultural perspectives of the regions where people live [2]. However, the control and continuity of scientific facts for the production of these designs ensure the efficient production of the manufactured product.

The machines, environmental conditions, and labor used in the production processes increase the quality of the product, the added value of the production facility, and, consequently, its visibility in the market. Maintaining all these facts at a high level is possible by transferring scientific knowledge to production facilities. This is possible by using all the information required by technology and production in the production process.

As in every production facility, the first parameter considered in ceramic materials is cost. The term production cost refers to the total cost incurred by a business to produce a certain amount of product, including labor, raw materials, or consumables. In economics, it is similar to the expenses incurred to obtain the cost factors of capital, labor, materials, and other processes involved in the production process. To produce a certain amount of output at a lower cost, companies must select inputs that minimize costs [3]. In ceramic material production, selecting the lowest-cost materials that can maintain the highest level of production efficiency is one of the most important factors in delivering the produced material to the user and ensuring that production facilities continue their production activities.

In this section, we will discuss the use of technology in ceramic product production, the benefits of studies on the efficient use of materials in the final product conversion process, and developments in industrial ceramic materials. In this book, which focuses on industrial ceramics, the process components that are important in the production of ceramic materials are considered. In this section, the impact of different developments and technological applications on ceramic materials, as they are presented to the market, will be discussed. The topics to be discussed will be illustrated with examples from both studies on the final product and technological developments made for production.

2. Experiences in Industrial Ceramics

Industrial ceramics include ceramic plates, porcelain, decorative items, all kinds of glassware, sanitary ware such as toilets, sinks, and shower trays, as well as floor and wall tiles used in everyday life [4]. Developments in these areas will be examined in their entirety in this section. Although there are process variations, the production method for each material is based on heat and raw materials. Relatively the same raw materials are used. What these raw materials are has been explained in previous sections. The final product and process development applications applied to the high-strength materials obtained by exposing these raw materials to heat after they have undergone certain processes can meet human needs in the marketing of products. The developments described in this context provide advantages for the industry in terms of both business and customer perspectives in the marketing of ceramic products with different characteristics.

2.1 Technological developments in production processes

The production of industrial ceramic materials is challenging, and in addition, the time spent on the production of some products is very long, the inputs in the production processes of some products are changed quickly according to customer needs, and labor is at the forefront in the production of all materials. The human labor required in ceramic product production has some negative effects on production processes. Technological developments are utilized to minimize these effects. Additionally, improvements made in processes that reduce production costs enable the use of production capital in new product production and create new employment opportunities.

When examining the production of ceramic materials across different products, it can be seen that the production of ceramic coating materials requires and uses more automation than the production of medical devices and tableware [5]. The use of automatic control systems has made it possible to collect data instantly during process control, thereby enabling continuous rheological control of different solid-containing suspensions. Digital rheology controllers, which operate in conjunction with computer systems via a fully digital communication network and continuously transfer data, enable the measurement of important rheological parameters such as density, viscosity, and temperature. The use of such devices enhances the ability to control and intervene in ceramic liquid suspensions, which are highly susceptible to changes in environmental conditions.

Digital control systems are used in the presses used in the production of ceramic coating materials to ensure that products are produced in standard sizes. These systems enable production stability by controlling certain data such as tile thickness, press pressure power, and particle size distribution of the ceramic powder used. Additionally, in recent years, devices capable of moldless shaping have been designed to enable the semi-finished shaping of floor and wall tiles of different thicknesses and sizes. These devices, which operate entirely using digital printing methods, are environmentally friendly due to the recovery of ceramic powders, energy-efficient in terms of energy consumption, and also provide flexibility in production. With these devices, the production of tiles up to 1200x2400 mm in length is possible [6].

High-volume production in ceramic tile manufacturing necessitates the storage of semi-finished products in the production area. Cassette-type racks are used as storage areas in production. However, transfers between these storage areas are carried out using robotic vehicles. These vehicles are devices that follow a specific path with predetermined routine stops and can carry

Materials Research Forum LLC
https://doi.org/21741/9781644903834

cassette holders containing ceramic semi-finished tiles. These devices have been used in ceramic tile production for years. However, in recent years, the integration of augmented reality systems has added many capabilities to these devices, such as the ability to find their own way and decide which stop to go to, thereby minimizing production errors caused by human error [6].

Decoration is a very important design argument in the production of ceramic coating materials. The most traditional decoration method used in the production of ceramic coating materials is screen printing. However, in recent years, decoration using digital printing has both enhanced the decorative designs of products and enabled more flexible production. Inkjet printing technology has become the leading technology in ceramic tile production. These systems, which have the ability to decorate quickly, offer advantages in terms of cost and aesthetics with features such as contactless decoration, intelligent ink management, the ability to print on textured surfaces at high resolution, more efficient decoration methods, and ease of control on the production line. These systems also enable the development of pigment-based liquid suspensions used in the printers. With the introduction of the first inkjet digital printing machines in 2000, it was understood that the particle size distribution ratios of the solid particle suspensions required for use with these machines had to be very fine and narrow, and that the rheological properties of the ink had to be continuously monitored. With the development of these suspensions, the development of digital printing machine spray heads was also achieved [7].

The use of digital printing inks in ceramic tile production requires serious engineering and control. This has led to a shift from simple measurement methods to more complex but more understandable and controllable methods for measuring the particle size, density, and viscosity of these suspensions. The development of ink formulations has revealed numerous engineering challenges, such as the development of a pigment, solvent, and print head necessary for this suspension to enable digital printing. As a result, important chemical, physical, and technological variables have been identified, and the value ranges of these properties have been developed according to the digital printing machines to be used [8].

In recent years, smart coordination systems that can work synchronously between processes have been developed to control semi-finished and final products in the production of ceramic coating materials. These types of systems facilitate the work of production employees and ensure the control and management of the line in all possible combinations. The digitization of lines during flexible production transitions has minimized the likelihood of errors in production planning by digitizing the entire system, including decoration and even slurry preparation [6].

The control of the final products of ceramic coating materials is also a challenging and time-sensitive stage, similar to the tracking of semi-finished products. The application of manual quality control processes in the production of ceramic products, which are produced in high quantities, are heavy, and have a continuous flow, increases the likelihood of the product reaching the customer in a different quality or with defective material. As a result of digitalization, digitalization has also been achieved in quality control, and systems have been developed that incorporate cameras from different angles, are equipped with LED lighting systems, and have the ability to capture instantaneous high-speed images as ceramic products pass through. These systems, which can detect various structural defects and decorative flaws with extremely high precision and perform automatic sorting, have prevented errors that could occur during the quality control process [6].

When examining the technological devices used in the production of ceramic sanitary ware, it is seen that there are not as many technologies in production facilities as there are in ceramic coating materials. One of the reasons for this is that the products have a three-dimensional design. Although automation is used in the production of these products, since it is a sector where labor and manpower are at the forefront, developments such as digitization of ceramic coating materials and process management cannot be used in the production of ceramic sanitary ware. However, with the development of robotic technologies in recent years, robots have been used in some production processes and have helped to reduce manpower.

The most difficult and labor-intensive process in the production of ceramic sanitary ware is the casting process. Although all processes in the ceramic sanitary ware industry, which is generally a labor-intensive sector, involve heavy work, the difficulties in the casting process are not only related to the heavy nature of the work. Challenges such as errors in the plaster molds used in traditional shaping methods, time losses due to ceramic suspensions being left in the molds, and the long waiting time required for the molds to dry and be ready for recasting affect production numbers. Additionally, the large size and space-consuming nature of traditional casting benches increase production areas and the energy and costs required to maintain the necessary environmental conditions for ceramic production. To address these issues, high-pressure casting benches are used. The polymer mold technology used in these machines and the high pressure applied to the suspension enable products to be shaped in a very short time. Although the initial investment costs for these machines are high, the casting life of the polymer molds used is much longer than that of plaster molds. While a maximum of 100-120 castings can be obtained from a plaster mold (and the error rate of products coming out of the molds increases towards the end of the casting life), polymer molds can produce up to 70,000 (100,000 in simple designs) and, with proper maintenance, up to 100,000 castings. This reduces costs in the long term and enables faster production of products with higher surface quality [9] .

The slurry suspensions used in pressure casting machines have the same formulations as those used in products produced using the plaster mold technique. However, using slurry suspensions with different rheological properties enables both more efficient use of the system and faster operation of the system [10]. Higher slurry temperatures than normal conditions and relatively high slurry densities increase the slurry's ability to thicken. The temperature-controlled slurry storage tanks used in the system and the ability to measure viscosity and density values with automation systems enable the system to operate fully automated, quickly, and reliably.

Pressure casting systems, which are fully automated systems used in the production of ceramic sanitary ware, have been developed in recent years with different molding techniques [11,12]. In the early years of their design, these systems were mainly used to produce flat sinks and simple reservoirs, but in recent years, shower trays and sinks have been produced using the suspended molding method, and ring-bonded toilets and single-piece floor-mounted toilets have been produced by completely automatically bonding two different ceramic products. The system's ability to automatically collect slurry suspensions, measure slurry rheology, and intervene, as well as provide full automation for removing products from the workbench via manipulator robots for surface processing, represents a significant advancement for production facilities that traditionally rely on labor-intensive methods. Another advantage of these systems is that they eliminate labor-related errors such as retouching mold joints and smoothing surface roughness on semi-finished products, as they produce products with higher-quality surfaces than plaster molds. These machines increase production efficiency and reduce production costs.

Another technological development used in ceramic sanitary ware is in glazing processes. Manual applications in glazing processes reduce labor errors and the ability to produce quickly. With the use of robots in glazing processes, fully automated system designs have been achieved, minimizing errors. In order to adapt these machine systems to ceramic sanitary ware production systems, it has become necessary to develop different glazing techniques. Although it is assumed that workers use different methods for different product designs in manual glazing, the possibility of glazing products with the same design differently must also be taken into account. Accordingly, determining the glazing path that the robot must follow during glazing is very important. At this stage, engineers who will provide automation teach these paths to the robots, ensuring that the system operates fully automatically (Figure 1). When applied correctly, the rheological values that are stabilized by mixing the glaze tanks help to create an environment in which the system can operate without errors [13].

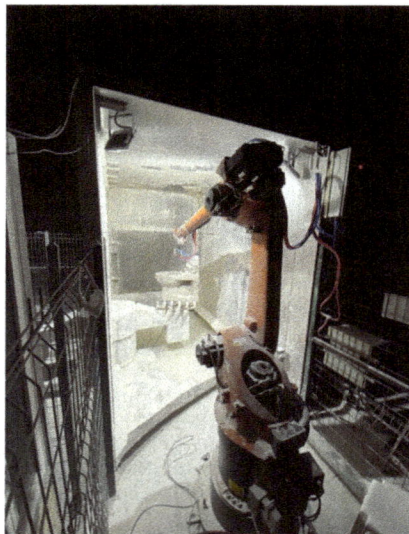

Figure 1. Robot Glazing Technology in Ceramic Sanitary Ware Production [14].

In ceramic healthcare products, robots are used not only as manipulators and glazing robots but also for unloading products from the kiln after sintering and during quality control stages. Similar to imaging systems used in quality control stages for ceramic coating materials, cameras are also used in these systems to transfer visual and structural defects on the product to a software program, enabling a fully automated solution for identifying these defects and separating quality control from human error.

In the production of ceramic tableware, as in the production of other industrial ceramic materials, certain technological advancements are utilized. Among these, the isostatic pressing method is particularly noteworthy. The isostatic pressing method enables the production of products in various sizes and complex shapes. Robotic systems adapted to these systems, which are used for production automation and the production of more complex products, enable the system to be fully automated. Automation systems that provide a high-quality and repeatable work process also increase the traceability of the product in the process [15].

As in ceramic healthware, high-pressure casting benches and manipulator robot systems connected to these systems are also used in ceramic tableware.

Glazing processes in tableware production are carried out using the dipping method. Glazing defects caused by labor errors in the dipping method lead to product losses due to labor-related issues. Systems capable of automatic dipping glazing have been developed to minimize these defects. Glazing automation, which has been applied in ceramic tableware for many years, has reduced production costs and increased production speed.

Digital decoration methods similar to those used in ceramic coating materials are also applied in ceramic tableware. The increasing complexity of the decorations customers want to see on tableware today, and the need to purchase colorful or high-resolution ceramic tableware rather than new designs, has led to an increase in the use of automation and the ability to respond to market demands in production facilities.

In general, the use of technology in industrial ceramic material production enables the minimization of labor-related errors in production facilities, the reduction of production costs, the ability to produce standard and stable products, the flexibility of production systems, and the reduction of labor.

2.2 Technological developments in final products

Technological developments applied in the production processes of industrial ceramics are applications that make production processes easier, increase controllability, and minimize labor and workmanship errors. With these applications, product tracking in production processes and reducing customer complaints after the products reach the customer can be ensured. Improvements in final products must be of a quality that meets customer expectations for ceramic products and ensure that the products are more durable. In this context, companies that produce ceramic products using industrial production methods are working to meet customer demands for different purposes and are developing their products accordingly.

Ceramic products are a material with a wide range of uses, from bricks and tiles to electronic and magnetic products. Industrial ceramics are generally silica and clay-based materials that are produced using advanced production techniques, but are also often produced using traditional production methods. Competition among manufacturers is achieved through more efficient and cost-effective product production [16]. These materials are used in many areas of our lives. The use of ceramic products is widespread in our homes and public spaces. In this regard, the use of ceramic materials with antibacterial surface properties is desired in many public spaces, especially in places where hygienic environments are required, such as hospitals. Hospital infections and various skin diseases can cause potential health problems. As a result, the use of antibacterial and antimicrobial ceramic materials and the demand for these products have increased in public spaces [17]. Many additives are used to create antibacterial surfaces. Antibacterial glaze is formed by adding active ingredients such as TiO_2, ZnO, Ag, and Cu to the glaze [18]. Scientific studies have shown that after these oxides are added to the glaze

Materials Research Forum LLC
https://doi.org/21741/9781644903834

composition, the formation of bacteria and microbes on the glaze surface (the final product after sintering) is prevented, and there is no antibacterial activity on the surface.

The use of ceramic materials with improved technical and aesthetic properties has gained significant attention in the market in recent years. Ceramic coating materials with high-resolution decorative designs that are thin and large in size go beyond traditional ceramic applications from an architectural perspective. Self-cleaning products have been produced using TiO2-containing components such as anatase and rutile in the glaze compositions of wall tiles used in exterior wall coverings [19]. This effect is defined as a photocatalytic effect. Photocatalytic materials are substances that break down harmful bacteria and microorganism formations under sunlight containing UV rays and possess the ability to self-clean their surfaces. Currently, TiO_2 is primarily used to create this effect. Among TiO_2 polymorphs, the anatase phase exhibits the most effective photocatalytic effect [20]. In this type of ceramic material, the electron pairs on the surface exposed to UV rays are excited by the UV rays, and these electrons react with O2 to form anions. Subsequently, these anions react with water in the electron vacancies to form OH ions, thereby causing organic components such as liquid and oil layers to be broken down on the surface [21].

In the daily use of ceramic healthware, antibacterial and, in some applications, photocatalytic properties of glaze components are utilized. Another desired feature in the market is surfaces that do not retain stains or dirt. These surfaces, which are also applied using hot processing, can also be achieved through cold coating applications. These hydrophobic surfaces are used in ceramic healthcare products and various industrial ceramic applications [22]. Ceramic materials with this property are referred to as hydrophobic or superhydrophobic surfaces, which have a high contact angle with water. The most beautiful example of superhydrophobicity in nature is the lotus flower. In nature, it has been observed that dust and pollutants settle on the surface of this flower, and when it rains, the superhydrophobic property causes the pollutants to be cleaned off [23].

In ceramic materials, especially in outdoor floor tiles, ceramic surfaces with high slip resistance are desired. Traditional slip-resistant treatments involve embedding high-hardness components such as quartz or alumina particles into a glassy matrix binder to create a rough surface on the tile. These treatments prevent the surface from being slippery and chemically erode the surface [24]. Materials with these properties are preferred for surfaces with heavy traffic or those prone to getting wet and stained with oil. They are particularly used to prevent slipping on wet or dirty surfaces [25]. In recent years, there has been a demand for this type of surface in some ceramic products, such as shower trays, which include ceramic sanitary ware. Since these surfaces are rough, they are highly susceptible to staining, but it is also desirable that materials with this surface property do not stain during production. In addition, the resistance of the surfaces to chemicals and cleaning agents is proven by tests performed on the final product.

In industrial ceramic products, especially ceramic coating materials, matte surfaces are preferred. These surfaces, which have also been preferred in ceramic sanitary ware in recent years, are desired not only for their matte appearance but also for their color and certain surface applications. The development of matte glazes for such applications requires research and development activities and the exploration of different alternative formulations. In literature reviews, matte glazes applied to ceramic coating materials are typically used for rapid firing. The prolonged firing times in ceramic healthcare products, coupled with pigment additives that vary depending on the desired color, create challenging effects on the production process. The necessity of using pigments with different oxide contents ensures that the color characteristics on the product surface vary in each sintering and glaze application. In cases of such errors or deviations from color standards, re-firing is performed to bring the color values of the products in line with standard deviation values.

3. Conclusion

Industrial ceramics are products manufactured under harsh conditions. They are materials that require intensive use of technological developments and science in production systems. The use of scientific data should facilitate the production of the material, enable the understanding of errors, and lead to their resolution, thereby increasing production efficiency and added value. The production efficiency of industrial ceramic products manufactured using traditional methods should be increased by transferring technological developments to production systems. Therefore, the production of industrial ceramic materials requires intensive engineering work and research and development activities that have a significant impact on the product.

The transfer of customer demands to production systems in the production of ceramic materials depends on the organizational efficiency of production facilities, which are living organisms. The ability of people at all levels working in production facilities to correctly perceive market demands through their work on product and system development and efficient production will increase the added value of production facilities and, consequently, the country.

As with the production of any product, the experience of engineers and other workers involved in industrial ceramic production should be of a quality that enhances the efficiency of material production. Increasing the efficiency of a product's production requires the efficient use of qualified knowledge along with experience. Industrial ceramic production depends on the efficiency of interdisciplinary work in many engineering fields. In this context, the collective work of engineers and other employees who influence production in companies depends on the qualifications of the employees. To ensure a qualified workforce, engineers who have received qualified education must conduct training and work to educate other engineers and employees and to improve the quality of their work. Such efforts will increase the added value of production facilities and contribute to the country's economy.

Referances

[1] Geneva (2021). Catching technological waves Innovation with equity.

[2] Maryam Ali Yasin, S. *et al.* (2018). Article ID: IJMET_09_05_114, The Ceramic Tableware Design Elements of Local Studio Ceramic Designers. *International Journal of Mechanical Engineering and Technology (IJMET.*

[3] Cetin, S. (2023). Production of sintered glass-ceramic composites from low-cost materials. Ceramics International. https://doi.org/10.1016/j.ceramint.2023.04.068

[4] Derin Coşkun", N. *et al.* (2023). Investigation of Defects Occurring in the Casting Shop Process of Ceramic Sanitaryware on the Final Product. *Black Sea Journal of Engineering and Science.* https://doi.org/10.34248/BSENGINEERING.1347161

[5] Gustavo Mallol Gasch, J.Con-45 (2007), Control and Automation In The Ceramic Industry. Evolution And Perspectives.

[6] Atılgan Türkmen, B. et al. (2021). Environmental impact assessment of ceramic tile manufacturing: a case study in Turkey. Clean Technologies and Environmental Policy.

https://doi.org/10.1007/s10098-021-02035-w

[7]   Dondi, M. *et al.* (2014). Ink technology for digital decoration of ceramic tiles: An overview.

[8]   Jaramillo Nieves, L.J. *et al.* (2020). Digital decoration for ceramic tiles: The effect of glazes particle size distribution on the inkjet decoration. *Boletín de la Sociedad Española de Cerámica y Vidrio.* https://doi.org/10.1016/J.BSECV.2019.06.005

[9]   Dadic, Z. *et al.* (2017). High pressure die casting mould repair technologies. *Mechanical Technology and Structural Materials.*

[10]   Kaygısız, N.N.T. and Kaygısız, H. (2021). ALÇI KALIP İLE SERAMİK OBJE ÜRETİMİNDE ÜÇ BOYUTLU YAZICI KULLANIMI. *International Journal of 3D Printing Technologies and Digital Industry.* https://doi.org/10.46519/IJ3DPTDI.880661

[11]   Arapoğlu, İ. and Kahraman, D. (2024). SERAMİK SANATINDA ALTERNATİF KALIP YÖNTEMLERİ* Alternative Mold Methods in Ceramic Art. *Sayı/Number.* https://doi.org/10.34189/asd.2024.21.008

[12]   Makalesi, A. (2016). Karaelmas Fen ve Mühendislik Dergisi. *Karaelmas Fen ve Müh. Derg.*

[13]   Qian, Z. *et al.* (2020). The Robot Intelligent Spraying Glazing System for Sanitary Ceramics Industry. *Journal of Physics: Conference Series.* https://doi.org/10.1088/1742-6596/1653/1/012028

[14]   Torres, R. and Ferreira, N. (2022). Robotic Manipulation in the Ceramic Industry. *Electronics (Switzerland).* https://doi.org/10.3390/ELECTRONICS11244180/S1

[15]   Makalesi, A. *et al.* (2022). Çeşitli Toz Sıkıştırma Yöntemlerinin Karşılaştırılması. *European Journal of Science and Technology Special Issue.* https://doi.org/10.31590/ejosat.1219539

[16]   Sudha, P.N. *et al.* (2018). Corrosion of ceramic materials. *Fundamental Biomaterials: Ceramics.* https://doi.org/10.1016/B978-0-08-102203-0.00009-3

[17]   Özcan, S. *et al.* (2017). F Formation of Antibacterial Effect On Ceramic Tile Surfaces. *Anadolu University Journal of Science and Technology A - Applied Sciences and Engineering.* https://doi.org/10.18038/AUBTDA.300424

[18]   Maryani, E. *et al.* (2020). The Effect of TiO2 additives on the antibacterial properties (Escherichia coli and Staphylococcus aureus) of glaze on ceramic tiles. *IOP Conference Series: Materials Science and Engineering.* https://doi.org/10.1088/1757-899X/980/1/012011

[19]   da Silva, A.L. *et al.* (2018). Photocatalytic ceramic tiles: Challenges and technological solutions. *Journal of the European Ceramic Society.* https://doi.org/10.1016/J.JEURCERAMSOC.2017.11.039

[20]   Etacheri, V. *et al.* (2015). Visible-light activation of TiO2 photocatalysts: Advances in theory and experiments. *Journal of Photochemistry and Photobiology C: Photochemistry Reviews.* https://doi.org/10.1016/J.JPHOTOCHEMREV.2015.08.003

[21]   Ayode Otitoju, T. et al. (2020). Advanced ceramic components: Materials, fabrication, and applications. Journal of Industrial and Engineering Chemistry. https://doi.org/10.1016/j.jiec.2020.02.002

[22]   Reinosa, J.J. *et al.* (2012). Copper based hydrophobic ceramic nanocoating. *Journal of the European Ceramic Society.* https://doi.org/10.1016/J.JEURCERAMSOC.2011.08.013

[23]   Aydin, T. and Kunduraci, N. (2017). KİR TUTMAYAN SERAMİK DUVAR KAROLARI. *Omer Halisdemir University Journal of Engineering Sciences.*

[24]   Javed, S. et al. (2025). Strategies and pathways to improve circularity in ceramic tile production. Journal of Cleaner Production. https://doi.org/10.1016/j.jclepro.2025.145788

[25]   Vermol, V. V. *et al.* (2011). A study on porcelain anti slip tile design. *2011 IEEE Colloquium on Humanities, Science and Engineering, CHUSER 2011.* https://doi.org/10.1109/CHUSER.2011.6163699

Chapter 18

Indispensable Materials in the Metallurgical Industry: Refractories Sustainability, Challenges, and Future Prospects

Recep Artir[1*], Ebru Halvaci[2], Cumhur Eren Isık[3], Iskender Isık[4], Fatih Sen[2*]

[1]Marmara University, Faculty of Engineering, Department of Metallurgical and Materials Engineering, Goztepe Campus, 34722, Kadikoy, Istanbul, Türkiye

[2]Sen Research Group, Department of Biochemistry, Kutahya Dumlupinar University, Kutahya 43000, Türkiye

[3]Department of Handicrafts, Kutahya Fine Arts Vocational School, Kutahya Dumlupınar University, Evliya Çelebi Campus, 43000 Kutahya, Türkiye

[4]Department of Materials Science & Engineering, Faculty of Engineering, Kutahya Dumlupınar University, Evliya Çelebi Campus, 43100 Kutahya, Türkiye

rartir@eng.marmara.edu.tr, fatihsen1980@gmail.com

Abstract

Refractory materials are essential for the metallurgical and ceramic industries, providing the necessary resistance to extreme heat, thermal shock, and chemical attack in furnaces, kilns, and reactors. Their reliability directly impacts process efficiency, equipment lifespan, and product quality. The aim of this study is to analyze the classification, properties, and applications of refractory materials, while addressing sustainability challenges and identifying future development directions. A literature-based analytical approach was applied, reviewing scientific data on chemical and physical characteristics, manufacturing processes, and performance behavior under varying industrial conditions. Comparative analysis was conducted for steelmaking, aluminum production, and copper smelting, alongside an examination of the technological and material linkages between the refractory and ceramic industries. The results indicate that refractories are categorized into acidic, basic, and neutral types, each suited for specific slag chemistries and operating environments. Despite advancements, the industry faces challenges such as thermal shock damage, corrosion, high production costs, and environmental concerns related to non-renewable raw materials, carbon emissions, and post-use waste. Emerging solutions include hybrid composite materials, integration of Industry 4.0 smart monitoring, circular economy practices, green manufacturing processes, and nanotechnology-based performance enhancements. It is concluded that refractories will remain indispensable for high-temperature processes, but their future will rely on balancing performance, cost-efficiency,

A Journey from Raw Materials to Ceramics
Materials Research Foundations 184 (2025)

Materials Research Forum LLC
https://doi.org/21741/9781644903834

and sustainability. Innovation, digitalization, and eco-friendly production methods will be critical to ensuring their role in the evolving landscape of metallurgical and ceramic manufacturing.

Keywords

Refractory Materials, Sustainability, Thermal Shock Resistance, Corrosion, Ceramic Industry

1. Introduction

Refractory materials are indispensable to both the metallurgical and ceramic industries, serving as the backbone for high-temperature processes in furnaces, kilns, reactors, and crucibles. Their unique ability to withstand extreme heat, thermal shock, and chemical corrosion makes them critical to the performance, safety, and longevity of industrial operations [1,2].

In recent years, growing global attention toward sustainability has introduced new challenges for the refractory sector. These include reducing environmental impact, minimizing reliance on non-renewable raw materials, and improving recyclability without compromising material performance. At the same time, advances in ceramic science and smart manufacturing offer promising pathways to develop next-generation refractories that are both efficient and environmentally responsible [3,4].

This study explores the vital role of refractories in the metallurgical industry, highlighting sustainability challenges and examining their evolving connection with the ceramic sector. Furthermore, it considers prospects driven by innovation, digitalization, and the transition toward greener industrial practices.

2. Classification of Refractory Materials

Refractory materials are a category of inorganic, non-metallic compounds specifically engineered to endure extreme conditions of heat, pressure, and chemical attack. These materials are primarily used to line up high-temperature industrial equipment, such as furnaces, reactors, and kilns, where structural integrity must be maintained despite prolonged exposure to temperatures exceeding $1000°C$. Their essential role lies not only in providing thermal insulation but also in protecting the structural components of industrial systems from thermal degradation and corrosion [5–8]. Figure 1 illustrates refractory waste recycling and waste treatment.

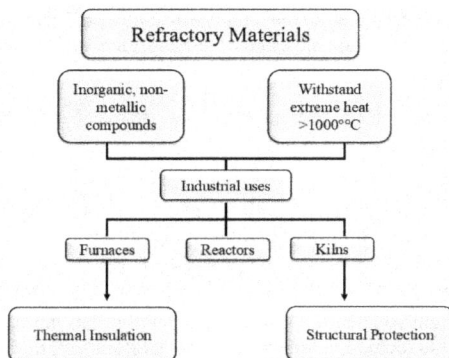

Figure 1. Classification of heat-resistant materials.

2.1 Classification, raw materials, and manufacturing methods

Refractory materials are traditionally classified based on their chemical behavior when exposed to slags and other reactive substances in high-temperature industrial environments. The three primary categories acidic, basic, and neutral each exhibit specific advantages depending on the process chemistry.

Acidic refractories, such as those based on silica and fireclay, are resistant to acidic slags and environments. They are commonly used in glass melting furnaces and coke ovens, where basic contaminants are minimal. However, their performance significantly deteriorates when exposed to basic slags, limiting their use in metallurgical applications involving alkaline components [9–13].

Basic refractories, composed primarily of magnesia (MgO) and dolomite ($CaMg(CO_3)_2$), show excellent resistance to basic slags, making them indispensable in steelmaking, particularly in basic oxygen furnaces and electric arc furnaces. Their resistance to alkali attack, coupled with high refractoriness, makes them suitable for environments with intense chemical corrosion and thermal loads [14–16].

Neutral refractories, such as alumina (Al_2O_3), chromite ($FeCr_2O_4$), and carbon-based materials, are chemically stable in both acidic and basic conditions. Their versatility allows for broader industrial application, particularly in processes where slag chemistry is variable or where material compatibility is uncertain [17,18].

The performance of these refractories is largely determined by the choice and purity of raw materials. Alumina, widely sourced from bauxite, offers excellent thermal and mechanical properties and is used in both shaped and unshaped refractories. Silica, derived from high-purity quartz, provides strong resistance to acidic environments and is common in lightweight insulating bricks. Magnesia, extracted from magnesite ore or synthesized through seawater processing, is the key ingredient in basic refractories due to its high melting point and low reactivity with basic slags. Zirconia (ZrO_2), though more expensive, is increasingly used in advanced refractory applications, especially where exceptional thermal shock resistance and phase stability are required such as in aerospace or high-performance metallurgical systems [8,19–23].

Manufacturing methods for refractories vary depending on the material type, application, and required performance characteristics. Pressing whether uniaxial or isostatic is commonly used to shape dense refractory bricks with controlled porosity and strength. Casting techniques are employed for monolithic refractories, allowing for in-situ application of materials in complex geometries. Thermal forming processes, including sintering and fusion casting, enhance grain bonding and structural integrity, resulting in materials that can withstand repeated thermal cycling without degradation [9,22,24].

Modern refractory production increasingly incorporates digital control, automation, and nano-scale additives to improve uniformity, reduce energy consumption, and extend service life. This integration of material science and process engineering reflects the industry's shift toward more sustainable, high-performance solutions tailored to the evolving demands of high-temperature manufacturing.

2.2 Chemical properties of refractory materials

The chemical properties of refractory materials are critical in determining their stability and performance in harsh industrial environments where they are exposed to reactive slags, gases, and molten metals. One of the most essential aspects is their chemical resistance that is, the ability to withstand corrosive attack without undergoing structural degradation or phase transformation.

This resistance is primarily dictated by the material's chemical composition and its reactivity with surrounding substances. Acidic refractories, such as silica, are resistant to acidic slags but deteriorate rapidly when exposed to basic environments. Conversely, basic refractories like magnesia and dolomite are ideally suited for steelmaking operations involving basic slags, showing excellent resistance to alkali-based corrosion while being vulnerable to acidic media. Neutral refractories, including alumina, chromite, and carbon-based materials, are chemically stable in both acidic and basic conditions, making them especially valuable in processes where the chemical environment may fluctuate or contain mixed slags [6,18,25–28].

In addition to slag interaction, chemical properties also encompass oxidation and reduction resistance, which are crucial in furnaces where oxygen levels vary or reducing agents are present. For example, carbon-containing refractories such as graphite exhibit excellent thermal stability but are highly susceptible to oxidation at elevated temperatures, often requiring protective coatings or operation under inert atmospheres [25,29–32].

Phase stability at high temperatures is another vital factor. Some materials undergo crystal structure changes at specific temperatures, which can lead to mechanical stress, cracking, or thermal distortion. Partially stabilized zirconia, for instance, maintains its crystal phase even beyond 2400°C, making it a preferred choice in extreme applications such as aerospace, nuclear reactors, and certain chemical reactors [9,33–36].

Recent advances in materials science have allowed for the development of complex refractory formulations with nano-additives or multi-phase ceramic systems that enhance chemical resistance without compromising physical strength. These innovations contribute to longer service life, reduced downtime, and improved sustainability in heavy industrial operations. Table

1 Summary of Section 2.2 illustrates the chemical classification, properties, and applications of refractory materials.

Table 1. Chemical classification, properties, and applications of refractory materials.

Category	Typical Composition	Chemical Behavior	Common Applications	Key Advantages	Limitations
Acidic	Silica (SiO_2), Fireclay	Resistant to acidic slags, deteriorates in basic slags	Glass melting furnaces, coke ovens	Good resistance to acidic environments, good thermal stability	Poor performance in basic/alkaline slags
Basic	Magnesia (MgO), Dolomite ($CaMg(CO_3)_2$)	Resistant to basic slags and alkali attack	Steelmaking (BOF, EAF), environments with high basicity	Excellent alkali resistance, high melting point, strong under thermal and chemical load	Weak resistance to acidic slags
Neutral	Alumina (Al_2O_3), Chromite (Cr_2O_3), Carbon-based materials	Chemically stable in both acidic and basic environments	Variable slag chemistry processes, general-purpose high-temperature linings	Versatile, wide industrial use, stable under varied chemical conditions	Performance depends heavily on purity and processing

2.3 Physical properties of refractory materials

The physical properties of refractory materials are fundamental to their ability to perform reliably under extreme industrial conditions. These properties govern how well a refractory can resist mechanical stress, temperature fluctuations, and prolonged thermal exposure without failure. Among the most critical is a high melting point, which allows the material to maintain its structure at operating temperatures often exceeding 1500°C. For example, pure alumina (Al_2O_3) exhibits a melting point of approximately 2050°C, making it widely used in steel furnace linings where thermal loads are intense and continuous [8,37–39].

Another key parameter is porosity, which refers to the volume of voids within the material's microstructure. Lower porosity generally enhances resistance to slag penetration and reduces thermal conductivity, contributing to better insulation. Zirconia-based refractories, for instance, are favored in applications requiring dense, low-porosity structures for optimal thermal isolation [40,41].

Thermal shock resistance is also a vital physical property, especially in industries where rapid heating and cooling cycles are common such as glass manufacturing or thermal treatment lines. The ability of a refractory to expand and contract repeatedly without cracking or spalling greatly influences its service life and operational reliability [37,42,43].

Cold crushing strength (CCS) indicates the mechanical strength of the material at room temperature and is essential during the installation and early operation phases. High CCS values ensure that the refractory structure can withstand physical impacts, handling stress, and load-bearing conditions without damage. Magnesia-based bricks, known for their high mechanical integrity, are often used in rotary kilns where physical stress is significant [38,44,45].

Thermal conductivity is another critical aspect that varies depending on the specific application. In scenarios where heat containment is essential, such as insulation layers in ladles or tundishes, low-conductivity materials are preferred. In contrast, components like silicon carbide tubes, which are used for direct heating, require high thermal conductivity to efficiently transfer energy [26,42,46,47].

In modern refractory engineering, these physical properties are not evaluated in isolation but rather in combination, according to the specific thermal and mechanical demands of the application. With the advancement of ceramic processing technologies, engineers are now able to fine-tune microstructures at the microscopic level to produce materials that offer both durability and energy efficiency. This has led to the development of highly specialized refractories tailored for each industrial context, enhancing performance while reducing maintenance and energy costs. Table 2 Summary of Section 2.3 illustrates the Advanced scientific of physical properties of refractory materials.

Table 2. Advanced scientific of physical properties of refractory materials.

Property	Scientific Description	Influencing Factors	Representative Materials	Industrial Relevance & Applications
High Melting Point	Retention of shape and mechanical integrity at extreme temperatures (>1500°C).	Chemical composition, purity, bonding phases.	Alumina (2050°C), Magnesia (2800°C).	Essential for steel furnace linings, smelters, high-temp reactors.
Porosity	Ratio of void volume to total volume; low porosity improves chemical resistance and insulation efficiency.	Particle size distribution, pressing pressure, sintering temperature.	Dense zirconia, high-fired alumina.	Glass tank linings, insulating refractories, slag-resistant furnace walls.
Thermal Shock Resistance	Ability to survive rapid temperature changes without cracking or spalling.	Thermal expansion coefficient, modulus of rupture, microstructural elasticity.	Alumina-silicate refractories, stabilized zirconia.	Glass manufacturing, thermal treatment kilns, casting molds.
Cold Crushing Strength (CCS)	Compressive strength at ambient temperature; critical for installation stability.	Grain bonding quality, binder type, firing cycle.	Magnesia-carbon bricks, high-alumina bricks.	Rotary kilns, steel ladles, areas with heavy mechanical loading.
Thermal Conductivity	Rate of heat transfer through the material; adjustable depending on application.	Grain composition, density, phase distribution.	Low: Alumina-silicates; High: Silicon carbide (SiC).	Low: Ladles, tundishes (insulation).High: Heating elements, heat exchangers.
Integrated Design Approach	Modern refractories balance all physical parameters to meet specific process demands.	Digital processing, nano-additives, hybrid ceramic-metal composites.	Multi-phase engineered refractories.	Customized solutions for energy-efficient, long-life furnace linings.

3. Applications for Refractory Materials in the Metallurgical Industry

Refractory materials are fundamental components in the infrastructure of the metallurgical industry, where they are used to line equipment exposed to extreme thermal and chemical conditions. Their role is crucial in ensuring process efficiency, protecting structural integrity, and reducing maintenance downtime. Figure 2 illustrates applications of refractory materials in the metallurgical industry.

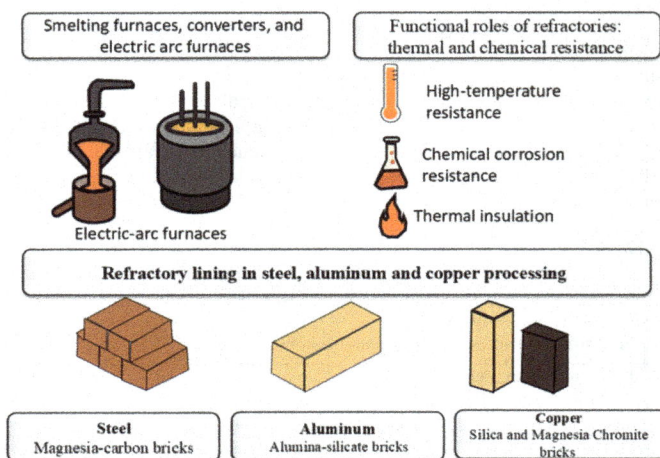

Figure 2. Applications of refractory materials in the metallurgical industry.

3.1 Smelting furnaces, converters, and electric arc furnaces

In metallurgical operations, refractories are used in high-temperature processing units, most notably in metal smelting furnaces. These furnaces require refractory materials with high melting points and excellent resistance to corrosive slags and direct contact with molten metals. Refractories are also used in converters, such as basic oxygen furnaces (BOFs), which operate under harsh conditions involving basic slags and rapid thermal cycling, necessitating linings with high chemical and thermal stability. Additionally, electric arc furnaces (EAFs), widely used in modern steel production, expose refractory linings to severe thermal gradients, frequent temperature fluctuations, and intense mechanical impacts from scrap charging, requiring durable refractories capable of withstanding such extreme conditions efficiently [6,25,48–51].

3.2 Functional roles of refractors: thermal and chemical resistance

Refractories serve multiple critical functions in metallurgical environments. They provide high-temperature resistance, ensuring structural stability above 1600°C and preventing deformation or failure of the lining during continuous operations. They also offer chemical corrosion resistance, which protects furnace linings from aggressive interactions with slags, gases, and molten metals, thereby extending service life and maintaining process purity. In addition, refractories contribute to thermal insulation, reducing heat losses, improving energy efficiency, and maintaining consistent process temperatures. Their mechanical durability enables them to withstand impact, abrasion, and dynamic loading resulting from charging, tapping, and deslagging operations [8,10,37,52].

A further critical role is thermal insulation, which minimizes conductive, convective, and radiative heat losses. This not only improves furnace thermal efficiency but also contributes to precise temperature control, a parameter directly linked to metallurgical reaction kinetics and product quality [12,37,53].

Additionally, refractories demonstrate notable mechanical durability, allowing them to withstand cyclic mechanical stresses arising from thermal gradients, impact loading during raw material charging, abrasive wear from slag and metal flow, and dynamic forces during tapping and deslagging. The integration of these properties ensures optimal furnace performance, reduces unscheduled downtime, and enhances the overall energy and cost efficiency of metallurgical operations [18,54,55].

Table 3. Advanced comparison of refractory materials in metallurgical industries.

Aspect	Steelmaking	Aluminum Production	Copper Smelting
Operating Temperature	Among the highest in industry (>1600°C), especially in Electric Arc Furnaces (EAF) and Basic Oxygen Furnaces (BOF)	Moderate to relatively low (~700–1000°C), but requires thermal stability to preserve aluminum quality	Moderate to high (~1100–1300°C), depending on furnace type (e.g., reverberatory, electric arc)
Slag and Chemical Environment	Basic slag rich in alkaline oxides; attacks acidic refractories, causes rapid corrosion; also involves frequent thermal shocks and intense mechanical stress	Reactive molten aluminum capable of penetrating and chemically interacting with refractories, posing risks of contamination and degradation	Acidic slags with sulfur-rich gases (e.g., SO_2); highly aggressive chemical conditions requiring corrosion-resistant refractories
Operational Challenges	- Frequent thermal shock- Erosion from basic slag- Mechanical wear due to charging and tapping	- Deep penetration of molten aluminum- Risk of chemical reaction and contamination- Need for non-wetting surfaces	- Chemical corrosion from sulfur gases- Reaction with acidic slag- Need for thermal stability under chemically shifting conditions
Refractory Types	- Magnesia-carbon (MgO-C) bricks- Doloma bricks- Sometimes zirconia-based bricks for critical areas	-Alumino-silicate refractories- non-wetting coatings such as **boron nitride**- Occasionally high-purity alumina bricks	- Silica bricks- Chrome-magnesite or chromite-containing refractories- Sometimes neutral refractories for chemical balance
Required Properties	- Resistance to basic slag- Excellent thermal shock resistance- Mechanical durability	- High resistance to wetting and penetration- Chemical inertness- good erosion resistance under moderate temperature	- Resistance to acidic and sulfurous corrosion- Thermal shock resistance- Long-term structural and chemical stability
Purpose of Refractory Use	To ensure uninterrupted operation, reduce premature lining failure, and minimize production losses due to wear and maintenance interruptions	To protect metal purity, minimize contamination, and extend refractory lifespan by reducing interaction and infiltration	To withstand aggressive chemical environments, maintain structural integrity, and reduce downtime through improved refractory durability

3.3 Comparison of refractories in steel, aluminum, and copper processing

Different types of refractories are used depending on the metal being processed, due to variations in operating temperature, slag chemistry, and reactivity. In steelmaking, basic refractories such as magnesia-carbon bricks are commonly used because of their high resistance to basic slags and thermal shock, making them suitable for converters and electric arc furnaces. In aluminum metallurgy, refractories must resist penetration and wetting by molten aluminum; therefore, alumina-silicate materials and non-wetting coatings are preferred to reduce contamination and erosion. For copper smelting, where acidic slags and sulfur-rich gases are present, acidic or neutral refractories like silica bricks or chromite-containing materials are used to withstand the aggressive chemical environment [56–58]. Table 3, Summary of Section 3, illustrates the applications of refractory materials in metallurgical processes.

4. The Relationship Between Refractories and the Ceramic Industry

Refractories and the ceramic industry share a close scientific and technological relationship, as both belong to the category of non-metallic, inorganic materials. They have many fundamental properties in common, such as heat resistance, chemical stability, and structural hardness. Although their applications may differ between the two industries, each relies on the other for development, knowledge exchange, and technological advancement.

4.1 The use of ceramic techniques in refractory manufacturing

Ceramic manufacturing techniques play a central role in the production of refractory materials, particularly in shaping, drying, and sintering processes. The refractory industry relies heavily on the principles of traditional ceramic science to control grain structure, reduce porosity, and achieve strong mechanical bonding between particles [7,42,59,60].

Techniques such as hydraulic pressing, isostatic pressing, and slip casting are borrowed from the ceramic industry and have been adopted in the production of refractories with complex shapes or specialized applications. Moreover, high-temperature sintering processes used to increase density and reduce internal voids are among the most critical operations shared by both industries [27,61,62].

4.2 The role of advanced ceramics in enhancing the efficiency and sustainability of refractories

The past decade has seen significant growth in the use of advanced ceramics to improve the performance of refractory materials, especially in industrial environments with demanding conditions. These advanced materials include high-purity alumina, stabilized zirconia, silicon carbide, and various nitrides [8,60,63].

Such materials exhibit exceptional properties, including outstanding resistance to thermal shock, tailored thermal conductivity (either high or low depending on application), and excellent chemical corrosion resistance even in highly acidic or basic environments. From a sustainability perspective, these materials help extend the service life of refractory linings, reduce energy consumption, and minimize waste associated with frequent replacements. For instance, the use of partially stabilized zirconia linings in nuclear metal processing furnaces reduces the need for regular maintenance operations [64–66].

4.3 Knowledge and Material Transfer Between the Two Industries

The relationship between the ceramic and refractory industries extends beyond material usage to include scientific knowledge exchange, joint development of composite materials, and collaborative research between universities and industrial centers.

Materials originally developed for use in technical ceramics such as zirconia or boron carbide have later been adopted in thermal applications due to their exceptional properties. Conversely, thermal endurance testing in the refractory industry has contributed to improving the design of ceramics used in engineering applications, such as engines and turbines [9,67,68].

Moreover, joint research programs in areas like ceramic 3D printing, nanomaterial fabrication, and composite materials have led to significant advancements in both fields, accelerating innovation and enhancing industrial efficiency [27,69,70]. Table 4, Summary of Section 4, illustrates the relationship between the refractory and ceramic industries.

Table 4. Relationship between the refractory and ceramic industries.

Section	Focus Area	Key Practices / Technologies	Representative Materials	Industrial / Scientific Benefits
Use of Ceramic Techniques in Refractory Manufacturing	Adaptation of traditional ceramic manufacturing methods for refractory production.	Hydraulic pressing, isostatic pressing, slip casting, high-temperature sintering, drying control.	Alumina-silicate bricks, complex-shaped monolithic.	Enhanced density, reduced porosity, improved grain bonding, precision in complex shapes.
Role of Advanced Ceramics in Efficiency & Sustainability	Integration of high-performance ceramic materials into refractory linings for extreme environments.	High-purity alumina, stabilized zirconia, silicon carbide (SiC), nitrides (Si_3N_4, BN).	Nuclear reactor linings (ZrO_2), EAF sidewalls (SiC), chemical reactors (Si_3N_4).	Extended service life, reduced energy loss, higher resistance to thermal shock and corrosion, lower maintenance frequency.
Knowledge & Material Transfer Between Industries	Cross-disciplinary R&D and technology transfer between ceramic and refractory sectors.	Zirconia, boron carbide, ceramic composites, nanomaterials.	Engine components, turbines, high-temperature kilns, aerospace applications.	Accelerated innovation, hybrid material development, ceramic 3D printing adoption, improved mechanical & thermal performance.

5. Technological and Operational Challenges

The refractory industry faces a series of technological and operational challenges that directly affect performance, cost-efficiency, and the overall sustainability of metallurgical and high-temperature processes. These challenges are particularly significant in industries where refractories operate under severe thermal, chemical, and mechanical stresses.

5.1 Performance Limitations

One of the most persistent limitations lies in the ability of refractory materials to maintain thermal shock resistance under extreme operating conditions. In many metallurgical units, rapid heating and cooling cycles such as those encountered during tapping, charging, or process interruptions can lead to cracking, spalling, and premature failure of the lining. Additionally, prolonged exposure to aggressive slags and molten metals results in chemical corrosion, which gradually degrades the microstructure and weakens the mechanical integrity of the refractory. High-velocity gas flows and particle-laden streams, common in converters and blast furnaces, also accelerate erosion, reducing service life and requiring more frequent repairs [22,53,71,72].

5.2 High costs of advanced refractories

The transition from traditional-shaped brick linings to advanced monolithic refractories, such as castable and gunning mixes, has improved installation flexibility and performance. However, these solutions often involve higher production costs due to the use of nano-additives, high-purity raw materials, and complex manufacturing processes. Stabilized zirconia, silicon carbide, and ultra-high-alumina systems, while offering superior performance, require significant capital investment, which may not be economically feasible for all operators—particularly in facilities with limited production scales or budget constraints [52,73–75].

5.3 Need for Improved Lifespan

Frequent shutdowns for refractory replacement present another major operational challenge. Each maintenance cycle not only interrupts production but also generates additional costs related to labor, material procurement, and lost operational time. As industrial processes push towards higher energy efficiency and reduced downtime, there is an increasing demand for refractories with extended service life and minimal performance degradation over prolonged use. This has driven research into hybrid materials, protective coatings, and microstructure engineering aimed at enhancing resistance to wear, corrosion, and thermal fatigue [12,40,76]. Table 5, summary of Section 5, illustrates the technological and operational challenges in refractory applications.

Table 5. Technological and operational challenges in refractory applications.

Challenge	Underlying Causes	Industrial Impact	Potential Solutions
Performance Limitations	- Rapid heating/cooling cycles causing thermal shock. - Chemical corrosion from slags and molten metals. - Erosion from high-velocity gas and particulate flow.	- Premature refractory failure. - Frequent unplanned maintenance. - Reduced process efficiency.	- Use of thermal shock–resistant compositions (e.g., stabilized zirconia, SiC). - Anti-corrosion coatings. - Optimized installation & firing procedures.
High Costs of Advanced Refractories	- Use of high-purity raw materials. - Nano-additives and advanced binders. - Complex manufacturing techniques (e.g., isostatic pressing).	- High capital and operational costs. - Limited adoption in small-scale plants. - Budget constraints for upgrades.	- Hybrid designs combining cost-effective base with premium wear zones. - Recycling and reusing refractory aggregates. - Process optimization to extend lining life.
Need for Improved Lifespan	- Wear and fatigue from prolonged high-temperature exposure. - Frequent shutdowns for replacement. - Mechanical damage during operation.	- Increased downtime. - Higher labor and material costs. - Reduced overall productivity.	- Development of wear-resistant hybrids. - Microstructure engineering for enhanced durability. - Predictive maintenance using AI and sensors.

6. Sustainability Challenges in the Refractory Industry

Despite the critical role refractories play in supporting high-temperature heavy industries, the production sector faces growing challenges considering global shifts toward environmental and economic sustainability. With increasing pressure to reduce emissions and improve resource efficiency, there is an urgent need to reassess the entire refractory value chain from raw material extraction to post-use management. Figure 3 illustrates refractory waste recycling and waste treatment.

Figure 3. Refractory waste recycling and waste treatment.

6.1 Dependence on non-renewable resources

The refractory industry relies heavily on naturally extracted minerals such as bauxite, magnesite, zircon, and chromite resources that are finite and non-renewable. With the continued growth in industrial demand, the future availability of these materials is at risk, particularly in the face of geopolitical challenges and vulnerabilities in global supply chains. This presents a direct threat to the stability of the industry and drives the need to explore synthetic alternatives or strategies for reusing spent refractory materials [77–80].

6.2 High carbon footprint during production

Refractory production is among the most energy-intensive industrial processes, particularly during the firing (sintering) stages, which require temperatures exceeding 1500°C. These high-temperature operations result in substantial carbon dioxide emissions, making the carbon footprint of the refractory industry relatively high compared to many other sectors. Such emissions are a growing global environmental concern, especially under increasingly strict environmental regulations that compel manufacturers to reduce their emissions or face carbon taxes [81–83].

6.3 post-use refractory waste

At the end of their service life, refractory materials are typically removed entirely from furnaces or reactors, resulting in the generation of large volumes of solid refractory waste. This waste is often disposed of in industrial landfills without treatment, despite containing materials that may be reusable or recyclable. Moreover, certain types of refractories contain heavy metals or oxide compounds that can pose environmental hazards if not properly managed [84,85].

6.4 Recycling and treatment challenges

Although recycling spent refractory materials is a promising approach to reducing resource depletion and minimizing industrial waste, its implementation faces several technical and economic challenges. Among the most significant obstacles are the difficulty of separating different components within used refractory materials, the degradation of their physical and chemical properties after service, the lack of adequate infrastructure for large-scale processing, and the absence of economic incentives for many manufacturers to adopt recycling practices. In light of these challenges, it has become essential to develop integrated solutions that include designing recyclable refractory materials, advancing treatment technologies, and establishing supportive regulatory policies to achieve true sustainability in this vital industrial sector [84,86,87].

7. Future Prospects

7.1 Material Innovations

Future development in refractories will focus on hybrid systems that combine the advantages of different materials. Integrating advanced ceramics such as stabilized zirconia, silicon carbide, and silicon nitride into conventional refractories will enhance thermal shock resistance, corrosion resistance, and lifespan. Tailored composite microstructures will allow optimization for specific industrial conditions, reducing downtime and maintenance costs [88,89].

7.2 Digitalization and Smart Manufacturing

Industry 4.0 technologies will enable smart monitoring of refractory linings through embedded sensors that collect real-time data on wear, temperature, and chemical attack. Artificial intelligence (AI) will process this data to predict maintenance needs, optimize furnace operation, and extend service life. Automated manufacturing, including robotic shaping and precision sintering, will improve quality consistency and reduce human error [90–92].

7.3 Circular Economy Integration

Recycling spent refractories will become a core strategy, with advanced separation and treatment technologies enabling the recovery of high-value raw materials such as alumina, magnesia, and zirconia. Reintroducing these recovered materials into production will reduce dependency on non-renewable resources and lower industrial waste volumes, supporting sustainable production cycles [71,87,90].

7.4 Green Refractory Production

Sustainability goals will drive the adoption of low-carbon manufacturing processes, including renewable energy-powered kilns and low-temperature sintering methods. Eco-friendly binders and additives will replace hazardous substances, ensuring compliance with environmental regulations and reducing the industry's carbon footprint [25,67,93,94].

7.5 Nanotechnology Applications

Nanotechnology will revolutionize refractory design by enhancing grain bonding, reducing micro-cracking, and increasing resistance to thermal shock and corrosion. Nano-coatings will form protective barriers against aggressive slags and molten metals, while engineered nanostructures will deliver unprecedented durability and performance under extreme industrial conditions [67,95].

8. Conclusion

Refractory materials are fundamental to the operational success of metallurgical and ceramic industries, ensuring stability, efficiency, and safety of high-temperature processes. This study examined their classification, chemical and physical properties, and diverse industrial applications, along with the strong technological link between refractories and the ceramic sector.

The analysis highlighted that refractories classified into acidic, basic, and neutral types offer tailored performance for specific slag chemistries and process conditions. While advancements in raw material selection, manufacturing methods, and material engineering have enhanced performance, significant challenges remain. These include vulnerability to thermal shock and corrosion, high costs of advanced compositions, and environmental issues linked to non-renewable resource dependency, energy-intensive production, and disposal of spent materials.

The study also identified promising future directions. Material innovations, such as hybrid composites and advanced ceramics, are expected to improve durability and efficiency. Digitalization, through Industry 4.0 technologies like smart sensors and AI-driven predictive maintenance, can optimize operational performance and extend service life. Circular economic strategies and green refractory production can minimize environmental impact, while nanotechnology offers advanced microstructural control for unprecedented thermal and chemical resistance.

In conclusion, refractory materials will remain indispensable to high-temperature industries, but their future success depends on integrating technical innovation with sustainable practices. Achieving this balance will not only improve performance and reduce costs but also ensure alignment with global environmental and energy-efficiency goals, securing their role in the evolving industrial landscape.

References

[1]   B.L. Krasny, N.A. Makarov, K.I. Ikonnikov, D.O. Lemeshev, D.D. Bernt, A.L. Galganova, A.S. Sizova, O.I. Rodimov, Calcium Zirconate: Methods of Synthesis and Applications of Ceramic and Refractory Materials Based Thereon—A Review. Part 2. Applications of Ceramic and Refractory Materials Based on Calcium Zirconate, Glas. Ceram. 81 (2024) 73–77.

[2]   B.L. Krasnyi, N.A. Makarov, K.I. Ikonnikov, D.O. Lemeshev, D.D. Bernt, A.L. Galganova, A.S. Cizova, O.I. Rodimov, Ceramic and Refractory Materials Based on it a Review. Part 1: Methods of the Synthesis of Calcium Zirconate, Glas. Ceram. 80 (2024) 535–541.

[3]   Z.J. Ting, X. Meng, Z. Yang, S.A. Jiskani, L. Hu, W. Dong, M. Zhao, Solid Recovered Fuel (SRF): A Comprehensive Review of Its Origins, Production, and Industrial Utilization, Energy and Fuels 39 (2025) 9726–9761. https://doi.org/10.1021/acs.energyfuels.5c01674

[4]   M. V Vasić, P.M. Velasco, N. Mijatović, M. Radormirović, Z. Radojević, Sustainable

approach to raw clays for ceramic and refractory applications: insights from updated traditional ternary diagrams, Clay Miner. 59 (2024) 202–212.

[5]   A. Spyridakos, D.E. Alexakis, I. Vryzidis, N. Tsotsolas, G. Varelidis, E. Kagiaras, Waste Classification of Spent Refractory Materials to Achieve Sustainable Development Goals Exploiting Multiple Criteria Decision Aiding Approach, Appl. Sci. 12 (2022) 3016. https://doi.org/10.3390/app12063016

[6]   R. Sarkar, Refractory technology: fundamentals and applications, CRC Press, 2023.

[7]   S. Seifert, S. Dittrich, J. Bach, Recovery of raw materials from ceramic waste materials for the refractory industry, Processes 9 (2021) 1–17. https://doi.org/10.3390/pr9020228

[8]   A. Zharmenov, S. Yefremova, B. Satbaev, N. Shalabaev, S. Satbaev, S. Yermishin, A. Kablanbekov, Production of Refractory Materials Using a Renewable Source of Silicon Dioxide, Minerals 12 (2022) 1010. https://doi.org/10.3390/min12081010

[9]   A. Ghosh, S. Sinhamahapatra, H.S. Tripathi, Refractories as Advanced Structural Materials for High Temperature Processing Industries, Futur. Landsc. Struct. Mater. India (2022) 279–292. https://doi.org/10.1007/978-981-16-8523-1_11

[10]   O.R. Ishola, Effect Of Additives On Refractory Properties Of Some Selected Clay Deposits In Niger State, Nigeria, (2022).

[11]   A.S.M. Mohamed, R.E.A. Ngida, R.S. Farag, M.F. Zawrah, Industrial Wastes for Production of Thermally Stable and Corrosion Resistant alumino-silicate Refractory Ceramic Bodies, J. Inorg. Organomet. Polym. Mater. 34 (2024) 4661–4673. https://doi.org/10.1007/s10904-024-03108-1

[12]   R. Pereira, A.A.L. Ferreira, S.M. Toffoli, An overview and the recent advancement of combustion applied to glass melting furnaces, Ceramica 71 (2025) eNEAN2178. https://doi.org/10.1590/NEAN2178

[13]   A. Fadel, Design of Industrial Furnace (End Port Glass Type), (2023). https://www.researchgate.net/publication/372526128

[14]   T. HARPER, Towards Circular Economy for Steel-Assessing the Efficiency of Yellow Gypsum Synthesis from BOF Slags, (2021).

[15]   Y. Mao, W. Liu, W. Liu, Y. Shen, A Novel Amino Acid Derivative Collector for Green Recovery of Low-Grade Magnesite Ore Resources: Purification Effect and Interface Adsorption Mechanism, Available SSRN 5316661 (n.d.).

[16]   J. Chu, L. Zhang, Y. Bao, N. Ali, C. Zhang, H. Zhou, Corrosion behaviour of MgO-based refractories by different existence states of manganese-containing volatile phases, J. Eur. Ceram. Soc. 43 (2023) 3755–3769. https://doi.org/10.1016/j.jeurceramsoc.2023.01.045

[17]   Z. Xiangchong, Z. Anzhong, L. Hongxia, X. Kuangdi, Refractories, Properties and Application of, in: ECPH Encycl. Min. Metall., Springer, 2023: pp. 1–5. https://doi.org/10.1007/978-981-19-0740-1_443-1

[18]   W. Jiguang, Nonoxide Refractories, Types and Applications of, in: ECPH Encycl. Min. Metall., Springer, 2024: pp. 1461–1463.

[19]   Georgitzikis K, Mancini L, Elia E, Vidal-Legaz B, Sustainability aspects of Bauxite and Aluminium Climate change, Environmental, Socio-Economic and Circular Economy considerations, 2021. https://rmis.jrc.ec.europa.eu

[20] M.K. Kar, M.A.R. Önal, C.R. Borra, Alumina recovery from bauxite residue: A concise review, Resour. Conserv. Recycl. 198 (2023) 107158. https://doi.org/10.1016/j.resconrec.2023.107158

[21] P.S. Reddy, N.G. Reddy, V.Z. Serjun, B. Mohanty, S.K. Das, K.R. Reddy, B.H. Rao, Properties and Assessment of Applications of Red Mud (Bauxite Residue): Current Status and Research Needs, Waste and Biomass Valorization 12 (2021) 1185–1217. https://doi.org/10.1007/s12649-020-01089-z

[22] F.R. Ghifari, F. Mubarok, Magnesia Refractory Bricks for Cement Rotary Kiln: A Systematic Review, Asian J. Eng. Soc. Heal. 4 (2025) 579–592.

[23] M.B. Veysel, B. Tuba, Investigation of the effects of precision-casting waste sands on the thermal shock resistance properties of fire clay refractory materials, Pamukkale Univ. J. Eng. Sci. 41 (2024) 1–12.

[24] S. Mandal, Investigation of Zinc Aluminate as a Refractory Material, (2022).

[25] S. Gan, Basic Refractories, Types and Properties of, in: ECPH Encycl. Min. Metall., Springer, 2024: pp. 127–128.

[26] N. Kratz, 10.4 Structural ceramics, Walter de Gruyter GmbH & Co KG Berlin, Germany, 2022.

[27] S. Bose, C. Das, Introduction to Ceramics: Fabrication, Characterizations, and Applications, CRC Press, 2024.

[28] N. Deva, I. Ibrahimi, Substantiation of refractory lining influence on the electric furnace efficiency for the production of ferronickel, Min. Miner. Depos. (2021).

[29] O.O. Otegbeye, MgO-C Refractory-Slag interaction: A Study on the effect of antioxidants and slag MgO content on MgO-C refractory-slag interactions in Si-killed steel refining, (2023).

[30] H. Wang, Y. Zhang, X. Shen, K. Cui, Q. Yue, Microstructure evolution, mechanical characteristics and erosion behavior of Cr2O3–Al2O3–MgO–ZrO2 refractories in molten reduction slag, Ceram. Int. 50 (2024) 19502–19514.

[31] J. Lv, H. Zhang, H. Gu, F. Liang, A review on the application of nanomaterials to boost the service performances of carbon-containing refractories, High-Temperature Mater. 1 (2024) 10005.

[32] Z. Chenrui, Y.A.N. Mingwei, L.I. Hongyu, N.I.U. Zhiwang, L. Baoqing, S. Junli, Recent progress on Al2O3-C refractories with low/ultra-low carbon content: a review, China's Refract. 31 (2022) 35.

[33] F. Findik, Review of high temperature materials, Herit. Sustain. Dev. 5 (2023) 213.

[34] B.A. Kahl, Ultra-High Temperature Ceramic Coatings for Hypersonic and Space Applications, (2023).

[35] B.W. Lamm, D.J. Mitchell, Chemical vapor deposition of zirconium compounds: A review, Coatings 13 (2023) 266.

[36] A. Lynam, A.R. Romero, F. Xu, R.W. Wellman, T. Hussain, Thermal spraying of ultra-high temperature ceramics: a review on processing routes and performance, J. Therm. Spray Technol. 31 (2022) 745–779.

[37] D. Vitiello, Thermo-physical properties of insulating refractory materials, (2021).

[38] L. Plioplys, Development and mechanical performance analysis of high-temperature resistant concrete composite, (2025).

[39] M. Song, Y. Zong, C. Wang, J. Zhang, H. Shi, X. Yang, Analysis of erosion characteristics of Al2O3-SiC-SiO2-C castable-poured in blast furnace hearth, Eng. Fail. Anal. (2025) 109903.

[40] C. Perez Velasquez, M. Montazerian, J.C. Mauro, Corrosion of zirconium-based refractories in glass-contact areas: Mechanisms and challenges, Int. J. Appl. Ceram. Technol. 22 (2025) e15064.

[41] L. Treccani, Introduction to ceramic materials, Surface-functionalized Ceram. Biotechnol. Environ. Appl. (2023) 1–46.

[42] J.G.P. da Fonseca, Refractory Ceramics in the Context of the Fabrication of Rocket Nozzles: Conventional Alumina and Industrial Waste Alumina-Silicate Refractory Compositions, (2023).

[43] J. Škamat, R. Boris, J. Malaiškienė, V. Antonovič, R. Stonys, A. Kudžma, Possibilities to Recycle Thermal Power Plant By-Products in Refractory Castables., Sustain. 16 (2024).

[44] W. Du, S. Jin, Discrete element modelling of cold crushing tests considering various interface property distributions in ordinary refractory ceramics, Materials (Basel). 15 (2022) 7650.

[45] W. Du, S. Jin, S. Emam, D. Gruber, H. Harmuth, Discrete element modelling of ordinary refractory ceramics under cold crushing testing: Influence of minimum element size, Ceram. Int. 48 (2022) 17934–17941.

[46] J. Djalilov, Introduction To The Teaching Of Physical Properties Of Ceramic Materials, Ment. Enlight. Sci. J. 4 (2023).

[47] Z. Zongtai, P. Xigao, Refractory Fiber Products, Types and Application of, in: ECPH Encycl. Min. Metall., Springer, 2024: pp. 1779–1780.

[48] X. Anjun, Q. Ying, BF-BOF Integrated Steelmaking Route, in: ECPH Encycl. Min. Metall., Springer, 2024: pp. 143–144.

[49] J. Demeter, B. Buľko, P. Demeter, M. Hrubovčáková, Evaluation of Factors Affecting the MgO–C Refractory Lining Degradation in a Basic Oxygen Furnace, Appl. Sci. 13 (2023) 12473.

[50] M. Karbowniczek, Electric Arc Furnace Steelmaking, CRC Press, 2021.

[51] J. Schwietz, B. Panic, M. Saternus, J. Pieprzyca, K. Janiszewski, Improvement of the Foaming Agent Feeding Process to an Electric Arc Furnace by Analyzing the Sound Generated by an Electric Arc and the Coefficient of Variation of Active Power Consumption, Materials (Basel). 17 (2024) 5860.

[52] G.A. Khater, M. Romero, A. López-Delgado, I. Padilla, A.A. El-Kheshen, M.M. Farag, M.S. Elmaghraby, N.H.S. Nasralla, Synthesis and characterization of ceramic refractories based on industrial wastes, Sci. Rep. 14 (2024) 25137.

[53] X. Yin, R. Li, X. Sun, Q. Feng, B. Zhang, S. Zhao, Prediction of composite performance of refractory and ceramic materials based on deep learning, JOM (2025) 1–14.

[54] J. Pötschke, Refractory Fundamentals in Metallurgical Practice, Springer, 2024.

[55] M. Mandal, D. Sarkar, Manufacturing excellence in ceramic industry, in: Ceram. Process.

Ind. Pract., CRC Press, 2019: pp. 1–36. https://doi.org/10.1201/9781315145808-1

[56] R. Kundu, R. Sarkar, MgO-C refractories: a detailed review of these irreplaceable refractories in steelmaking, Interceram-International Ceram. Rev. 70 (2021) 46–55.

[57] F. Barandehfard, Refractories corrosion resistance improvement against molten aluminum alloys by aluminum nitride suspension plasma synthesized coatings, (2022).

[58] A. Yurkov, Refractories for the Metallurgy of Copper, in: Copper-From Miner. to Final Appl., IntechOpen, 2022.

[59] I. Jastrzębska, J. Szczerba, Design, manufacturing and properties of refractory materials, Materials (Basel). 17 (2024) 1673.

[60] G.A. Khater, M. Romero, A. López-Delgado, I. Padilla, A.A. El-Kheshen, M.M. Farag, M.S. Elmaghraby, H. Shendy, N.H.S. Nasralla, Utilizing Ceramic Factory Waste to Produce Low-Cost Refractory Ceramics, Recycling 9 (2024) 98.

[61] P. Sengupta, I. Manna, Advanced high-temperature structural materials in petrochemical, metallurgical, power, and aerospace sectors—An overview, Futur. Landsc. Struct. Mater. India (2022) 79–131.

[62] M. Bustillo Revuelta, Ceramic Products, in: Constr. Mater. Geol. Prod. Appl., Springer, 2021: pp. 339–374.

[63] M.F. Zawrah, M.A. Taha, R.A. Youness, Advanced ceramics: stages of development, in: Adv. Ceram., Springer, 2023: pp. 1–46.

[64] L. Liya, Reaction Sintering, in: ECPH Encycl. Min. Metall., Springer, 2024: pp. 1764–1765.

[65] R. Mu, G. Gu, Y. Xu, Y. Xi, G. Wei, Photo-stimulating dual-network microminiature hydrogels reinforced with silver sulfide nanoparticles for regulated cancer photothermal therapy, Chem. Eng. J. 510 (2025) 161836. https://doi.org/10.1016/j.cej.2025.161836

[66] P.P. Fedorov, E.G. Yarotskaya, Zirconium dioxide. Review, 23 (2021) 170–188.

[67] I.I. Ekanema, A.E. Ikpe, A systematic review of the trends in ceramic materials and its viability in industrial applications, J. Mater. Charact. Appl 2 (n.d.) 63–78.

[68] A.K. Prajapati, S. Aryan, S. Singh, D. Dwivedi, N. Serikova, B.K. Purohit, Application of Functional Ceramics in Oil and Gas Industries: Properties and Current Status, in: Funct. Mater. Oil Gas Ind., CRC Press, 2023: pp. 39–50.

[69] C. Gómez Rodríguez, Synthesis and sintering of advanced ceramic materials, (2024).

[70] Y. Chaolu, Z. Fuchuan, C. Qianren, S.H.I. Gang, M.K. Masiko, Development and Application of Modern Building Ceramic Materials, Res. Appl. Mater. Sci. 5 (2024).

[71] A.M. Muhaba, Multi-objective optimization and performance evaluation of rotary furnace refractory linings using locally sourced materials, Adv. Mech. Eng. 17 (2025) 16878132251346230.

[72] D. Ozkan, G. Binal, Y. Ozgurluk, O. Odabas, S.E. Yilmaz, M.S. Gok, Y.Y. Ozbek, E. Cevlik, A.C. Karaoglanli, Characterization and Tribology Performance of Refractory Materials Coatings with Blast Furnace Slag (BFS) Fabricated by Thermal Spraying, JOM 77 (2025) 336–352.

[73] C. Ren, R. Enneti, G. Ouyang, Refractory materials for corrosive or high-temperature

319

environments, JOM 74 (2022) 4305–4306.

[74] E. Omowumi, E. Akinbolaji, E. Oluwasehun, Evaluation of Termite Hill as Refractory Material for High Temperature Applications, Int. J. Res. Innov. Appl. Sci. 8 (2023) 62–71.

[75] R. Liutyi, D. Liuta, I. Petryk, Structural construction of binders based on orthophosphoric acid and refractory materials, Adv. Mater. Sci. Eng. 2021 (2021) 6667769.

[76] R. Patel, M.L. Chaudhary, A.F. Martins, R.K. Gupta, Mastering Material Insights: Advanced Characterization Techniques, Ind. Eng. Chem. Res. 64 (2025) 8987–9023. https://doi.org/10.1021/acs.iecr.5c00447

[77] A. Zghari, H. Amar, M.L. El Hachimi, Reuse of Lead Mine Processing Tailings for Ceramic Tiles Production: Zeïda Mine Case Study, Morocco, Int. Res. J. Multidiscip. Scope 06 (2025) 1142–1156. https://doi.org/10.47857/irjms.2025.v06i01.02823

[78] M. Singh, M. Singh, J. Singh, M. Singh, H. Singh, Recent Advances in Green Energy Materials: A Review, Digit. Manuf. Perform. An Environ. Perspect. (2026) 51–72.

[79] V. Ramdas, S.G. Njokweni, P. Letsoalo, S. Motaung, S.O. Ramchuran, Bio-Coal Briquetting as a Potential Sustainable Valorization Strategy for Fine Coal: A South African Perspective in a Global Context, Energies 18 (2025) 3746.

[80] P. Orosco, O. Barrios, F. Tunez, L. Barbosa, Sustainable Recycling of Silicon from End-of-Life Photovoltaic Panels for the Synthesis of Porous Cordierite Via Bischofite-Assisted Chlorination, Silicon 17 (2025) 889–903.

[81] R. Pramanik, U. Sengupta, D. Sarkar, Al2O3–SiC–C castable for high performance and lower carbon footprint, J. Mater. Sci. (2025) 1–12.

[82] Y. Wang, Y. Shangguan, C. Wang, X. Zhou, H. Liu, Y. Cao, X. Liu, Y. Guo, G. Yan, P. Kang, Life Cycle Emissions and Driving Forces of Air Pollutants and CO2 from Refractory Manufacturing Industry in China Based on LMDI Model, Toxics 13 (2025) 533.

[83] O. Jankovský, A. Jiříčková, M. Záleská, M. Pavlíková, Z. Pavlík, A. Pivák, C.G. Aneziris, A.-M. Lauermannová, Utilization of carbon-bonded magnesia refractory waste in MOC-based composites: Towards CO2-neutral building materials, Open Ceram. 18 (2024) 100592.

[84] C. Hao, L. Yu, X. Chen, Y. Chen, C. Li, P. Qi, J. Sun, H. Zhou, Research on the development of recycling technology for magnesium refractory materials, Mater. Res. Express 11 (2024) 115504.

[85] N. Tessier-Doyen, E. Thune, M. Huger°, C. Xing°, Coupled Numerical Simulation And Post-Mortem Analysis To Evaluate Tundish Lining Refractories Lifetime And The Impact Of Anchoring Systems, (2024).

[86] M. Panigrahi, R.I. Ganguly, R.R. Dash, Applications, challenges and opportunities of industrial waste resources ceramics, High Electr. Resist. Mater. Ferrochrome Slag Resour. Ceram. (2024) 241–256.

[87] S. Sarner, O. Guillon, A perspective on ceramic recycling, J. Eur. Ceram. Soc. 45 (2025) 117511. https://doi.org/10.1016/j.jeurceramsoc.2025.117511

[88] J. Zhang, R.S. Uwanyuze, S. Shuster, J.E. Kanyo, B. Yavas, S. Schaffōner, L.D. Frame, S. Alpay, S.L. Suib, Interfacial Reactions Between Alloy 718 Melts and Investment Mold

Refractories, J. Mater. Eng. Perform. (2025) 1–9.

[89] B. Matović, J. Maletaškić, V. V Srdić, Book of Abstracts/8th Conference of The Serbian Society for Ceramic Materials, 8CSCS-2025, in: 8th Conf. Serbian Soc. Ceram. Mater. 8CSCS-2025, Institut za multidisciplinarna istraživanja Kneza Višeslava 1, 11000 …, 2025: pp. 1–133.

[90] A.A. Saddiq, A.M. Baghdadi, Y. Algamal, Antimicrobial Activity of Some Refractory Ceramic Bodies (RCBs) Prepared from Petroleum Waste Sludge (PWS) and Local Bauxite Mineral, ACS Omega (2025).

[91] A. Kumar, S. Yadav, S. Kujur, Multifunctional Materials in Engineering and Processing Engineering of Multifunctional Materials, Multifunct. Mater. Eng. Biol. Appl. (2025) 273–294.

[92] S. Antony Jose, Z. Lapierre, T. Williams, C. Hope, T. Jardin, R. Rodriguez, P.L. Menezes, Wear-and Corrosion-Resistant Coatings for Extreme Environments: Advances, Challenges, and Future Perspectives, Coatings 15 (2025) 878.

[93] E.H. Koren, S. Senanu, Refractories for Hydrogen Use in Metal Production: A Review, in: TMS Annu. Meet. Exhib., Springer, 2025: pp. 117–130.

[94] S. Gan, Neutral Refractories, Types of, in: ECPH Encycl. Min. Metall., Springer, 2024: p. 1424.

[95] N. Senthilnathan, R. Namdeti, U.R. Meka, Nanocomposites and Environmental Pollution Control, in: Nanomater. Environ. Remediat., CRC Press, n.d.: pp. 231–250.

Materials Research Forum LLC
https://doi.org/21741/9781644903834

Chapter 19

The Construction of Civilization Through Art; "Turkish Tile and Ceramic Art"

Gökhan Akca[1*], Cumhur Eren Isık[2], Ebru Halvaci[3], Selcuk Erdogan[3], Iskender Isık[4], Fatih Sen[3*]

[1]Faculty of Fine Arts, Department of Ceramics and Glass, Kutahya Dumlupinar University, Kutahya 43000, Türkiy

[2]Department of Handicrafts, Kutahya Fine Arts Vocational School, Kutahya Dumlupınar University, Evliya Çelebi Campus, 43000 Kutahya, Türkiy

[3]Sen Research Group, Department of Biochemistry, Kutahya Dumlupinar University, Kutahya 43000, Türkiy

[4]Department of Materials Science & Engineering, Faculty of Engineering, Kutahya Dumlupınar University, Evliya Çelebi Campus, 43100 Kutahya, Türkiy

gokhan.akca@dpu.edu.tr, fatihsen1980@gmail.com

Abstract

Turkish tile and ceramic art, which flourished in Anatolia and reached its zenith during the Ottoman era, serves as a powerful testament to how art functions as a means of civilization-building. This chapter examines this art form within a framework that explores its role in expressing Islamic identity, social values, and political power. It traces the historical evolution of the craft from its Seljuk origins to its golden age in 16th century Iznik, highlighting its distinctive aesthetic and technical characteristics, including its unique quartz-based body, vibrant color palette, and intricate floral and calligraphic motifs. The analysis extends to the profound symbolism embedded within the art, its extensive use in imperial architecture to project cultural unity, and its wide-ranging influence across the Islamic world and Europe. Ultimately, the paper argues that Turkish tiles were not merely decorative but constituted a vital visual language that articulated the Ottoman worldview and values, leaving a legacy that continues to be recognized as a significant cultural heritage today.

Keywords

Civilization, Art, Culture, Turkish Art, Ceramic, Tile

1. Introduction

Since the dawn of civilizations, art has played a pivotal role in building identity and shaping the features of societies. Works of art in their various forms, from architecture to sculpture to ceramics, reflect a society's values and culture and contribute to transmitting these values across generations. Perhaps decorative arts are a rich visual language carrying profound cultural and spiritual meanings. They are not merely aesthetic elements for decoration, but rather tools for communication and expressing the depth of a nation's cultural identity. In the Islamic context, the decorative arts have gained a prominent position as one of the pillars of expressing a civilization's vision and ideals, blending sensual beauty with spiritual symbolism within an integrated aesthetic framework [1,2].

Within this framework, Turkish tile and ceramic art which developed in Anatolia and reached its peak during the Ottoman era stands as a shining example of how art contributed to building civilization. This art, which specialized in decorating walls and buildings with glazed ceramic pieces (tiles) and producing decorative ceramic vessels, combined practical function with visual beauty, fine craftsmanship with profound symbolism. It reflected the cultural diversity of the Ottoman Empire and was open to diverse influences, making it a true repository of the values and aspirations of Ottoman Islamic identity [3–5].

In this study, we will examine Turkish tile and ceramic art within the framework of "building civilization through art," through the following themes: the role of art in expressing identity and social and political values; the historical background of this art from the Seljuks to the Ottomans; its aesthetic and technical characteristics; its use in architecture (especially mosques and palaces); its symbolism and civilizational connotations; its influence and spread in Islamic and global arts; and concluding with a summary that illustrates its importance and civilizational role throughout history.

2. Art as a Means of Civilization Building: The Role of Art in Expressing Identity and Social and Political Values

Art has always been an essential means of building civilizations and establishing their identity in the collective consciousness. It is sometimes a silent and sometimes a blatant expression of prevailing social and political values, and a mirror that reflects the aspirations and ideals of society. In the Ottoman civilization, for example, architectural and decorative art were used as an effective tool to broadcast the state's identity and strengthen its presence. Major architectural projects decorated with tiles and ceramics formed the cornerstone of the Ottoman identity and contributed to spreading a distinctive Ottoman style among the subjects of the state and in its various regions. Mosques and palaces decorated with tiles and ceramics were not only beautiful buildings, but also a visual statement of the state's power and cultural unity. By the mid-16th century - at the height of Ottoman power - tiles adorned the facades of these architectural projects, a clear indication of the establishment of a unified artistic style representing the state [3,6,7].

Art also creates a common language between members of society and fosters a sense of belonging to a unified civilization. Works of art often carry symbols and connotations that are understood by members of the same civilization and reflect their beliefs, until it was rightly said that they are "a mirror that reflects the traditions, beliefs, and cultural values of different societies." In the Ottoman context, decorative tiles and ceramics formed part of the visual discourse that expressed the Islamic identity of the state and its diverse civilization. Religious symbols (such as the verses of the Qur'an and the names of God) were intertwined with geometric and geometric elements; thus, the message of faith was combined with symbols of power and organization, consolidating the values of society in attractive artistic forms that were easy for the public to absorb and relate to [8–10].

Moreover, Ottoman art had an important political and social dimension. The sultans recognized the value of art in embodying power and prestige and used it as a soft means to consolidate their legitimacy and demonstrate their civilizational superiority. The Ottoman court's patronage of ceramic and tile art was a clear example of this: Sultans, led by Suleiman the Magnificent, lavished support on craftsmen and artists and commissioned them to decorate mosques and edifices in the most beautiful ceramic suits. The sultan's patronage of the arts carried the implicit message that the empire combined a deep belief in Islamic values with an artistic renaissance. Indeed, the arts flourished under the umbrella of the state and became one of the means of building civilization, not only architecturally but also symbolically and morally. Thus, art emerged as a tool of social unification and political expression, linking the members of society to a single identity and cementing the image of their civilization and its achievements in subsequent generations [11,12].

Table 1 summarizes Section 2 and illustrates the role of art in reinforcing Ottoman identity and values.

Table 1. The role of art in reinforcing Ottoman identity and values.

Theme	Focus on your argument	Media & Forms	Mechanism (how art "builds" civilization)	Ottoman-specific notes	Outcomes
Civilization identity & state branding	Art as a means of establishing a civilization's identity in collective consciousness	Architecture; decorative arts (tiles, ceramics)	Creates a unified visual style that signals who "we" are	Mid-16th-century surge in tiled façades on mosques and palaces; a distinctive "Ottoman style" spread across regions	Publicly legible statement of state power and cultural unity
Shared language & social belonging	Art as a common language that encodes beliefs, traditions, values	Qur'anic inscriptions; divine names; geometric/arabesque programs	Symbols understood by members of the civilization foster cohesion and belonging	Religious text intertwined with geometric order—faith + discipline/power in easily absorbed visual forms	Consolidates social values; makes identity intuitive and everyday
Political legitimation & soft power	Patronage as a tool for prestige and authority	Court-sponsored tilework/ceramics; major architectural commissions	Shows rulers' refinement and piety; turns aesthetics into legitimacy	Sultans especially Süleyman the Magnificent funded craftsmen and projects to clothe monuments in tiles	Reinforces dynastic legitimacy; projects civilizational superiority
Symbolic legacy & intergenerational memory	Art as a mirror of ideals that endure beyond the moment	Monumental ensembles; decorative programs	Encodes ideals in durable, admired forms that are revisited across generations	State-umbrella flourishing of arts made monuments both architectural and moral touchstones	Sustained social unification and a lasting image of Ottoman civilization

3. The Historical Background of Turkish Tile and Porcelain Art: From Seljuk to Ottoman Times

Turkish tile and ceramic art have a history spanning more than a thousand years, and its roots can be traced back to the earliest Turkish Muslim states in Central Asia. The Qurhanids (10th-11th century AD) are known to have had an early interest in ceramics and tile making, showing that ceramic art developed early as a craft among the Muslim Turks. This art flourished under the Great Seljuks (rulers of Persia and Iraq in the 11th-12th centuries AD) and moved with them to Anatolia with the emergence of the Seljuq Rum state. Inspired by older Islamic artistic traditions (such as Abbasid and Fatimid ceramics) and taking advantage of their position as a link between the Islamic East and Central Asia, the Seljuks made extensive use of glazed tiles to decorate their mosques, madrasas, and palaces. Anatolian Seljuk works were often characterized by the "ceramic mosaic" style, in which colored ceramic pieces are cut and set to form geometric and floral motifs on the walls. They also mastered the techniques of glazing and sometimes metallic glazes, which added vivid colors and luster to their ceramic pieces [9,13–15].

One of the finest examples of Seljuk ceramic art in Anatolia are the tiles uncovered in the Kubadabad Palace on the shore of Lake Bishehir near Konya. The palace was built during the reign of Seljuk Sultan Alaeddin Kiqabad I (reigned 1220-1237 AD) and was decorated with magnificent ceramic paintings. Tiles with octagonal star-shaped tiles and cross-shaped tiles were found in the palace, made by glazing and metallic luster techniques, respectively. The star tiles are richly decorated with figurative motifs depicting scenes from daily life at the Seljuk court, portraits of the sultans and their courtiers, as well as mythical and mythological creatures (such as sphinxes, mermaids, peacocks, lions, and two-headed eagles) that were believed to protect the palace and symbolize the prestige of the sultan. Crusader tiles were decorated with intricate floral motifs and produced using a fine metallic luster technique. One of the most prominent symbols that appeared on the tiles of this palace was the two-headed osprey, which was painted covering the entire surface of the tile and inscribed on its body with the words "The Exalted Sultan" in Kufic script. This symbol carried deep political and spiritual connotations, as it was seen as an emblem of power, authority, and protection, and was adopted by the Seljuks as a symbol of their statehood. These rare examples reflect the level of sophistication that Seljuk ceramic art reached in the 13th century in terms of techniques, decorations and symbols, paving the way for what followed in later eras [16–19].

With the demise of the Anatolian Seljuks in the late 13th century, the emergence of local Turkish principalities, and the rise of the Ottoman Empire in the 14th century, a new era of tile and ceramic art began in Anatolia. The Ottomans inherited the tile-making traditions of their Seljuk predecessors and capitalized on the expertise of local craftsmen and the influence of Ilkhanid and Timurid art. The first distinctive examples of Ottoman tiles appeared in the early 1500s, in the city of Bursa, the first capital of the Ottomans. The Yeşil Jama Mosque (Green Mosque), completed by Sultan Mehmet Çelebi in 1421 AD, and its annex, are two of the first buildings in which colored tiles were widely used in the decoration of the interior walls. The Muradiye Mosque in Bursa (completed in 1426) stands out as a site rich in early ceramic tiles, as does the Muradiye Mosque in Edirne (1433) in the early Ottoman period. In Istanbul, Sultan Mehmet the Conqueror built a building known as the Glazed Kiosk (Çinili Köşk) in 1472, a pavilion in the Topkapi Palace whose interior walls are covered with beautiful, glazed tiles, reflecting the early Ottoman influence on Iranian and Mamluk court styles. These early Ottoman works were often executed in ceramic mosaic technique or painting over glaze (such as the metallic luster tiles in

Bursa), with limited colors such as indigo (dark blue), blue, and turquoise green with touches of black and white. The decorative patterns were simple geometric, or floral compared to what would come later. During this period, the term "qashani" was used to refer to wall cladding tiles, a term derived from the Persian city of Kashan, famous for its glazed tiles, while the term "ifani" was used for porcelain vessels such as plates and jugs. This shows the extent of the cultural connection between the Islamic Levant and the early Ottomans in the field of ceramics, and that the Ottomans benefited from the accumulation of previous technical expertise [9,10,20–22].

The golden age of Turkish court art and ceramics came with the rise of the Ottoman Empire in the 16th century. The reign of Sultan Suleiman the Magnificent (1520-1566 AD) is seen as the peak of Ottoman artistic and cultural production, as the empire witnessed an unprecedented boom in the arts under the patronage of the sultan and the ruling elite. During this era, a distinctive Ottoman school of ceramics known as Iznik ceramics (after the town of Iznik in northwestern Anatolia) crystallized, which became the main center to produce fine tiles and ceramic vessels. Historical sources indicate that the earliest clear evidence of Iznik pottery production dates to the reign of Sultan Mehmet the Conqueror (ruled 1451-1481 AD). It is said that the conqueror was so impressed by the fine blue and white Chinese pottery (Ming Dynasty porcelain) that he encouraged local potters to imitate these fine products. As a result, around the late 15th century, ceramic workshops in Iznik began producing ceramic pieces influenced by Chinese models, mostly decorated vessels. Then, by the mid-10th century, with increasing support and demand from the Ottoman court, Iznik tiles appeared heavily on the facades of mosques and palaces built by the Ottomans during this period. In other words, ceramics, which began as Chinese-inspired vessels and artifacts, quickly evolved into a major decorative element of Ottoman architecture during the reign of Suleiman the Magnificent and beyond [11,23–25].

In terms of design and color, the sixteenth century also saw quantum leaps. After about 1520, Ottoman workshops (especially in Iznik) began to develop new decorative styles that combined local Islamic decorative traditions with influences from the Far East and Europe. For example, botanical arabesques appeared in more complex and graceful compositions, incorporating elements from Chinese art (such as Chinese cloud motifs and lotus motifs) but with a distinctly Ottoman flavor. The use of twisted lines and saz - long, serrated leaves - was a common decorative element attributed to the famous Ottoman inscriptional designer Qara Mimi at the court of Sultan Suleiman. In terms of color, dark cobalt blue and sky blue (turquoise) continued to be used as primary colors inspired by Chinese porcelain in the first half of the century, but starting around 1550, a striking new color, crimson red Iznik, was added. This distinctive red color first appeared in the decoration of the tiles of the Sulaymaniyah Mosque in Istanbul (1550-1557 AD), possibly derived from iron-oxide-rich soil brought from Armenia. This third color (along with blue and turquoise) added vibrancy and striking color contrast to the designs and became a prominent feature of the Ottoman court in the second half of the 16th century. By the end of the century, the color palette had expanded to include emerald green and dark violet as well, further enriching the overall appearance of the tiles. This period witnessed the production of large numbers of ceramic masterpieces that combined artistic creativity with technical prowess, thanks to the direct patronage of the Ottoman court. It is no exaggeration to say that the sultans, their families, and courtiers were the most important customers for Iznik ceramics throughout the 16th century, which ensured a thriving market for the artisans, great financial and moral support, and a favorable climate for the development of this art to the fullest extent [26–29].

However, the heyday of Ottoman court art in the 16th century was followed by a period of gradual decline at the beginning of the 17th century. As the Ottoman state entered a period of relative political and economic stagnation, the tile industry faced challenges that affected its viability. Sources state that starting in the late 16th century, tensions arose between Iznik artisans and the Ottoman court: As Iznik ceramics became increasingly popular beyond the state's borders, artisans began to accept private orders from wealthy Europeans and others, diverting much of their energies from fulfilling the orders of the sultan's palace. Ottoman authorities responded by issuing strict edicts forbidding court makers from fulfilling any foreign orders until the Sultan's orders were completed, but many craftsmen grumbled and did not fully comply. At the same time, the state was suffering from inflation and economic issues in the early seventeenth century, and officials did not respond to demands to raise the wages of potters' products in line with rising costs. As a result of these administrative and economic factors, as well as natural disasters and epidemics, the quality of Iznik production gradually deteriorated. Historians report that the early 17th century saw the emergence of defects in the tiles, such as cracked glazes, color mixing, and low purity in the drawings. Although design innovation continued during this period, technical standards declined significantly. By the end of the 1600s, the production of luxury tiles in Iznik had ceased altogether, signaling the end of a golden era in the history of art [24,26,30,31].

After the decline of Iznik as a ceramic center, Kutahya (Central Anatolia) emerged as an alternative player in Ottoman tile-making. Interestingly, Kutahya had been producing some tiles since the late 14th century, but its real breakthrough came in the second half of the 16th century, when Iznik was at the height of its prosperity. The authorities at the time sought to establish production workshops in Kutahya to meet the growing demand for tiles in Istanbul and the major cities, especially as production in Iznik began to decline. By the 18th century, Iznik's tile industry had all but disappeared, leaving Kutahya as the only center in the country to continue the craft. During the eighteenth and nineteenth centuries, Kutahya provided the last of the Ottoman potters' tiles, although they were considered inferior to the golden masterpieces of the sixteenth century. One of the last great Ottoman craftsmen in this field was Haji Mehmet Emin Efendi of Kutahya, whose hands produced beautiful examples that were considered the end of the bright chain of traditional Ottoman tile art [6,32–34].

Table 2 summarizes Section 3 and illustrates: the Historical background of Turkish tile and porcelain art: From the Seljuks to the Ottomans.

Table 2. *Historical background of Turkish tile and porcelain art: From the Seljuks to the Ottomans.*

Period/Polity	Approx. Dates	Centers	Techniques/Colors	Motifs & Symbols	Notable Examples	Notes/Impact
Qarakhanids ("Qurhanids")	10th–11th c. CE	Transoxiana / Central Asia	Early interest in ceramic and tile-making as crafts	—	—	Shows the early emergence of ceramic arts among Muslim Turks
Seljuks (Great Seljuks → Seljuks of Rum)	11th–early 13th c. CE	Persia & Iraq → Anatolia	Glazed revetment; "ceramic mosaic"; occasional metallic luster	Geometric/vegetal, drawing on earlier Islamic traditions	Mosques, madrasas, palaces	Transfer of expertise to Anatolia; broad architectural use
Kubadabad Palace (near Lake Beyşehir/Konya)	1220–1237 CE	Anatolia (Seljuks of Rum)	Glazed octagonal star tiles; cross-shaped luster tiles	Courtly scenes; portraits; mythic creatures (sphinxes, mermaids, peacocks, lions, double-headed eagles); Kufic "The Exalted Sultan"	Excavated palace tiles	Peak 13th-c. technical/symbolic maturity; double-headed eagle as emblem of power/protection
Early Ottoman period	15th c. (into early 16th)	Bursa, Edirne, Istanbul	Ceramic mosaic; overglaze painting; metallic luster; limited palette: indigo/blue/turquoise with black & white	Simple geometric and floral	Green Mosque (1421) & annexes; Muradiye Mosque, Bursa (1426); Muradiye Mosque, Edirne (1433); Çinili Köşk/Glazed Kiosk (1472)	"Qashani" for wall tiles; "ifani" for porcelain vessels; Iranian & Mamluk court influences
Formation of the İznik school	Late 15th → mid-16th c.	Iznik	Initially Chinese-inspired blue-and-white wares (decorated vessels), then heavy shift to architectural tiles	Chinese cloud and lotus motifs within local frameworks	—	Driven by the Ottoman court; swift move from tableware to a major architectural element
Golden Age (Suleiman the Magnificent & after)	1520–late 16th c.	Iznik (tile workshops)	Major stylistic advances: elegant vegetal arabesque; serrated *saz* leaves (attributed to Kara Memi); palette: cobalt/turquoise, then İznik crimson red (~1550, iron-oxide rich clay), later emerald green & dark violet	Distinct Ottoman synthesis of Islamic traditions with Chinese/European influences	Tiles of the Süleymaniye Mosque (1550–1557), among others	Direct patronage by sultans and elite; thriving market and craftsmanship

Early decline	Early 17th c.	Iznik	Technical issues: crazed glazes, color bleeding, reduced drawing purity	—	—	Workshop–court tensions; edicts prioritizing palace orders over foreign ones; inflation/wage disputes; disasters/epidemics; luxury production ends by late 17th c.
Shift to Kütahya	18th– 19th c. (roots late 14th c.)	Kutahya	Continued tile production at quality below the 16th-c. peak	Late Ottoman ornament	Works of Hacı Mehmed Emin Efendi (of Kütahya)	Final link in the Ottoman tile tradition before its fade

4. Aesthetic and technical characteristics: Materials used, decorations, colors and artistic styles

Turkish tile and ceramic art (especially Ottoman Iznik ceramics) is known for a unique set of aesthetic and technical characteristics that have earned it a special place among the Islamic arts. The first of these characteristics relates to materials and manufacturing techniques: Iznik potters created a special formula for ceramic paste based on a high percentage of powdered quartz (rock crystal) mixed with clay and other additives, giving the product the texture of stone frit, which is known for its bright whiteness and hardness. This combination gave the tiles a soft-touch, ultra-white base, like fine Chinese porcelain, perfect for color painting. After the tile was shaped and decorated with mineral pigments, it was covered with a transparent glaze and melted in high-temperature furnaces. This glaze gave the surface of the tile a distinctive shine and enhanced the durability of the colors, as the paintings are completely united with the glaze. Thanks to this advanced technique - painting under the glaze - a perfect harmony between form and decoration was achieved; Iznik tiles had a smooth, glossy surface and pure, consistent colors, making them technically superior to many of their counterparts in the Islamic world at the time. This technical transition (from mosaic to freehand painting under glaze) revolutionized Ottoman ceramic art during the 16th century and allowed artists to unleash their creativity in painting and coloring without being limited by the limitations of small mosaic pieces [35–38].

The decorative and aesthetic characteristics of Turkish tile art are manifested through several prominent elements. The first of these is color: Ottoman tiles traditionally relied on a basic color palette of dark cobalt blue and turquoise blue along with a bright white ground. These two colors - cobalt blue and turquoise - were the dominant color palette for the first half of the 16th century, as blue in Islamic culture symbolizes spiritual serenity and the connection between earth and heaven. Then, with the introduction of scarlet red (Iznik red) around the mid-16th century, the Ottoman artist had a color triad that allowed for rich combinations: blue/turquoise/red, with occasional touches of emerald, green and black to accentuate details. Iznik tiles were thus characterized by a strong yet balanced color palette - in many designs, cool colors (blues and shades of blue) framed spaces and gave them depth, while warm reds created vibrant, eye-catching focal points. This color harmony was not random but was carefully considered by the craftsmen to achieve the desired visual effect and symbolic significance (for example, the pairing of blue and white was seen as an expression of spiritual purity, while red represented life energy and strength) [18,19,39].

Another essential element is the motifs used in the decoration. Ottoman artists drew inspiration for their patterns from both lush nature and the classical Islamic decorative heritage. Botanical motifs were widespread on tiles, with graceful tulips with their petals blooming upwards, carnations with their branching leaves swaying beside the tulips, wild roses and iris flowers in a variety of colors, and long, serrated saz leaves wrapping around the flowers. These botanical elements were painted in a simplistic yet elegant style; the leaves elongate in graceful wavy lines, and the flowers are simplified into relatively flat forms with balanced repetition. The choice of this vocabulary was not spontaneous but was associated with deep belief symbols and connotations for the Ottomans - as we will detail in the section on symbolism - the tulip, for example, embodied the symbol of divine beauty and spiritual unity, the carnation symbolized renewed life, and so on. In addition to plants, the Ottomans maintained the presence of traditional geometric motifs in their ceramic works, albeit to a less prominent degree than in other Islamic arts. Stars, rhombuses, and interlocking polygons were used as frames or dividing lines between areas of botanical motifs, striking a balance between the organic freedom of botanical forms and strict geometric order. Moreover, the principle of repetition was central to the decorative composition: Elements such as tulips and carnations were often arranged in symmetrical rows or circles, creating a visual rhythm that suggests continuity and infinity - an aesthetic as well as spiritual value in Islamic art [18].

Arabic calligraphy was also prominent as a decorative element in some Turkish tiles, particularly in religious contexts. Quranic verses, hadiths, and the names of God and the Prophet Muhammad (peace and blessings be upon him) were inscribed in the finest Thuluth and Kufic scripts on ceramic panels placed in mosques and lodges to evoke sacred meanings. A famous example is the group of calligraphic tiles bearing the names of God (Allah), the Prophet Muhammad (peace and blessings be upon him), the Rightly Guided Caliphs, and the Prophet's grandsons (Hasan and Husayn), may God be pleased with them. All these tiles were written in white Thuluth script on a blue background within decorative circles. These tiles were created in the 16th century to decorate a triangular section under the dome of an important Ottoman building and were likely used in major mosques such as the Sulaymaniyah Mosque and Rüstem Pasha Mosque in Istanbul, or the Selimiye Mosque in Edirne. Undoubtedly, the presence of such sacred inscriptions within the decoration added a powerful spiritual dimension to the space, in addition to their calligraphic aesthetic value [8,40–42].

It is worth noting that Ottoman artists also embraced foreign artistic influences and creatively incorporated them into their decorations. Among the most prominent of these influences was the Chinese ceramics, which reached the Ottoman court through trade and gifts. Iznik potters initially attempted to imitate the blue and white Chinese ceramics that appealed to the conquering Sultan, borrowing some elements such as Chinese cloud designs, lotus flowers, and mythical dragons. However, they soon repurposed these elements in a local spirit that was in harmony with the rest of the Islamic vocabulary. The result was a unique Ottoman style that blended traditional Islamic arabesques with classic Eastern decorative styles in a harmonious synthesis. The influence of Iranian art is also evident in Ottoman decorations, especially in the early stages. For example, we see the continued use of the rock and wave pattern (a decorative border of Persian origin representing intertwined rocks and waves) on some of the borders of plates and tiles. We also note the striking similarity between the decorative elements of Iznik ceramics and their Safavid counterparts in Tabriz and Isfahan albeit in different colors. This cultural interaction undoubtedly enriched Turkish court art, giving it broader aesthetic dimensions, and making it an art with a universal spirit, even if it had an Ottoman identity [30,43–48].

5. The Symbolism and Cultural Connotations of Turkish Tile and Ceramic Art

Turkish tile and ceramic art are not merely an aesthetic creation for visual enjoyment; it is more like a visual text encoded with profound cultural symbols and connotations. These tiles and ceramic pieces carry within their compositions a cultural and spiritual message that parallels their artistic value. The Ottomans realized this and transformed their art into a "silent language" that spoke of their history and values. For example, art historians view Ottoman tiles as an indirect expression of the empire's achievements, victories, and interactions with other peoples, while simultaneously embodying its spiritual and intellectual values. They were not merely decorative wall coverings; rather, they were a cultural discourse through which the Ottomans sought to integrate multiple dimensions: artistic, scientific, and spiritual, in what they called an effort to "combine divine harmony with earthly beauty" in their works. Thus, the Ottoman tiles can be read as a historical and symbolic source that reveals the Ottomans' worldview and the faith that motivated their creativity [3,18,38].

The first thing that draws attention to the symbolism of this art is the common plant motifs and the meanings they carry, derived from Islamic culture, particularly Sufism. The tulip (laleh), for example, which was repeated on thousands of Ottoman tiles and plates, was loved not only for its beautiful appearance but also for its profound spiritual significance. The Ottomans considered the tulip a symbol of divine unity and the divine essence, because the letters of the Ottoman word "laleh" (in the Arabic) are the same as the letters of the word "Allah." The single tulip, which stands out among the leaves on its thin stem, was for them an embodiment of the Creator's oneness, as it grows alone, its head tilted toward the sky in submission and reverence. Sufis loved this flower and saw it as a symbol of spiritual contemplation, even decorating their underwear with it as a kind of talisman and constant remembrance of God. Another beautiful connotation is that when the tulip is fully in bloom, it bows its head to the ground. This was seen as a symbol of humility before the Creator's majesty. Hence, the tulip earned the nickname "Flower of God" or "Flower of Divine Love" in the Ottoman imagination [49,50].

Another floral motif is the hyacinth, a flower that inspired both Ottoman and Sufi poets. In literature, poets likened the hyacinth spikes to the intertwining strands of a beloved's hair, expressing their captivating beauty and intricacy. Pioneering Sufis, including Mevlana Jalal ad-Din Rumi, founder of the famous Mevlevi order of the Seljuk period, borrowed the image of the spike in their eulogies of the Prophet. In one of Rumi's poems, he praises the Prophet Muhammad (peace and blessings be upon him), saying, "You are the rose of the garden of the Sharia, the sublime spike!" equating the Prophet's status to the exalted spike. Consequently, the spike became a common symbol of the Prophet (peace and blessings be upon him) in Ottoman culture, symbolizing his exalted status and spiritual influence. This symbolism also spread to art, where some tiles and plates were decorated with images of spikes swaying in the breeze, perhaps implicitly alluding to prophetic love or spiritual longing.

The plum blossom (or peach blossom) also appears in the Karamemi style of decorations from the second half of the 16th century. This five-petaled flower appears at the edges of the external decorative motifs and has been interpreted as a symbol of fertility and the abundance of heavenly blessings. This is evidenced by its use in specific areas of funerary architecture: it adorned the entrances to the tombs of important Ottoman figures (such as the tomb of Hürrem Sultan, wife of Suleiman the Magnificent, and the tomb of Grand Vizier Rüstem Pasha). It also adorns the niches and minarets of some mosques (such as the Süleymaniye Mosque of Sultan Selim II and the Hürrem Sultan Mosque in Istanbul). The plum blossom's presence in these locations is believed to symbolize the eternal gardens and their omens it suggests unfading blossoming branches and everlasting fruit, alluding to a paradise where "nothing is forbidden or restricted." In this sense, this flower can be considered a frame that embraces the other flowers (tulips, carnations, and irises) within a complete paradise scene [51–53].

In addition to the decorative symbols themselves, there are the spatial and cultural connotations of Turkish tile art. The selection of specific locations for extensive tile use was itself significant. For example, when Sultan Suleiman the Magnificent decided to restore the Hagia Sophia (converted into a mosque by his predecessor, Mehmed the Conqueror) in Istanbul and decorate parts of it with Iznik tiles in the mid-16th century, this was not merely an aesthetic improvement; it carried great cultural symbolism. This move was seen as a fusion of two imperial legacies: the Byzantine and the Ottoman Islamic, and a declaration that the Ottoman Caliphate was the heir to both glories. The addition of Islamic tiles within a towering Byzantine edifice like Hagia Sophia, a symbol of the Eastern Roman Empire reflected the idea of a unified civilization under the banner of Ottoman Islam. Some historians even see the tile decorations in Hagia Sophia (particularly those bearing the names of the Rightly Guided Caliphs and the floral surahs of the Qur'an) as a message that Islam encompassed the previous Byzantine civilization without destroying it, but rather unified it with its own civilization in a single form. This is an important indication of the Ottomans' approach to assimilating peoples and cultures within their empire, as art here expresses the concept of cultural communication and religious tolerance through visual symbolism [12,54].

6. Influence and Diffusion: The Relationship of Turkish Court Art to Islamic and World Arts

Turkish tile and ceramic art were not solely a product of its local environment; it is part of a system of Islamic arts that has influenced and been influenced by other forms of art throughout history. On the one hand, the Ottomans inherited and developed this art form from their Muslim ancestors, while on the other hand, their art left a clear imprint on other regions, even in Europe. This interactive process can be observed on more than one level:

Within the Islamic world: The Ottomans derived many of their glazed ceramic traditions from centers that had flourished before them in the Levant. The word "qishani," commonly used in Arabic references to describe this type of tile, for example, originates from the Persian city of Kashan, a prominent center to produce glazed tiles and ceramics from the Seljuk period onward. Ottoman ceramics were indeed influenced by Iranian styles in their early stages, both in the metallic luster technique (mastered by Kashan craftsmen in the 13th century) and in some of the motifs and colors. During the Mamluk period (1250–1517), Egypt and the Levant preceded the Ottomans in the production of glazed tiles for the decoration of mosques and palaces. The Ottomans were familiar with these products, and some were later decorated in mosques in Damascus and Cairo during their rule. However, once their Iznik School took root in the 16th century, the reverse influence began to spread from Istanbul to the rest of the Islamic world. Iznik tiles found their way to decorating numerous mosques and shrines in various Ottoman provinces, from the Balkans to the Hijaz. For example, during his restoration of the Dome of the

Rock in Jerusalem (c. 1545 CE), Sultan Suleiman the Magnificent re-covered the exterior of the dome with blue ceramic tiles brought from Iznik and Kütahya, giving the holy building a new look with a striking Ottoman-Islamic character. The use of Ottoman tiles also spread to decorate mosques in Damascus and Aleppo during the 16th and 17th centuries, to the point that some of these mosques (such as the Adiliyeh Mosque and the Murad Pasha Mosque in Damascus) now bear a clear Iznik imprint. We should also not forget the influence of the Ottoman school on Persia and India. Although the Safavids in Iran and Mughal India developed their own court styles (such as the seven-colored tiles of Isfahan or the glazed tiles of the Taj Mahal), some historians see similarities and interplay between these schools, given the artistic contacts and competition that existed in the 16th century and beyond. Thus, it can be argued that the Ottoman court in its time became a neoclassical form of Islamic art, inspired by other centers or adapting its vocabulary into their own creations [7,46,55,56].

Beyond the Islamic world: Europe was also influenced perhaps unintentionally by Islamic ceramics in general, and Ottoman ceramics in particular. Historically, Islamic decorative arts were among Europe's most valuable exports during the Middle Ages. These exports included textiles, carpets, and ceramics, providing Europeans with a glimpse into the rich aesthetics of the East. European churches and institutions preserved some fine Islamic ceramics (Syrian, Egyptian, and Iranian) in what were called Renaissance cabinets of curiosities. With the fall of Andalusia and the advent of the Age of Discoveries, Europe's contact with the arts of the Islamic world increased. In the 15th century, Italian and Spanish artists were influenced by some Islamic decorative elements especially arabesques and Arabic calligraphy and this was evident in stucco and ceramic decorations that imitated their Andalusian counterparts. Then, in the 16th century, when Ottoman ceramics reached their peak, their products spread through trade and gifts to Europe. Magnificent Iznik dishes were presented to kings and nobles, and Venetian and Venetian merchants carried them to Italy. Historians point out that around the late 16th century, some Iznik craftsmen began accepting special orders for wealthy European clients, demonstrating the Europeans' intense fascination with these products. Ottoman documents indicate that sultanic decrees were issued at the time prohibiting craftsmen from devoting their efforts to satisfying European demand before fulfilling the state's needs. This suggests that the European rush was so influential that it disrupted the balance of domestically directed production. There is no doubt that this interaction bore fruit in the long run in Europe. Ceramic experts believe that the emergence of fine ceramics in Europe such as Italian maiolica in the Renaissance and later Dutch Delft porcelain in the 17th century owes much to Islamic influences. Delft artists, for example, imitated Chinese blue and white porcelain but also encountered Ottoman colored porcelain and perhaps drew inspiration for some of its decorative elements. By the 18th century, with the invention of European porcelain, Europeans were able to produce products that rival Eastern imports, but this did not diminish their continued fascination with Eastern artifacts.

In the subsequent centuries (19th–20th AD), Western interest in the Ottoman ceramic heritage grew as part of the wave of Orientalism and artistic trends that viewed the East as a source of inspiration. Collectors acquired many ancient Ottoman tiles and vessels, and several of them entered European museums. For example, the collections of the Louvre in Paris and the Victoria and Albert Museum in London contain 16th-century Ottoman tiles, displayed as examples of the splendor of Islamic art. Similarly, in the United States, the Art Institute of Chicago and the Metropolitan Museum in New York, among others, include examples of Iznik plates and tiles in their exhibitions of Islamic art. This subsequent Western interest aroused the curiosity and pride of the Turks themselves in this heritage, leading to what can be called a modern renaissance of

Turkish ceramic art. Beginning in the mid-20th century, intensive efforts began in Turkey to revive the traditions of tilemaking using the old method. Craftsmen, artists, and researchers conducted studies of historical recipes for pastes and dyes, and training courses were organized to pass on skills to a new generation. Workshops in the town of Iznik itself have reopened (after an interruption of approximately 300 years) with the support of government agencies and international experts, and these workshops have been able to restore many ancient techniques of shaping, painting, and firing. The ceramics industry has also flourished in Kutahya, which is still known today for its handmade ceramics (albeit mostly for gifts and the tourist market). These efforts have borne fruit in recent decades, with Turkish tiles making a strong comeback: modern mosques in Turkey and abroad have revived the tradition of Ottoman-style decorative tile wall coverings, and many Turkish architects are using handmade tiles in the interior decoration of contemporary buildings to revive their heritage. Visitors can now visit Iznik to witness firsthand skilled craftsmen painting and glazing tiles in the same traditional method their ancestors did five centuries ago. Museums and specialized shops have also been established, showcasing both traditional and modern works inspired by the ancient style [57–60].

This local and international momentum culminated in an important global recognition: in 2016, UNESCO inscribed the Turkish art of Iznik ceramics on its Representative List of the Intangible Cultural Heritage of Humanity. This was considered a significant cultural achievement, recognizing the historical and human value of this art and the need to preserve it as part of the world's heritage. It also encouraged the Turkish government to support artisans and institutions involved in this field. Today, Iznik tiles and other traditional Turkish ceramics remain in demand and admired by art and heritage enthusiasts around the world. They are acquired to adorn modern homes and lavish halls, just as they once adorned the palaces and mosques of sultans. Through this new spread, we can say that Turkish tile art has completed its full civilizational cycle: from a humble local origin to a global imperial heyday, to its decline with the decline of that empire, and finally its final resurgence as a shared human cultural heritage that transcends the boundaries of time and space [61–63].

7. Conclusion

We can consider the art of Turkish tiles and ceramics a profound cultural lesson in how artistic creativity can transform into the building blocks of civilization. It demonstrates how nations leave their mark not only through wars and wealth, but also through the artifacts they craft with taste and thought. Every decorated Ottoman tile is a story of faith, love of beauty, and pride in identity, and every wall adorned with it is a page in the history of that civilization. Preserving and reviving this art form is nothing less than preserving these bright pages of human history. In our contemporary world, which seeks harmony and peace, this art remains an ambassador of shared aesthetic values and a bridge between East and West, between past and future, in accordance with the saying: "The arts contribute significantly to shaping the cultural identity of nations. They are a powerful tool for documenting the past, expressing the present, and shaping the future." So, a salute to all those who have created and preserved this heritage. Let us always reflect on the arts that God has bestowed upon us, which speak of their magic, bearing witness that we have built a civilization when we crafted beauty and deposited it in pottery and stone to tell our story to eternity.

References

[1]   E. Mundt, Art, form, and civilization, Univ of California Press, 2022.

[2]   O.S. Berker, Architecture and Ceramics, Procedia - Soc. Behav. Sci. 191 (2015) 291–295. https://doi.org/10.1016/j.sbspro.2015.04.645

[3]   S. Yılmaz, Traditional Objects That Sources to Contemporary Turkish Ceramic Art, Art-Sanat Derg. (2020) 417–441.

[4]   İ. Yardımcı, The glazed tile techniques of the Seljuk and Beylik periods, J. Lit. Art Stud. 3 (2013) 42–51.

[5]   H. Sazci, Development Of The Ceramic Tile And Tile Industry In The Republic Of Turkey, in: Qualicer 2000. VI World Congr. Ceram. Tile Qual., 2000.

[6]   D. Demirarslan, O. Demirarslan, An Overview Of Tile Art In Turkish Architecture., J. Int. Soc. Res. 14 (2021).

[7]   S. Sarıhan, E. Akoğlu, A.A. Kaymaz, Synthesis of Art in Greek and Turkish Cultures: A Comparative Analysis of Mycenaean and Tile Art, J. Turkology 35 (2025) 259–277.

[8]   A.N. Karakozak, The relationship between art and religion: Turkish-Islamic artworks, Antakiyat 4 (2021) 361–376.

[9]   B.A. Kutty, A Reflection on the Culture and Tradition of Islam on Art and Architecture, (n.d.).

[10]   S.E. Osim, Islamic art and architecture: A reflection of the culture and tradition of Islam, Soc. J. Soc. Sci. Humanit. 1 (2021) 174–185.

[11]   G. Necipoğlu, The Aesthetics of Empire: Arts, Politics and Commerce in the Construction of Sultan Süleyman's Magnifijicence, in: Battle Cent. Eur., Brill, 2019: pp. 115–159.

[12]   W.B. Denny, Islamic Visual Art, in: Oxford Res. Encycl. Relig., 2024.

[13]   T.D. Apaydin, The Types of Monochrome Glazed Ceramics in the Becin Castle Excavation (2014-2021)/Becin Kalesi Kazisinda Ortaya Cikarilan (2014-2021) Tek Renk Sirli Seramik Kaplarin Form Ozellikleri., J. Art Hist. (2022) 57–83.

[14]   T.D. Apaydın, Beçin Kalesi kazısında ortaya çıkarılan (2014-2021) tek renk sırlı seramik kapların form özellikleri, Sanat Tarihi Yıllığı (2022) 57–82.

[15]   S. Karakitsos, The Lustre Pottery Techniques Continuum Through the Silk Roads, in: Int. Conf. Silk Road Sustain. Tour. Dev. Cult. Herit., Springer, 2021: pp. 177–183.

[16]   I. Gürgen, Palace architecture and its rhetoric in Seljuk Anatolia: The conceptualization of Kubadabad, (2019).

[17]   Ç. Öztürk, G. Şimşek Franci, İ.M. Kuşoğlu, An archaeometric assessment study of Seljuk period glazed tiles from Kılıçarslan Square (Konya, Turkey), Herit. Sci. 10 (2022) 174.

[18]   K. Özkul, M.R. Başar, Graphic Design Analysis of The Color and Shape Features in Kubad Abad Palace Tiles, J. Art Time 2 (2022) 34–45

[19]   K. Özkul, M.R. Başar, Graphic design analysis of the color and shape features in Kubad Abad Palace Tiles, Art Time (2023) 34–45.

[20]   J. Živković, V. Bikić, M. Georgakopoulou, J.C. Carvajal López, Archaeology of craft and artisans in the Ottoman Empire: a case of ceramic production in Belgrade during the sixteenth and seventeenth centuries, Archaeol. Anthropol. Sci. 13 (2021) 63.

https://doi.org/10.1007/s12520-021-01306-3

[21] M. EskandariMarjin, R. Nouri Shadmahani, A Look at the Pottery Style of Abu Zayd Kashani, Iran. J. Archaeol. Stud. 12 (2022) 59–72.

[22] W. Shaded, Palestinian Traditional Architecture and Methods in Designing Dwellings: Case Study Old City of Hebron, ARCHive-SR (2023) 48–65.

[23] J.S. Putra, Z. Rahmi, The Economic Progress of the Ottoman Empire: An Analysis of Advancement During the Reign of Sultan Suleiman the Magnificent, J. Philol. Hist. Rev. 2 (2024) 138–150.

[24] E. Gökçe, Iznik ceramics: History and present-day, Athens J. Humanit. Arts 5 (2018) 225–242.

[25] B. Zidan, Cross-Cultural Exchange Between the Islamic World and Europe (Iznik ceramic and Italian maiolica as a case study), J. Gen. Union Arab Archaeol. 4 (2019) 1–46.

[26] S. Faroqhi, Making things to serve sultans, viziers and army commanders (1450-1800), Mediev. Hist. J. 21 (2018) 69–99.

[27] B. Hickman, a Forgotten FiFteentH-century ottoman mosque and its inscriptions, Muqarnas Online 36 (2019) 209–219.

[28] G. Simsek, O. Unsalan, K. Bayraktar, P. Colomban, On-site pXRF analysis of glaze composition and colouring agents of "Iznik" tiles at Edirne mosques (15th and 16th-centuries), Ceram. Int. 45 (2019) 595–605. https://doi.org/10.1016/j.ceramint.2018.09.213

[29] P. Colomban, B. Kırmızı, G.S. Franci, Cobalt and associated impurities in blue (And green) glass, glaze and enamel: Relationships between raw materials, processing, composition, phases and international trade, Minerals 11 (2021) 633. https://doi.org/10.3390/min11060633

[30] H.C. Evans, C. Alchermes, I.B. McCabe, A. Ballian, S.R. Canby, K. Colburn, Y. Crowe, M.D. Findikyan, R. Goshgarian, G. Grigoryan, Armenia: Art, Religion, and Trade in the Middle Ages, Metropolitan Museum of Art, 2018.

[31] N. Shafir, In an Ottoman Holy Land: The hajj and the road from Damascus, 1500–1800, Hist. Relig. 60 (2020) 1–36.

[32] S. Erdogan, Multiple Narratives of Display and Heritage in Museums: Iznik Ceramics in Comparison, (2020).

[33] H. Suleiman, B. Gülgün, K. Yazici, O 9. Building A Historical Place Bridge With The Concept Of Cultural Landscape; Tile Art And Mosaics, Proceeding B. (n.d.) 37.

[34] F. Zahra, S. Shahir, Aesthetical aspects and historical significance of architectural ornament Iznik: Hermeneutical study, (2024) 165–175.

[35] P. Colomban, G. Simsek Franci, Timurid, Ottoman, Safavid and Qajar ceramics: Raman and composition classification of the different types of glaze and pigments, Minerals 13 (2023) 977.

[36] C. Altundal, Characterization of Glazes of the 11-14th-Centurt Potsherds and Tiles from Comana Pontica in Tokat, Turkey, (2023).

[37] M.B. Er, an Archaeometrical Investigation on Provenance and Technological Properties of Seljuk Period Pottery From Komana (Tokat), (2020) 276.

[38]   C. Gökçe, E. Feyzoğlu, A General Overview Of Portrait Works On Ceramic Surfaces, Turkish Online J. Des. Art Commun. 14 (2024) 763–784.

[39]   P. Blessing, The blue-and-white tiles of the Muradiye in Edirne: Architectural decoration between Tabriz, Damascus, and Cairo, Muqarnas Online 36 (2019) 101–129.

[40]   J. Alhamar, Islamic Style Botanical Surface Decorations, (2019).

[41]   G.Ó. Hansson, Productive syncretism in Saami inflectional morphology, in: 2007: pp. 91–135. https://doi.org/10.1075/cilt.288.05han

[42]   L. Saleh, Damascus in the Ottoman Period (Sixteenth Century-Eighteenth Century): A Comparative Study, (2025).

[43]   H.E. Al Khalifa, M.W. Lafi, Ornamentation and Islamic identity in contemporary European mosques: An analysis of cultural, aesthetic, and functional dimensions and modern influences, Buildings 15 (2025) 1302.

[44]   N. Khatcherian, Armenian crafts in the Ottoman Empire: Armenian identity and cultural exchange, Natl. Identities 21 (2019) 485–505.

[45]   G. Collaço, The Image as Commodity: The Commercial Market for Single-Folio Paintings in Ottoman Istanbul, 17th–18th C, Harvard University, 2020.

[46]   V. Bruccoleri, Crossed Flowers: Interconnection of Decorative Patterns on Blue-and-White Ceramics in Central Asia, Iran, and Turkey (15th-17th Centuries), Connect. Common Legacies Cent. Asia (2024)

[47]   A.S. Levey, J. Coresh, K. Bolton, B. Culleton, K.S. Harvey, T.A. Ikizler, C.A. Johnson, A. Kausz, P.L. Kimmel, J. Kusek, A. Levin, K.L. Minaker, R. Nelson, H. Rennke, M. Steffes, B. Witten, R.J. Hogg, S. Furth, K. V. Lemley, R.J. Portman, G. Schwartz, J. Lau, E. Balk, R.D. Perrone, T. Karim, L. Rayan, I. Al-Massry, P. Chew, B.C. Astor, D. De Vine, G. Eknoyan, N. Levin, S. Burrows-Hudson, W. Keane, A. Kliger, D. Latos, D. Mapes, E. Oberley, K. Willis, G. Bailie, G. Becker, J. Burrowes, D. Churchill, A. Collins, W. Couser, D. DeZeeuw, A. Garber, T. Golper, F. Gotch, A. Gotto, J.W. Greer, R. Grimm, R.G. Hannah, J.H. Acosta, R. Hogg, L. Hunsicker, M. Klag, S. Klahr, C. Lewis, E. Lowrie, A. Matas, S. McCulloch, M. Michael, J. V. Nally, J.M. Newmann, A. Nissenson, K. Norris, W. Owen, T.G. Patel, G. Payne, R.A. Rivera-Mizzoni, D. Smith, R. Star, T. Steinman, F. Valderrabano, J. Walls, J.P. Wauters, N. Wenger, J. Briggs, K/DOQI clinical practice guidelines for chronic kidney disease: Evaluation, classification, and stratification, Am. J. Kidney Dis. 39 (2002).

[48]   L. Mehling, The Irony of Imperial Decorative Styles: The Role of Diversity in the Production of Unified Cultural Identities in the Ottoman Empire and France, c. 1500–1700, Diyâr 3 (2022) 190–241.

[49]   M. Manninen, Creative power of the flower of life, (2022).

[50]   B. Clemens, Mutants: Ceramic Objects, (2023).

[51]   S. Hamadeh, Ç. Kafescioğlu, A Companion to Early Modern Istanbul, BRILL, 2021. https://doi.org/10.1163/9789004468566

[52]   E. Atak, Importance of the Tulip in Turkish Culture and as Art from Central Asia to Anatolia, Cent. Asiat. J. 65 (2022) 151-+.

[53]   R. Tillinghast, Journeys into the Mind of the World: A Book of Places, University of

Tennessee Press, 2024.

[54] K.S. Bayraktar, Edirne Wall Paintings, (2022).

[55] M.S. Graves, Arts of allusion: Object, ornament, and architecture in medieval Islam, Oxford University Press, 2018.

[56] F. McGill, An Unusual Sawankhalok Ceramic at the Asian Art Museum, and Its International Connections, J. Siam Soc. 113 (2025) 187–198.

[57] N. Ramirez, Trans-Pacific Connections: Chinese Porcelain and Aztec Motifs in Colonial Mexico, (2025).

[58] J.P. Sires, G.J. Linares Matás, "Those who were bound together": illuminating Almoravid imperial entanglements, Azania Archaeol. Res. Africa (2025) 1–21.

[59] J.A. Burrison, Beautiful Clay: A Traditional Craft as Art, Indiana University Press, 2025.

[60] H.P. Gournis, A. Apostolakis, The case study of traditional pottery-making, Responsible Consum. Prod. Creat. Cult. Ind. Actions, Policies, Strateg. a Sustain. Futur. (2025) 94.

[61] Ö.D. Çakmaklı, Roman Glass Artefacts from the Northern Necropolis of Aizanoi (2012-2017): An Assessment in Social and Economic Context, Arkhaia Anatolika 8 (2025) 1–31.

[62] S. Mendeş, Managing Intangible Cultural Heritage in Turkey: Four Cases Originating in Antiquity, (2021).

[63] H. Ersoy, Spaces of Commonality and Change: Re-constructing Ruins Amongst Cultural Values in Iznik, Turkey, (2022).

Chapter 20

Environmental Footprint of the Ceramic Industry: Production Processes, Life Cycle Analysis (LCA) Practices and Sustainability Strategies

Merve Paksoy Balci[1*], Ebru Halvacı[2], Cenk Karakurt[3], Selcuk Erdogan[2], Fatih Sen[2*]

[1]Natural Stone Technologies Application and Research Center, Bilecik Seyh Edebali University, 11100, Bilecik, Türkiy

[2]Sen Research Group, Department of Biochemistry, Kutahya Dumlupinar University, Kutahya 43000, Türkiy

[3]Department of Civil Engineering, Bilecik Seyh Edebali University, 11100, Bilecik, Türkiy

merve.paksoy@bilecik.edu.tr, fatihsen1980@gmail.com

Abstract

Ceramics industry's energy-intensive production processes and meeting raw material demand stand out as the primary sources of environmental impacts. The ceramics industry's environmental impacts can be reduced through energy consumption, sustainable raw material use, waste management, and the integration of renewable energy systems. In this context, Life Cycle Assessment (LCA), innovative methods, policy regulations, and economic incentives are crucial. This study presents a comprehensive assessment of the ceramics industry's environmental impacts, addressing production, raw material supply, waste management and recyclability, LCA, innovative methods, policy regulations, and economic incentives. It has been stated that the most significant environmental burden, as commonly cited in literature studies, arises during the high-temperature drying and sintering stages. Raw material supply and transportation activities, in turn, contribute to environmental impacts such as greenhouse gas emissions, fossil fuel consumption, and acidification. The recycling of waste generated in ceramic production has been shown to have positive contributions to ecotoxicity and human health. LCA studies offer alternative solutions by separately considering all processes in ceramic production, from raw material supply to waste management. Common recommendations from the studies include the use of alternative raw materials, the adoption of energy-efficient technologies, the transition from fossil fuels to renewable energy, the development of local production and short supply chains, and the improvement of water and wastewater recycling systems. It was also reported that digitalization, smart technologies, and the expansion of policies and incentives will significantly contribute to the ceramics sector's environmental sustainability.

Keywords

Ceramic Production, Sustainability, Energy Efficiency, Life Cycle Assessment (LCA)

1. Introduction

Ceramics are among the oldest materials in the history of mankind; today, they have gained different areas of use in both traditional and advanced technology applications. Its historical development started from pottery to glazed products and became widespread in various fields with the advancement of technology. It is known that humanity's acquaintance with ceramics dates back to as early as 18,000-15,000 BC. [1], [2]. Today, traditional ceramics such as bricks, pottery, and tiles, white ceramics, and advanced ceramics such as technical ceramics (biomedical implants, electronic components) and composite ceramics, which are a combination of ceramics and metals, are used in a wide variety of fields [3].

Ceramics are inorganic materials obtained by sintering at high temperatures, consisting of clay minerals and other natural raw material mixtures that are abundant in nature. They have crystalline, semi-crystalline, or amorphous structures formed by oxygen and metal elements and are characterized by chemical resistance, mechanical hardness, high temperature resistance, and electrical insulating properties [4]. The estimated lifetime of a ceramic product is more than 50 years. This is because ceramics are highly resistant to factors such as humidity, temperature, chemicals, UV rays, and extreme weather conditions [5]. In this study, the environmental impacts of the ceramics industry, energy consumption, sustainable use of raw materials, waste management, and integration of renewable energy systems are discussed. In this context, the potential applications of the sector to contribute to sustainability, such as life cycle analysis (LCA) of ceramic production processes, innovative methods and policy regulations, and economic incentives, were evaluated.

2. Definition and Industrial Importance of Ceramics

Ceramic products were first used for purposes to meet the needs of daily life, usually in the form of terracotta pots. With the industrial revolution, the Chinese, Egyptian, and Mesopotamian civilizations pioneered the development of ceramics, resulting in a variety of ceramic materials with different firing temperatures and composition optimizations [6]. While China was a pioneer in terms of porcelain production and kiln technologies, Mesopotamian and Egyptian civilizations drew attention to the cultural and commercial importance of ceramics. In Europe, mass production started with the Industrial Revolution. With the development of electronics and space technologies, technical ceramics were developed in the 20th century. Due to population growth and industrialization, the environmental impact of ceramic production has become an increasingly important research topic [5]. Bricks and tiles are widely used in construction, pottery, and plates as household items, in the electronics industry as semiconductors and insulators, in the automotive industry as wear-resistant materials, in the medical field as implants, and in the space and defense industries as they are heat resistant. [3], [7].

In this context, ceramic products have become products with an increasing market share for industry and living spaces in terms of durability, aesthetics, ease of maintenance, and resistance to environmental conditions. [8]. According to MECS-Acimac Research Center's "Ceramic Tile Market Forecast Analysis: Trends 2024-2028" report, a gradual increase in ceramic tile production and demand is expected over the next five years. Production is projected to grow at an average annual rate of 2.2% and demand at 2.5% [9]. According to the sectoral analysis report titled "Pottery Ceramics Market Size, Share & Trends Report, 2022-2030" published by Grand View Research, it was reported that the global porcelain ceramics market was USD 10.72 billion in 2021 and is estimated to reach USD 15.00 billion by 2030 [10].

It is known that ceramic production was initially handcrafted by potters, but over time, it gained a production structure based on process development, institutional understanding, division of labor, and industrialization and started to be used in various fields [6]. The growth in the sector brings with it significant problems such as high energy consumption, resource utilization, and environmental impacts. In addition, fired wastes, raw wastes, scrapped products, and production wastes generated during ceramic production are also factors that increase environmental impacts [5]. Considering all these factors, it is important for a greener world that ceramic production processes are handled in line with the global economy and sustainable development goals.

3. Structural Properties and Mechanical Behavior

The chemical composition of ceramics mainly consists of oxides such as silica (SiO_2), alumina (Al_2O_3), clay minerals, and feldspar, as well as carbides (SiC), nitrides (Si_3N_4), and borides. The composition of traditional ceramics generally consists of clay (~20-30 %), quartz (~20-30 %), feldspar (~20-30 %), water, and other additives [2]. Technical ceramics are obtained from materials such as high-purity alumina, zirconia, silicon carbide, binders, and additives.

The composition of ceramics directly affects their physical and mechanical properties. Quartz (SiO_2) imparts hardness and mechanical strength. Feldspar facilitates the firing process by supporting liquid phase formation during production. Ceramics with high alumina (Al_2O_3) content have higher mechanical strength, while kaolin gives the ceramic whiteness and hardness at high temperatures. Zirconium is an element that gives the ceramic material a white and excellent color and has an opacifying effect on the glaze [10]. Kaolin, used as a ceramic raw material, is a type of clay that gives the product a light color. The glaze raw material also contains kaolin. Although anorthite-based products provide environmental advantages with lower sintering temperatures, their industrial use is limited due to the narrow range of sintering temperatures [8]. Mullite-based porcelain is still preferred in the sector due to its wider production tolerance and high mechanical strength [11]. The physical structure of ceramics is generally porous, hard, and brittle, while crystalline grains are accompanied by an amorphous glass structure in their microstructural properties. The crystals in their structure play a decisive role in their mechanical strength and thermal properties. Physico-mechanical properties of ceramics, such as thermal expansion, electrical properties, and chemical resistance, can be improved by using various additives [10].

4. The Fundamental Progresses and Process Dynamics in Ceramic Production

Ceramic production involves multi-stage processes starting from raw material procurement and ending with packaging. Each production stage has its own energy and water consumption and differs in terms of environmental impact [5]. In the general ceramic production process, raw materials are subjected to a grinding process in order to obtain the chemical and mineralogical composition suitable for the predetermined recipe. The powdered raw material goes through the stages of shaping, drying, polishing, decorating, and baking (sintering) to produce the final product. After the quality control process, the products are packaged, made ready for distribution, and offered for sale [12], [13]. The stages of ceramic production are given in Figure 1.

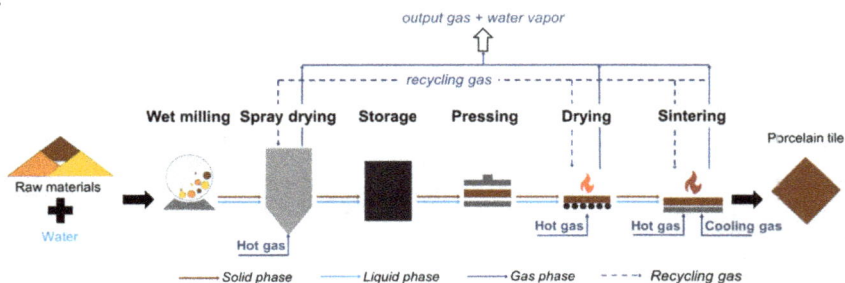

Figure 1. *Production steps of porcelain tile from raw materials to packaging [14], reprinted with permission from MDPI.*

The ceramic production process is not only limited to raw material preparation, but also consists of many critical process steps such as physical shaping, heat treatment, surface treatment, and quality control stages. Each stage of the production process must be carefully planned and controlled to produce ceramic products with the desired mechanical, physical, and chemical properties. The main stages in ceramic production are categorized as raw material procurement, preparation, forming, shaping, drying, firing, and surface treatment, and these basic stages and the dynamics of these processes are discussed from a technical point of view.

4.1 Raw Material Supply

The first step in ceramic production is raw material selection, extraction, and processing. Raw materials such as clay, quartz, and feldspar are extracted by the open pit method and used directly. Secondary environmental impacts such as habitat loss, soil destruction, and water use occur during the extraction of raw materials. Raw material procurement is the stage with direct environmental impacts. Raw material transportation also leads to significant energy consumption and emissions. Two different methods are used in ceramic production: wet and dry methods. These two production techniques differ in terms of raw material handling during the forming stage. The wet method has been developed instead of the existing dry method to produce products with superior performance, such as high quality and large dimensions [3].

4.2 Preparation step

The preparation stage consists of grinding raw materials, mixing, and creating a composition suitable for the production purpose. Dosing is carried out using two methods: by weight and by volume. In the mixing and grinding stage, the dosed raw materials and additives are mixed with water until they become homogeneous, and then they are ground in a ball mill. The grinding process is carried out as wet grinding to maintain the moisture content of the raw material [5], [10]. The grinding continues until the raw materials are in a suspension of about 35% water and 65% solids. After grinding, the sludge is spray-dried into fine droplets and dried with heated air to produce a ready-to-use granular product with a residual moisture content of approximately 7% [9].

4.3 Forming step

At this step, the free-flowing powder, which is transferred to various molds depending on the use of the product, is compacted with hydraulic pressure. The compression load for ceramic tiles is reported to be approximately 2500 tons [5]. One of the shaping methods, such as pressure molding, casting, injection, or extrusion, can be preferred.

4.4 Drying step

In the drying process, moisture is removed from the shaped ceramic product. This process is carried out slowly for several days by maintaining the relative humidity in a way that prevents deformations on the sample surface [5].

4.5 Baking step

To impart mechanical strength to the ceramic product, it is fired in high-temperature kilns. The sintering process removes water and organic compounds from the clay, and the inorganic particles bind together without changing their chemical composition to form a load-bearing solid [2]. Firing ovens are usually electric, gas, and wood-fired. After shaping, the biscuit is baked at a temperature of about 1200 °C [3]. When wet grinding is used during sample preparation, the firing process is completed using roller kilns for 60 minutes and at temperatures of 1150-1200°C. The next stage is the glazing process. During the firing phase, the clay is subjected to changes in chemical and crystalline structure under the influence of high temperature. In this case, the ceramic product gains properties such as expansion, mechanical strength, and porosity [15]. The temperature of the product is then reduced to 60 °C by controlled cooling. The final process is the packaging step [9].

4.6 Surface Finishing

It is the stage where mechanical processing, polishing, and glazing processes are carried out. Glazing is applied in order to make ceramic biscuits resistant to water and other factors. In the glazing stage, raw material supply, glaze-frit preparation, grinding, glaze application, and sintering processes are carried out. The glaze mixture, which is calculated in certain dosages, is mixed and ground in dry or wet mills and applied on the biscuit, then fired at temperatures ranging from approximately 800 °C to 1700 °C [3]. This process is carried out under 1 bar pressure and for one hour in a graphite furnace. This results in a smooth and uniform, as well as shiny surface [5]. Glaze material is prepared by mixing raw materials such as feldspar, color pigment, zirconium, various chemicals, and frit in a ball mill [16]. Depending on the size, shape, quantity, and desired surface properties of the ceramic product, dipping, spraying, bell, screen, disk, dripping, and electrostatic methods can be used in the glaze process. After glazing, ceramic products are quality controlled and packaged. The final stage of the process is waste management, which involves the recycling and disposal of end-of-life materials [12], [13].

5. Energy Consumption and Environmental Impacts in Ceramic Production

Ceramic production consists of energy-intensive processes with high energy and resource consumption requirements. In addition, the dependence on raw materials such as clay, feldspar, and quartz also increases energy consumption [13]. The production process relies mostly on various energy sources, such as electricity and thermal energy (Figure 2a). Electricity is mainly used as a driving force, while thermal energy is mainly supplied by natural gas, liquefied petroleum gas (LPG), and coal. Natural gas is the primary fossil fuel for the ceramics industry, accounting for more than 60-80% of the thermal energy mix, depending on the region. LPG and coal make up a smaller proportion, but still have a significant share in the energy profile of the industry [17]. According to literature studies, small ceramic plants consume a total of 253.72 MJ of energy, which is about twice the energy consumption of medium-sized plants. The process step with the highest energy consumption is glaze firing (Figure 2b). At this stage, 79.77% of the total energy is consumed in small plants and 35.41% in medium plants [5]. Therefore, strong support is needed to identify and implement sustainable solutions that can reduce the environmental, economic, and social impacts of the sector.

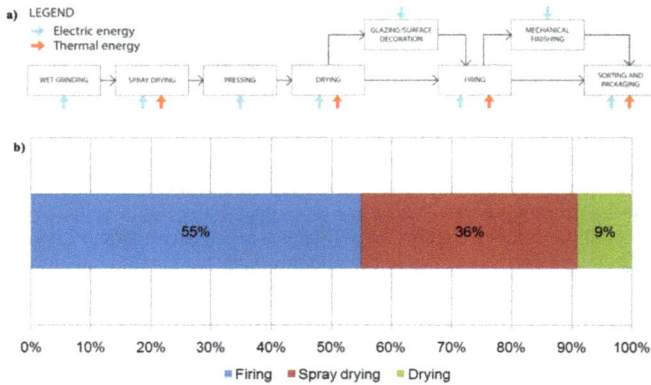

Figure 2. *a) Distribution of energy types in the ceramic tile production process, b) Distribution of thermal energy consumption [18], reprinted with permission from MDPI.*

All sub-sectors of the ceramic industry perform drying and sintering at high temperatures ranging from 800 to 2000 °C in their production processes. The biscuit and glaze firing stage of ceramic production, which is carried out at high temperatures, has the highest environmental, economic energy impact [19]. Therefore, production processes are energy-intensive, resulting in significant CO_2 emissions [7]. Approximately 90% of energy consumption is related to the drying stage, which requires the use of thermal energy [20]. According to [21], the sintering stage in ceramic production is the most critical step in the production process. [22], calculated the energy consumption in the ceramic industry based on experimental data on the ceramic tile production process and found that the largest part of the energy consumption belongs to the firing and sintering processes, with about 56%. In this context, among the methods to reduce the energy consumption of the ceramic industry, kilns and dryers will be the focus [23]. Successful operation of industrial decarbonization strategies can be achieved through furnace modernization in production processes. Expanding economic incentives for challenges such as high investment costs and technological adaptation processes would also be beneficial. Kilns used in the ceramic industry are responsible for more than 80% of sector-wide emissions due to their high level of energy consumption. Therefore, in terms of environmental impact and sustainability, it is necessary to reduce the impacts from the cooking stage. Developing environmentally friendly and advanced technologies in production methods can reduce all these environmental impacts [11].

Several efforts are underway to overcome the limitations of high-energy production stages. It is known that methods such as reducing cooking time, providing heat and humidity control in ovens and dryers with automation, equipping ovens with good insulation practices, waste heat recovery, digitalization, energy optimization in heating and cooling systems, and improving temperature distribution with the use of pusher fans are quite successful [23]. With the heat recirculation system, which will provide a 4.5% reduction in the energy required in the heating stage, the method requiring less energy at the cooking temperature has been achieved [19]. Furnace quality and raw material properties are important parameters affecting energy consumption. Since both plants in both scales obtain up to 80% of their energy from non-renewable sources, a shift towards renewable energy sources will greatly reduce environmental impacts.

To summarize, it will be possible to reduce energy consumption through technological and operational improvements such as the use of high-efficiency heating technologies in ceramic production processes, determination of optimum firing cycles and process optimization through digital simulation techniques, reduction of heat losses through the use of insulating materials, and equipping kilns with heat recovery systems. In addition, the use of renewable electricity, green hydrogen, biofuels, and synthetic gases instead of fossil fuels will be in line to neutralize the carbon footprint.

6. Waste and Products in Ceramic Production

Industrial waste and natural resource utilization are effective strategies for sustainable development and environmental protection. This necessitates the disposal of waste with high environmental impact [24]. Given the current situation, total waste generation is projected to increase by more than 60% by 2050. Globally, only 62% of the current waste is collected in a controlled manner, while the rest is released into nature. Only 19% of the waste collected can be recycled. All this waste is the most important source of greenhouse gas emissions, which are the main cause of climate change [24]. Billions of tons of agricultural, industrial, and human waste are disposed of every year, causing global destruction of the ecosystem [25].

With the increase in environmental awareness, the tendency towards environmentally friendly products has increased, and environmentally friendly product designs have become important in this situation. Although ceramics are made from natural materials, they are known for their significant energy consumption and pollutant emissions during production [8]. In ceramic production, different types of waste are generated at each stage of the production process (Figure 3). Production processes, from raw material procurement to the final product, pose a threat to human and environmental health [5].

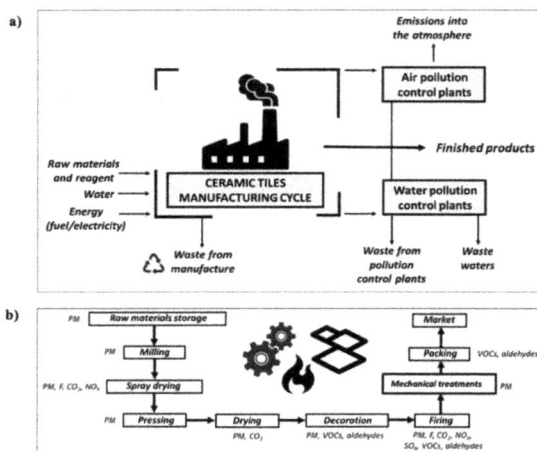

Figure 3. a) *Material, waste and emission outputs generated during the ceramic tile production process, **b)** Porcelain tile production steps and the main environmental pollutants produced in these steps (PM: particulate matter; F: fluorine, CO₂: carbon dioxide; NOₓ; nitrogen dioxides; VOCs: volatile organic compounds; SOₓ: sulfur dioxide) [26], reprinted with permission from MDPI.*

Ceramics that are thrown into the environment indiscriminately remain undissolved for hundreds or even thousands of years. In this case, toxic substances continue to be emitted, and the value-added components in the waste cannot be utilized. In addition, waste disposal causes the formation of dust and trace elements and adversely affects water and soil quality [24]. Wastes generated in ceramic production can be classified as raw wastes, which are materials such as broken, unformable, or unused clay mixtures, mold shards, etc. generated during production, hard ceramic pieces that are broken or become residual material after production, and wastes that may contain toxic or harmful chemicals from glazing and painting processes. While raw material wastes have a high recyclability potential, post-use wastes are important for their recovery potential.

While the final product is obtained in ceramic production, by-products are also generated. In this case, the disposal of the resulting wastes becomes mandatory. Raw material wastes include broken, cracked, and faulty products during the production phase, as well as clay, kaolin, and feldspar wastes spilled before and after processing. The complex composition and potential environmental risks of raw material wastes make it imperative to develop sustainable utilization methods that aim to minimize ecological impacts in the disposal and reuse of these wastes [27]. Production defects, excess and broken parts from use, construction, and demolition waste are non-biodegradable inorganic materials known as waste ceramics. Disposal of these wastes brings problems such as increased landfill and land degradation, while waste disposal increases energy expenditures [24]. Wastes generated during shaping and cutting, deformed products that crack during drying and firing, and dust wastes during the grinding process occur during the production process. Glaze materials that are over-applied or spilled during glazing and polishing are also considered waste. Dust accumulated in the filters of firing furnaces, flue gas and particle emissions, and waste heat are known as energy and filtration wastes. These particles and harmful waste gases, such as hydrofluoric acid, harm air quality. Materials such as cardboard, pallets, and plastics used in the packaging of the final product are also considered packaging waste. In the quality control phase, unsuitable products are sorted and discarded. Nano-particles released during sintering can contain oxides of Zn, Cr, Al, and Fe, which adversely affect lung health when inhaled. Particles emitted into the air during the production of nanoTiO$_2$ glazed porcelain containing nano titanium dioxide (TiO$_2$) particles have a 32.33% negative impact on human health, 34.34% on climate change, 4.07% on ecosystem quality, while their contribution to freshwater toxicity is 5% [5]. In ceramic production, an average of 20 liters of water is consumed for each square meter of tile production. Especially sludge preparation and glazing processes stand out with their intensive water consumption. The wastewater released during production is among the causes of water pollution due to its heavy metal content and chemical composition.

At the end of the production processes in the ceramic industry, various waste products such as clay, chamotte, soil, and liquid wastes can be disposed of directly or reprocessed and used in different areas. Wastes such as sludge, fluid residues, and broken band pieces that are released during the production process are classified as hazardous waste. Using such wastes instead of cement, fine and coarse aggregates in green concrete production can be an effective method for material recovery and reducing carbon emissions. Grinding broken ceramics into filler and drying waste sludge into briquettes also serve the goals of effective waste management and sustainability. Concretes containing ceramic waste offer technical and environmental advantages while contributing to sustainable construction practices through reduced cost, increased performance, and energy efficiency [24]. During the polishing stage of ceramic production, ceramic dust is produced as a by-product, and this waste causes air, water, and soil pollution. It is possible to use ceramic powder waste as a by-product as a substitute product in concrete [28].

To summarize the recycling processes of ceramic wastes, with regrinding, fired ceramic wastes can be ground and used as raw material reinforcement in new productions. Ceramic shards can be used as filler material in the production of waste-added concrete and building materials. Some ceramic wastes can be used in energy recovery thanks to their high calorific values. In addition, minimizing the amount of waste at the production stage, installing waste separation systems in facilities, legal regulations, and incentive mechanisms can be considered as important strategies for a holistic approach in waste management. By recycling waste, the ecological footprint shrinks by reducing the consumption of natural resources, the cost of raw materials is reduced, and the cost of waste disposal is also reduced. Re-utilization of both production waste and end-of-life ceramic products contributes to environmental sustainability through the integration of the circular economy approach, which aims to keep resources in the production cycle for the longest possible time, into the ceramics sector. In addition, both the minimization of production waste and the effective recycling of post-consumer ceramic products will be possible in the ceramic sector.

7. Environmental Impacts of Ceramic Production

Increasing population and industrialization bring with them increasing energy demand, waste discharge, and serious environmental problems. In particular, the widespread use of fossil fuel-based energy sources in industry leads to increased environmental impact [29]. Reducing the contribution of these sources to global greenhouse gas emissions has made it imperative to develop alternative solutions to achieve the Paris Agreement target and limit global warming to 1.5-2 °C [30]. Ceramics are the most widely used building materials in commercial and residential buildings. The ceramics industry is expanding its production volume at a rate of 6% per year, while consumption of ceramic products is growing steadily on a global basis [5].

Ceramic production is one of the most energy-intensive areas of the construction industry due to its high energy requirements and use of fossil fuels such as natural gas and coal, which leads to increased carbon dioxide emissions and thus global warming [15], [31]. Approximately 90% of the energy used in production is thermal energy, of which 55% is used directly in the firing process [8], [20]. Due to the fact that thermal energy is mostly derived from fossil fuels such as coal and natural gas, the sector generates high levels of carbon dioxide (CO_2) and other greenhouse gas emissions. Total energy consumption in the entire ceramics sector, which causes serious environmental pollution due to the large consumption of energy and raw materials during production, accounts for 30% of total production costs [32]. Ceramic production facilities are generally of two types: small and medium. In small plants, the raw material supply is outsourced, while in medium-sized plants, the raw material source is prepared on-site. The main greenhouse gas pollutant in plants is CO_2 emission (Figure 4). The emissions causing ozone depletion were calculated as 1.14×10^{-5} kg CFC11-eq in small-scale plants and 2.38×10^{-7} kg CFC11-eq in medium-scale plants, and it was determined that these values would cause serious acidification. It was reported that the small production facility emits 15.89 kg CO_2 equivalent greenhouse gas emissions per square meter. When the figures are analyzed, it is seen that small-scale facilities have a higher environmental impact for each parameter. It can be said that the fact that medium-sized facilities cause less pollution is due to their use of advanced technologies and equipment [5].

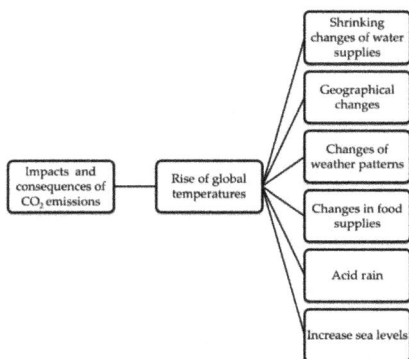

Figure 4. *Environmental impacts and possible consequences of greenhouse gas (CO_2) emissions [33], reprinted with permission from MDPI.*

The ceramic sector is an important focus of efforts to reduce carbon emissions. With around 19 million tons of CO_2 emissions per year, the European ceramics sector accounts for 1% of industrial emissions under the European Emissions Trading System (ETS) and 10% by number of installations [7]. The ceramic industry is known to emit approximately 1.84 million tons of carbon dioxide annually worldwide. The majority of these emissions come from energy sources and technologies used in kiln firing, drying, and shaping processes [17]. Firing and sintering processes account for more than 80% of total energy consumption and generate high thermal energy demand due to their high temperature requirements. The highest energy consumption and CO_2 emissions occur during drying and baking. In particular, the use of natural gas and coal as kiln fuel is known to increase greenhouse gas emissions. The ceramic production process consumes large amounts of non-renewable resources and generates significant levels of pollutants such as exhaust gases, wastewater, and solid waste [34]. Ceramic products made from different sources, in different plants or countries, have different environmental impacts.

According to the China Building Materials Federation, the ceramics industry caused 37.58 million tons of carbon emissions in 2020 alone due to its energy-intensive structure [35]. The primary sources of greenhouse gas emissions are agriculture and deforestation, with fossil fuels accounting for 53% of energy production. The estimated global warming potential of ceramic products is 14.4 kg CO_2-eq per square meter, and CO_2 emissions account for approximately 92.1% of the total environmental impact [23]. The European ceramics industry is estimated to emit around 19 million tons of CO_2 per year. This is equivalent to around 1% of Europe's industrial emissions [7]. In the production process, carbon emission sources were found to be 17% during mineral transformations, 64% with the use of fossil fuels, and 19% with the use of electricity [36]. The typical value of GWP for ceramic tile production is in the range 11-23 kg CO_2-eq/m^2. According to Italian data analysis, this value is reported to be ~13.8 kgCO_2-eq/m^2 on average [21], [37], [38]. Photochemical oxidation potential and ecotoxicity are significantly impacted by wastewater, dust, and chemicals [10].

The most important factors in terms of environmental sustainability in the production cycle in ceramic production can be listed as gas emissions, wastewater discharges, wastes, water consumption, and energy consumption [35]. It is essential to investigate the environmental impact of ceramic production due to the use of many chemicals in addition to the high-energy-demanding processes [16]. In addition to high energy and fossil fuels, large amounts of water and wastewater used during production are also the main causes of soil pollution [31]. All stages of the production process cause some pollutants to be released directly or indirectly into the environment. In addition, open mining operations during the supply of raw materials such as clay and quartz can also cause serious damage to the environment and ecosystem. Dust emissions during the grinding and cooking stages also adversely affect air quality and worker health [31]. Particle and silica emissions in the atmosphere may increase as a result of the addition of glaze, paint, and matting agents to the product during glaze application and decoration.

Air pollution is one of the main environmental impacts of ceramic waste. This leads to various ecological problems such as ozone depletion, global warming, acid rain, and eutrophication. Eutrophication is an environmental problem caused by the excessive accumulation of nutrients, especially nitrogen and phosphorus, in water environments such as lakes, rivers, and seas. These substances cause serious environmental problems, especially when they enter water resources through industrial discharges. Comparing ceramic production in terms of eutrophication, 0.34 kg NO_3-eq was calculated in small plants, while this value was calculated as 0.021 kg NO_3-eq in medium-sized plants [5]. The waste gases that cause air pollution during production are generated in five main process stages. Existing studies show that the most intensive stage in terms of CO_2 emissions and energy consumption in ceramic production is the firing stage. The pollutant gases produced in this process are dust particulate matter, sulfur dioxide (SO_2), and nitrogen oxides (NO_x). These gases cause serious damage to human health and ecosystem sustainability. In particular, they cause respiratory diseases and are also responsible for disrupting the ecological balance [34]. Although these wastes contain healthy natural materials such as clay, feldspar, and quartz, glaze, frit, and pigments contain toxic additives. Ultrafine and nanoscale air particles released during production have also been reported to threaten the environment and human health [5]. The ceramic dust is composed of particles that, when inhaled, can cause respiratory damage such as bronchitis, asthma, pneumoconiosis, and silicosis [24].

The danger of depletion of natural resources and climate change, and the ecological problems it causes are of concern for the world. Environmentally friendly productions play a major role in addressing these concerns. It can be said that environmentally friendly productions are possible with the dissemination of sustainable practices in ceramic production [39]. Considering all these reasons, it has become inevitable to assess the environmental sustainability of the ceramic industry. Existing literature studies confirm that there are not enough studies on the hazards caused by the ceramic industry [5]. In this context, comprehensively assessing and reducing the environmental impact of the ceramic industry is crucial for achieving sustainable production goals. Ceramic products are among the materials that stand out for offering environmentally friendly and sustainable solutions. It is possible to reduce environmental impact by improving production processes. With appropriate production methods and properly designed products, energy consumption can be reduced and environmental protection can be ensured.

8. Measures to Reduce Environmental Impact

Various methods have been developed to further reduce the ecological footprint of ceramics and increase their energy efficiency and sustainability [2]. Reducing the environmental impact of ceramic production can be achieved by increasing energy efficiency throughout the production process, optimizing raw material use, and reducing greenhouse gas emissions. In this context, recycling fired and unfired waste generated in ceramic production is considered one of the most effective methods for reducing environmental impact. Emissions from ceramic production can be divided into three main groups: direct emissions due to fuel use, process emissions due to the mineral transformation of clay, and indirect emissions due to electricity consumption. Generally, methods used to reduce these emissions include reducing the carbon content of additives, optimizing clay mixtures without increasing transportation emissions, reducing raw material use, and employing decarbonization technologies [7].

The use of materials and wastes obtained from industrial processes is a potential application in ceramic production to reduce raw material costs and environmental impacts [40]. Recycling waste ceramics contributes to sustainable construction and waste management by protecting natural resources, reducing energy consumption, and supporting a circular economy [24]. Using recycled ceramic waste to produce geopolymers, replacing aggregates and cement in concrete, will not only reduce CO_2 emissions but also reduce water consumption and acidification potential. This presents an effective method for reducing environmental impact and improving resource efficiency. This will also improve human health and environmental performance [41],

[42]. When ceramic waste, river sand, or concrete aggregate is used, the mechanical strength of the concrete increases, as does its resistance to fire and freezing. This also reduces the use of cement, which is a major contributor to CO_2 emissions in the construction industry [7]. Waste recovery and recycling of process water by purification are also among the recommended methods for reducing environmental impact [10].

In the ceramic industry, the process of densifying ceramic powders at high temperatures through atomic diffusion to solidify them is known as sintering. The sintering temperature of ceramics ranges from approximately 2250 to 2350°C. To transition to energy-efficient sintering methods, instead of heating the entire chamber, systems that heat ceramic products locally using electric, magnetic, microwave, or photonic stimulation, such as photonic sintering, microwave sintering, and flash sintering, could be expanded [2]. In line with new-generation furnace technologies and thermal efficiency targets, induction and intermittent furnaces in particular can reduce methane emissions. Literature studies indicate that furnace modernization can reduce environmental impacts by 9% to 62%, according to Life Cycle Analysis (LCA) data [10]. Energy efficiency can be achieved in modern furnaces through heat recovery, insulation, and waste heat utilization. Furthermore, to reduce dependence on fossil fuels, the use of photovoltaic panels, biomass-based energy systems, UV curing, and electric furnaces should be expanded. While the use of microwave-assisted furnaces can significantly reduce energy consumption by up to 99%, energy savings are estimated to be approximately 65% for hybrid furnaces [5].

In a study conducted by [34], ceramic production systems were evaluated using the Energy Analysis Method (EMA), an energy-based system analysis, using different inputs as a common unit (UEV). The study analyzed natural resources, energy, labor, and environmental impacts holistically. This allows for objectively demonstrating the systems' efficiency, environmental impact, and sustainability through energy indicators. A new ceramics factory in Wuhan, Hubei Province, China, was selected as the research area. The study determined that, according to energy contribution, 57.7% of the total energy comes from non-renewable resources. This is followed by purchased inputs (18.6%), labor and services (11.8%), comprehensive energy (7.49%), industrial pollutant emissions (3.86%), and renewable energy sources (0.60%). Increasing the rate of renewable energy input and the use of recycled materials were recommended as measures to increase sustainability.

It is possible to reduce the environmental impact of production by integrating environmentally friendly and advanced technologies such as the use of renewable resources in ceramic production, 3D printing methods, microwave firing, and low-temperature firing techniques, and recycling of waste [2]. 3D printing technology is among the innovative approaches to reducing the environmental impact of ceramic products and is quite remarkable [40]. 3D printing technology, including additive manufacturing, is an important tool for industrial production, especially in the construction sector, as it reduces resource and labor costs as well as greenhouse gas emissions [7]. As an alternative to traditional production methods, it is possible with the development of the additive manufacturing method, also known as 3D printing [43]. With this technology, material waste is reduced while a high level of specification is also possible [44]. It is important to reduce the high costs in ceramic production, increase the solid waste utilization rate, and improve the performance of the ceramic product in line with the application requirements. To this end, the raw materials used should be designed to be compatible with each other in terms of composition and plasticity [45]. Ceramics production is a sector responsible for significant environmental burdens in terms of energy and chemical consumption. Optimizing chemical reagent doses is one recommended measure to avoid unnecessary reagent consumption [27]. In line with sustainability goals, the development of alternative raw materials and glaze compositions that do not contain toxic additives will also have a positive impact on reducing the environmental impacts caused by the ceramics industry. While providing an affordable fuel supply is challenging due to facility type and resource availability, some production units are transitioning to carbon-free electricity. The ceramics industry's commitment to reducing carbon emissions is encouraging. However, to achieve carbon neutrality, all production facilities must transition to renewable energy sources [11].

Key solutions to achieving climate neutrality goals can be achieved through innovative solutions developed in areas such as renewable energy, hydrogen technologies, and the circular economy. However, achieving these goals is not solely driven by industry efforts. A holistic approach, supported by external factors such as a suitable regulatory framework, equal competitiveness, financing opportunities, and the provision of carbon-free energy sources, is required [7]. The fact that developing environmentally friendly and advanced technologies in production methods can reduce all these environmental impacts will also contribute to the sustainability goal.

9. Basic Components and Operation of the Life Cycle Analysis (LCA) Methodology

The Life Cycle Assessment (LCA) method generally aims at ecosystem health, human health, and the protection of natural resources by analyzing environmental impacts such as global warming, ozone depletion, acidification, eutrophication, photochemical oxidation, and human toxicity [5]. In this context, all stages and production processes within a product's life cycle are evaluated in terms of environmental performance, and performance improvements are identified [23]. LCA is a 4-stage system determined within the framework of international standards (ISO 14040 and ISO 14044). The first stage is the "definition" phase, which defines the analysis scope and system boundaries. Then, an "inventory analysis" is conducted, where inputs and outputs such as energy, materials, and emissions are quantitatively collected. The third stage involves an "impact assessment" analysis, which categorizes environmental impacts such as global warming, acidification, and water consumption. In the final stage, the analysis results are evaluated, and the findings are used as decision-support tools. Life Cycle Impact Assessment (LCIA) is used to convert inventory results into environmental impact scores. This contributes to the identification of critical points and potential improvements in the life cycle [9].

10. The Importance of Life Cycle Analysis (LCA) Applications in Ceramic Production

Pollution occurring in the process of ceramic production, starting from mining activities during raw material supply to the final product, negatively affects the ecosystem and causes destruction. Global impacts caused by the process can be assessed using the Life Cycle Assessment (LCA) method, Environmental Development of Industrial Products (EDIP), Eco-indicator 95, Eco-indicator,r 99, and Best Available Techniques (BAT) methods. While the Eco-Indicator 95 and Eco-Indicator 99 methods focus on human health, ecosystem quality, and resources, Best Available Technique (BAT) is an approach used to evaluate methods that can be applied to reduce environmental impacts in ceramic production. The method focuses on energy efficiency and air pollution. Before starting the BAT process, the environmental impacts of the existing system are determined using the Life Cycle Assessment (LCA) method [5], [21]. Among the existing methods, the most widely used evaluation method is LCA. Non-technical ceramics, such as concrete and stone, which are physically and chemically similar to the Earth's crust, have the lowest total ecological footprint compared to other structural materials. However, technical ceramics, including alumina, silicon carbide, and zirconia, and their waste, disrupt the ecological balance [2]. The environmental impacts of all processes in ceramic production, from raw material procurement and production to usage and waste management, are analyzed using the LCA method. An LCA assessment, defined from "cradle to grave," can improve decision-making processes in the sector, ensure the adoption of more sustainable practices, and contribute to the promotion of environmental innovation [3]. Although ecological monitoring in LCA analysis is complicated by uncertainties in the transportation and usage stages, indicators such as energy and water quantity, CO_2 emissions, and market value can be approximately estimated throughout the process [2]. For accurate and reliable LCA work, all inputs and outputs of materials, water, energy, emissions to air, emissions to water and soil, and solid waste must be specified specifically and reliably throughout the entire process from raw material supply of ceramics to the end of the life of the final product [21]. Controlling and reducing emissions is critical to the sustainability of the ceramics industry. In this context, LCA stands out as an important method that allows for comprehensive analysis of the environmental and energy impacts of production processes [3].

Life Cycle Assessment (LCA), the most widely used method in global impact analysis, aims to measure and evaluate the carbon footprint and energy consumption in ceramic production processes from the extraction of raw materials to the disposal stage at the end of life [15]. Because ceramic waste doesn't naturally decompose, it's difficult to dispose of and, if not managed properly, can cause serious environmental problems. Therefore, understanding and reducing the environmental impact of ceramic production is crucial for sustainability. In the study conducted by [10], the Life Cycle Assessment (LCA) method was applied to a ceramic sanitary ware production facility located in the Emilia Romagna region of Italy. The analysis adopted a "cradle-to-grave" approach, assessing all processes from raw material procurement to final product production in terms of environmental impacts within system boundaries. Additionally, the recovery of both raw and fired waste, as well as the disposal of glazes, slimes, and non-recyclable waste, was included in the analysis. Methane and ethane gases released into the environment during the firing process were influential in the ozone layer depletion (OLD) and global warming (GW) categories. The primary chemicals released in the release of acidic substances into the environment are nitric oxides, nitric acid, phosphates, and ammonia. The final product selection control stage affected freshwater aquatic ecotoxicity (FWAE) by 26.7% due to electricity consumption and particulate and silica dioxide emissions. The ozone layer depletion (OLD) value due to the use of natural gas in the drying stage, which requires high energy, was determined as 9.1%. Figure 5 shows a flow chart of the life cycle stages of porcelain tiles.

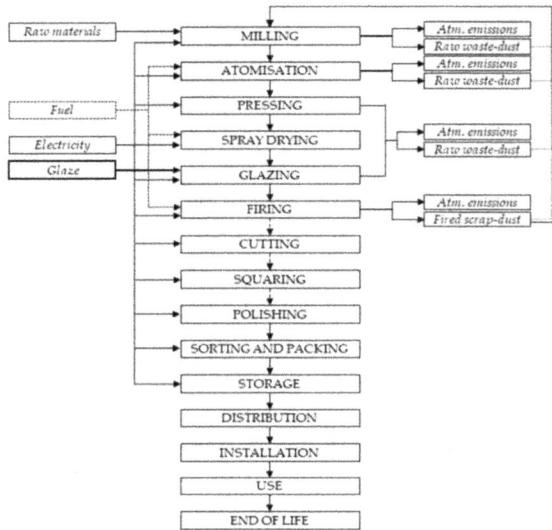

Figure 5. *flow chart of the life cycle stages of porcelain tiles [46], reprinted with permission from MDPI.*

The findings of the LCA analysis indicate that energy consumption and materials used during ceramic production are decisive factors in determining environmental impact. Practices such as water recycling and the integration of renewable energy systems have been suggested to reduce environmental impacts. Improving waste management systems will reduce the environmental burden by treating harmful gases such as sulfur oxides, hydrochloric acid, and hydrofluoric acid produced during the production process. Photovoltaic systems and energy recovery from high-temperature furnaces and chimneys have been proposed as additional improvements in many environmental impact categories. It has been stated that improving water and waste recycling, as well as increasing the use of renewable energy, will significantly contribute to the transition of the ceramics industry to a more sustainable production model [10]. In another LCA study in the literature LCA (Life Cycle Assessment) and LCC (Life Cycle Cost) methods were discussed together with a case study on ceramic production to measure the environmental and economic impacts of a ceramic tile production facility in China. The cost burden of the production process is mainly due to raw materials, and the total economic cost is calculated as 2.77 USD/m^2. The results revealed that among the environmental categories, the most affected were climate change, terrestrial ecotoxicity, marine ecotoxicity, human toxicity, and fossil fuel depletion. The environmental burden of raw materials consisting of inorganic chemicals is reported as 12.9% and the economic burden as 39.6%. The study recommends reducing the use of coal, electricity, and inorganic chemicals and optimizing raw material transportation to reduce environmental impacts and costs. Using alternative resources like coal and hydropower during production will reduce the impact on climate change and marine ecotoxicity by 98.4% and 96.4%, respectively. Furthermore, it has been stated that waste control at the facility eliminates most emissions, reducing environmental impact by providing significant reductions in key categories, particularly marine ecotoxicity, terrestrial acidification, and particulate matter formation [32].

[23], prepared a comprehensive life cycle inventory for the ceramics and construction sectors in Türkiye. The LCA method was used to assess the environmental impacts arising from glazed ceramic tile production, by ISO 14.040/14.044 guidelines. In total, 11 environmental impacts were estimated using the CML 2001 method, and sensitivity analysis was performed to test the reliability of the results. According to the LCA analysis results, the most important loads contributing to the eutrophication potential (EP) impact were determined to be phosphate (61.8%) emissions to freshwater and nitrogen oxide (10.5%) emissions to air.

The largest contribution to the global warming potential (GWP) came from the high energy consumption in the drying stages, which accounted for 73.4% of the total GWP. The second largest contributor was consumption from raw material extraction and processing loads, accounting for 12.9% of total GWP. The largest contributor to the ozone depletion potential (ODP) was non-methane volatile organic compounds (VOCs) released into the air, such as halon 1301 (55.6%) and halon 1211 (24.3%), from natural gas consumption. Among nearly all environmental impact categories, drying was identified as the phase with the largest impact due to fossil fuel-derived energy consumption. The study suggested that developing alternative materials or compositions containing waste from other industries instead of traditional raw materials would be effective in reducing environmental impacts at the raw material supply stage [23]. [15], conducted a one-year LCA study of 1 m^2 of ceramics produced at its ceramics facility in Brazil. The impact categories used for quantitative assessment of the environmental profiles examined during the study were determined as abiotic depletion elements (AD), abiotic depletion of fossil fuels (ADF), acidification (A), eutrophication (E), global warming (GW), photochemical oxidation (PO), and human toxicity (HT). The analyzed processes are grouped as follows: excavation for clay extraction (P1); extracted raw material (P2); well water used with a water pump (P3); mixing with an excavator (P4); firewood for drying (P5); firewood (P6); electricity supply- medium voltage- Northeast Region (P7); and transportation for tile distribution (P8). When the results were evaluated, the most important overall environmental impacts observed were identified at the raw material extraction stage, the fuelwood for incineration stage, and the product distribution stage. In another study conducted by [16], the environmental impacts of ceramics produced by a ceramics manufacturing company in Thailand were assessed using the EDIP and Eco-indicator 99 methodologies. According to the environmental impact assessment conducted using the EDIP methodology, the production activities were found to have significant global warming and human toxicity risks. Due to the facility's location relative to raw material sources, transportation distances are a key contributor to environmental impacts. Energy consumed in transportation accounts for approximately 60% of total energy consumption. This led to particularly high CO_2-eq values. In the assessment using the Eco-indicator 99 methodology, the transportation phase yielded the highest overall impact, while the cooking phase was the second-highest impact parameter. Reusing waste heat in the drying phase reduced the environmental impact. In light of the findings, for more environmentally friendly and sustainable ceramic production, it was suggested that a single firing technology should be adopted, the production facility should be located close to raw material sources, materials suitable for firing and sintering at lower temperatures should be developed, the use of chemicals containing heavy metals should be avoided, and local manufacturing companies should be preferred to reduce the transportation distance of the final product. In another case study conducted by [47], the environmental impacts of the vitrified ceramic floor tile supply chain were analyzed using Umberto NXT software using an LCA based on the ReCiPe method. The study analyzed all processes from raw material extraction to waste disposal as the system boundary and assessed the environmental impacts. The results showed that impacts such as climate change, human toxicity, depletion of fossil and metal resources are caused by red oxides used in glazing, electricity consumption, packaging production, concrete application, and transportation activities. According to LCA assessments conducted by [11], reducing the distance between raw material deposits, production facilities, and markets will reduce transportation costs and, consequently, CO_2 emissions. Additionally, adopting clean energy sources such as green hydrogen, biofuels, and renewable electricity has been reported to be effective in reducing greenhouse gas emissions. In a study conducted by [46] a case study was conducted for a representative example of Italian porcelain tile production, combining LCA, Life Cycle Costing (LCC), and Social Life Cycle Assessment (S-LCA) tools. Based on the findings, a Life Cycle Sustainability Assessment (LCSA) was conducted for the ceramic tile production facility, presenting the ceramic supply chain with a holistic approach that encompasses not only environmental impacts but also economic and social dimensions. It was reported that, in addition to the production process, raw material procurement and transportation processes are also critical for environmental impact categories. [9], conducted a Life Cycle Assessment (LCA) on the ceramic tile industry in Brazil, Colombia, and Mexico. The results showed that the Colombian ceramic tile industry had the lowest emissions among the three countries. Results for Mexico demonstrate significant impacts on global warming potential, fine particulate matter formation, and fossil resource scarcity. Initiatives in Brazil, such as tile thickness reduction and renewable energy use, have led to measurable improvements in emissions intensity. According to the study findings, it was recommended to invest in renewable energy sources, develop energy-efficient technologies for processes requiring high energy, and research innovations to optimize material and energy use without compromising product quality. As a result, the strategic application of advanced technologies can significantly reduce the carbon footprint of the ceramic tile industry in these countries. This is expected to contribute to the goals of limiting global temperature rise set by the Paris Agreement by reducing greenhouse gas emissions.

11. Sectoral Policy Supports

National and international regulations, incentive mechanisms, and financial support play a critical role in fostering the adoption of energy efficiency and low-carbon technologies. Reducing carbon emissions has become essential for the ceramics industry within the framework of the European Union's climate targets. In this context, adopting environmentally friendly standards through collaboration between manufacturers and policymakers is among the most effective methods for reducing environmental impacts. The European Union's goal of achieving climate neutrality by 2050 has been legislated within the framework of the European Green Deal and the European Climate Law of 2021. Climate policies aim to reduce greenhouse gas emissions by 55% by 2030 while also encouraging sustainable industrial transformation. The zero-emission environmental target, which encompasses all sectors and is expected to be effective in the ceramics industry by 2050. The ceramics industry is among the most energy-intensive sectors due to its production processes. In EU countries, the Industrial Emissions Directive (IED) and BAT (Best Available Techniques) reference values define greenhouse gas emission targets for the ceramics sector, thereby shaping sustainability policies. In addition to climate law, the European Emissions Trading System (EU ETS) limits manufacturers' carbon allowance for each ton of CO_2 emissions. Ceramic production accounts for 10% of the industrial facilities covered by the ETS. Expanding ISO 14001 certification, in addition to existing systems, is another major step toward eliminating the significant environmental problems caused by the ceramics sector.

12. Green Financing Tools and Incentive Mechanisms

The goal of a carbon-neutral sector requires high-cost investments. Technological advancements, as well as appropriate regulatory frameworks and financing mechanisms, are crucial for the successful implementation of the energy transition, the dissemination of green technologies, and the realization of a sustainable transformation process. Incentive programs, particularly for high-cost investments such as furnace modernization, the integration of renewable energy sources, the transition to electrification, and the use of green hydrogen, can accelerate the sector's carbon-neutrality goal. The EU aims to accelerate this process through Horizon Europe, the LIFE Program, and Green Deal funds.

13. Integration of Digital Transformation and Smart Manufacturing Technologies

The role of digitalization and smart production technologies in industrial processes is crucial. The Industry 4.0 approach is one approach that encourages the digitalization of ceramic production processes. Data collection, process monitoring, and AI-supported analysis should be expanded to optimize energy, water, and raw material usage. Furthermore, simulation and digital twin technologies allow for pre-testing process efficiency. In the short term, energy efficiency is one of the most effective and rapid ways to reduce carbon emissions. Thanks to digitalization and AI-supported systems in production, energy management can be achieved in a more controlled and efficient manner.

14. Conclusion

Considering the existing literature studies, the common findings reported by the authors can be listed as follows:

- In all studies, the most significant environmental impact was identified as the high-energy consumption of the production process, particularly during high-temperature stages such as drying and sintering. In this context, energy consumption was identified as the primary environmental burden.

- Raw material supply and transportation processes have been reported to cause serious impacts in categories such as greenhouse gas emissions, fossil fuel consumption, and acidification.

- The recovery of raw and fired wastes in the ceramic production process has important effects in terms of ecotoxicity and human health.

- Expanding the use of renewable energy sources such as solar, green hydrogen, and biofuels instead of fossil fuels in ceramic production is among the common recommendations of researchers.

- The literature review indicated that, based on LCA analysis results, each production stage in the ceramics industry has its own distinct impact. It was noted that when assessing environmental impact, not only the ceramic production process but also processes such as packaging, transportation, and waste disposal should be carefully evaluated.

- Common recommendations of all authors include the dissemination of energy-efficient technologies in production systems, transition to renewable energy systems, use of alternative raw materials, establishment of local production and short supply chains, and development of water and wastewater recycling systems.

References

[1]   X. Wu *et al.*, "Early Pottery at 20,000 Years Ago in Xianrendong Cave, China," *Science (1979)*, vol. 336, no. 6089, pp. 1696–1700, Jun. 2012. https://doi.org/10.1126/science.1218643

[2]   M. Mirkhalaf, "Sustainable structural materials inspired by nature," *Journal of the American Ceramic Society*, May 2025. https://doi.org/10.1111/jace.20710

[3]   A. W. Vieira *et al.*, "Life cycle assessment in the ceramic tile industry: a review," *Journal of Materials Research and Technology*, vol. 23, pp. 3904–3915, Mar. 2023. https://doi.org/10.1016/j.jmrt.2023.02.023

[4]   A. Imgirne *et al.*, "Potential use of mint waste and bentonite clay in sustainable porous ceramic manufacturing," *EuroMediterr J Environ Integr*, vol. 10, no. 3, pp. 1407–1423, Jun. 2025. https://doi.org/10.1007/s41207-024-00716-8

[5]   M. Muthukannan and A. S. Chithambar Ganesh, "The Environmental Impact Caused By The Ceramic Industries And Assessment Methodologies" *International Journal for Quality Research*, vol. 13, no. 2, pp. 315–334, Jun. 2019. https://doi.org/10.24874/IJQR13.02-05

[6]   R. B. Heimann, "On the nature of ceramics technology: from Empedocles to Dawkins," *Archaeometry*, vol. 67, no. 1, pp. 55–71, Feb. 2025. https://doi.org/10.1111/arcm.12997

[7]   U. Abdullaev, U. Dzhusuev, S. Asanova, Z. Matniyazov, and S. Pavlovskyi, "Research into Modern Methods of Producing Energy-Efficient Building Materials," *Architecture Image Studies*, vol. 6, no. 1, pp. 126–141, Mar. 2025. https://doi.org/10.62754/ais.v6i1.112

[8]   C. Lourenco Alves, J. de Oliveira Martins Müller, A. de Noni, and S. Heinrich, "Challenges and opportunities for increase sustainability and energy efficiency in ceramic tile industry," *Int J Appl Ceram Technol*, vol. 22, no. 4, Jul. 2025. https://doi.org/10.1111/ijac.15097

[9]   E. L. Saavedra and J. F. Osma, "Evaluating decarbonization technologies and optimizing pathways in ceramic tile production through life cycle assessment in Latin America," *J Clean Prod*, vol. 518, p. 145800, Aug. 2025. https://doi.org/10.1016/j.jclepro.2025.145800

[10]   M. P. Desole, L. Fedele, A. Gisario, and M. Barletta, "Life Cycle Assessment (LCA) of ceramic sanitaryware: focus on the production process and analysis of scenario," *International Journal of Environmental Science and Technology*, vol. 21, no. 2, pp. 1649–1670, Jan. 2024. https://doi.org/10.1007/s13762-023-05074-6

[11]   K. Li, E. de S. Cordeiro, and A. De Noni, "Comparison between Mullite-Based and Anorthite-Based Porcelain Tiles: A Review," *Eng*, vol. 4, no. 3, pp. 2153–2166, Aug. 2023. https://doi.org/10.3390/eng4030123

[12]   F. E. Garcia-Muiña, R. González-Sánchez, A. M. Ferrari, and D. Settembre-Blundo, "The Paradigms of Industry 4.0 and Circular Economy as Enabling Drivers for the Competitiveness of Businesses and Territories: The Case of an Italian Ceramic Tiles Manufacturing Company," *Soc Sci*, vol. 7, no. 12, p. 255, Dec. 2018. https://doi.org/10.3390/socsci7120255

[13]   G. Contini, F. Grandi, and M. Peruzzini, "Human-Centric Green Design for automatic production lines: Using virtual and augmented reality to integrate industrial data and promote sustainability," *J Ind Inf Integr*, vol. 44, p. 100801, Mar. 2025. https://doi.org/10.1016/j.jii.2025.100801

[14]   C. L. Alves, V. Skorych, A. De Noni Jr., D. Hotza, S. Y. G. González, and S. Heinrich, "Application of Flowsheet Simulation Methodology to Improve Productivity and Sustainability of Porcelain Tile Manufacturing," *Machines*, vol. 11, no. 2, p. 137, Jan. 2023. https://doi.org/10.3390/machines11020137

[15]   P. Viana, J. Bastos, D. Silva, A. Cabral, A. Ledoux, and E. Mesquita, "Mapping the environmental footprint: tracing ceramic tile production's impact on sustainable goals," *Journal of Building Pathology and Rehabilitation*, vol. 10, no. 1, p. 2, Jun. 2025. https://doi.org/10.1007/s41024-024-00510-7

[16]   N. Tikul and P. Srichandr, "Assessing the environmental impact of ceramic tile production in Thailand," *Journal of the Ceramic Society of Japan*, vol. 118, no. 1382, pp. 887–894, 2010. https://doi.org/10.2109/jcersj2.118.887

[17]   P. Quinteiro, M. I. Almeida, J. Serra, L. Arroja, and A. C. Dias, "Life cycle assessment of ceramic roof tiles: A temporal perspective," *J Clean Prod*, vol. 363, p. 132568, Aug. 2022. https://doi.org/10.1016/j.jclepro.2022.132568

[18]   M. Ancona *et al.*, "Energy and Environmental Assessment of Cogeneration in Ceramic Tiles Industry," *Energies (Basel)*, vol. 16, no. 1, p. 182, Dec. 2022. https://doi.org/10.3390/en16010182

[19]   N. Soussi, W. Kriaa, H. Mhiri, and P. Bournot, "Reduction of the energy consumption of a tunnel kiln by optimization of the recovered air mass flow from the cooling zone to the firing zone," *Appl Therm Eng*, vol. 124, pp. 1382–1391, Sep. 2017. https://doi.org/10.1016/j.applthermaleng.2017.06.111

[20]   S. Chitwaree, J. Tiansuwan, N. Thavarungkul, and L. Punsukumtana, "Energy saving in sintering of porcelain stoneware tile manufacturing by using recycled glass and pottery stone as substitute materials," *Case Studies in Thermal Engineering*, vol. 11, pp. 81–88, Mar. 2018. https://doi.org/10.1016/j.csite.2018.01.002

[21]   V. Ibáñez-Forés, M.-D. Bovea, and A. Simó, "Life cycle assessment of ceramic tiles. Environmental and statistical analysis," *Int J Life Cycle Assess*, vol. 16, no. 9, pp. 916–928, Nov. 2011. https://doi.org/10.1007/s11367-011-0322-6

[22]   E. F. S. Ciacco, J. R. Rocha, and A. R. Coutinho, "The energy consumption in the ceramic tile industry in Brazil," *Appl Therm Eng*, vol. 113, pp. 1283–1289, Feb. 2017. https://doi.org/10.1016/j.applthermaleng.2016.11.068

[23]   B. Atılgan Türkmen, T. Budak Duhbacı, and Ş. Karahan Özbilen, "Environmental impact assessment of ceramic tile manufacturing: a case study in Turkey," *Clean Technol Environ Policy*, vol. 23, no. 4, pp. 1295–1310, May 2021. https://doi.org/10.1007/s10098-021-02035-w

[24] G. F. Huseien *et al.*, "Sustainability of Recycling Waste Ceramic Tiles in the Green Concrete Industry: A Comprehensive Review," *Buildings*, vol. 15, no. 14, p. 2406, Jul. 2025. https://doi.org/10.3390/buildings15142406

[25] C. Zanelli, S. Conte, C. Molinari, R. Soldati, and M. Dondi, "Waste recycling in ceramic tiles: a technological outlook," *Resour Conserv Recycl*, vol. 168, p. 105289, May 2021. https://doi.org/10.1016/j.resconrec.2020.105289

[26] G. Boschi, G. Masi, G. Bonvicini, and M. C. Bignozzi, "Sustainability in Italian Ceramic Tile Production: Evaluation of the Environmental Impact," *Applied Sciences*, vol. 10, no. 24, p. 9063, Dec. 2020. https://doi.org/10.3390/app10249063

[27] L. Uribe, C. Salazar, and Y. Casas-Ledón, "Ceramic Foams as an Alternative for the Mining Tailing Valorization: A Life Cycle Analysis Perspective," *Journal of Sustainable Metallurgy*, vol. 11, no. 2, pp. 1604–1615, Jun. 2025. https://doi.org/10.1007/s40831-025-01069-4

[28] Y. O. Özkılıç *et al.*, "Influence of ceramic waste powder on shear performance of environmentally friendly reinforced concrete beams," *Sci Rep*, vol. 14, no. 1, p. 10401, May 2024. https://doi.org/10.1038/s41598-024-59825-7

[29] C. A. Horowitz, "Paris Agreement," *International Legal Materials*, vol. 55, no. 4, pp. 740–755, Aug. 2016. https://doi.org/10.1017/S0020782900004253

[30] K. Röper, N. Kunz, and L. Gast, "Renewable hydrogen in industrial production: A bibliometric analysis of current and future applications," *Int J Hydrogen Energy*, vol. 98, pp. 687–696, Jan. 2025. https://doi.org/10.1016/j.ijhydene.2024.12.034

[31] Y. Li, A. H. Sharaai, S. Ma, W. Wafa, Z. He, and L. A. Ghani, "Quantification of Carbon Emission and Solid Waste from Pottery Production by Using Life-Cycle Assessment (LCA) Method in Yunnan, China," *Processes*, vol. 10, no. 5, p. 926, May 2022. https://doi.org/10.3390/pr10050926

[32] L. Ye, J. Hong, X. Ma, C. Qi, and D. Yang, "Life cycle environmental and economic assessment of ceramic tile production: A case study in China," *J Clean Prod*, vol. 189, pp. 432–441, Jul. 2018. https://doi.org/10.1016/j.jclepro.2018.04.112

[33] K. Ahmed Ali, M. I. Ahmad, and Y. Yusup, "Issues, Impacts, and Mitigations of Carbon Dioxide Emissions in the Building Sector," *Sustainability*, vol. 12, no. 18, p. 7427, Sep. 2020. https://doi.org/10.3390/su12187427

[34] J. Zhang, H. Zhang, A. T. Asutosh, N. Sun, and X. Li, "The environmental sustainability assessment of building ceramic manufacturing based on the LCA-emergy approach," *Environ Dev Sustain*, vol. 27, no. 3, pp. 5999–6024, Nov. 2023. https://doi.org/10.1007/s10668-023-04114-y

[35] K. Ding, A. Li, J. Lv, and F. Gu, "Decarbonizing ceramic industry: Technological routes and cost assessment," *J Clean Prod*, vol. 419, p. 138278, Sep. 2023. https://doi.org/10.1016/j.jclepro.2023.138278

[36] A. Mezquita, J. Boix, E. Monfort, and G. Mallol, "Energy saving in ceramic tile kilns: Cooling gas heat recovery," *Appl Therm Eng*, vol. 65, no. 1–2, pp. 102–110, Apr. 2014. https://doi.org/10.1016/j.applthermaleng.2014.01.002

[37]   M. Pini, A. M. Ferrari, R. Gamberini, P. Neri, and B. Rimini, "Life cycle assessment of a large, thin ceramic tile with advantageous technological properties," *Int J Life Cycle Assess*, vol. 19, no. 9, pp. 1567–1580, Sep. 2014. https://doi.org/10.1007/s11367-014-0764-8

[38]   G. M. Nicoletti, B. Notarnicola, and G. Tassielli, "Comparative Life Cycle Assessment of flooring materials: ceramic versus marble tiles," *J Clean Prod*, vol. 10, no. 3, pp. 283–296, Jun. 2002. https://doi.org/10.1016/S0959-6526(01)00028-2

[39]   A. Villa, P. Gianchandani, and F. Baino, "Sustainable Approaches for the Additive Manufacturing of Ceramic Materials," *Ceramics*, vol. 7, no. 1, pp. 291–309, Feb. 2024. https://doi.org/10.3390/ceramics7010019

[40]   E. Mingione, M. Marconi, G. Rubino, and D. Salvi, "Reduction of energy consumption and environmental impact in ceramic sanitary ware production through fluxing agents: a case study," *The International Journal of Advanced Manufacturing Technology*, Jul. 2025. https://doi.org/10.1007/s00170-025-16069-7

[41]   Z. H. Joudah *et al.*, "Effects of Waste Glass Bottle Nanoparticles and High Volume of Waste Ceramic Tiles on Concrete Performance When Exposed to Elevated Temperatures: Experimental and Theoretical Evaluations," *Fire*, vol. 7, no. 12, p. 426, Nov. 2024. https://doi.org/10.3390/fire7120426

[42]   B. Zhang, H. Zhu, Y. Cheng, G. F. Huseien, and K. W. Shah, "Shrinkage mechanisms and shrinkage-mitigating strategies of alkali-activated slag composites: A critical review," *Constr Build Mater*, vol. 318, p. 125993, Feb. 2022. https://doi.org/10.1016/j.conbuildmat.2021.125993

[43]   E. Fiume, B. Coppola, L. Montanaro, and P. Palmero, "Vat-photopolymerization of ceramic materials: exploring current applications in advanced multidisciplinary fields," *Front Mater*, vol. 10, Oct. 2023. https://doi.org/10.3389/fmats.2023.1242480

[44]   T. Peng, K. Kellens, R. Tang, C. Chen, and G. Chen, "Sustainability of additive manufacturing: An overview on its energy demand and environmental impact," *Addit Manuf*, vol. 21, pp. 694–704, May 2018. https://doi.org/10.1016/j.addma.2018.04.022

[45]   Y. Liu, J. Tang, M. Li, Q. Zhang, and W. Zhang, "Production of high-strength eco-conscious ceramics exclusively from municipal solid waste," *Ceram Int*, vol. 50, no. 22, pp. 47851–47863, Nov. 2024. https://doi.org/10.1016/j.ceramint.2024.09.130

[46]   A. M. Ferrari, L. Volpi, M. Pini, C. Siligardi, F. E. García-Muiña, and D. Settembre-Blundo, "Building a Sustainability Benchmarking Framework of Ceramic Tiles Based on Life Cycle Sustainability Assessment (LCSA)," *Resources*, vol. 8, no. 1, p. 11, Jan. 2019. https://doi.org/10.3390/resources8010011

[47]   K. S. Sangwan, K. Choudhary, and C. Batra, "Environmental impact assessment of a ceramic tile supply chain – a case study," *International Journal of Sustainable Engineering*, vol. 11, no. 3, pp. 211–216, May 2018. https://doi.org/10.1080/19397038.2017.1394398

* 9 7 8 1 6 4 4 9 0 3 8 2 7 *